CW00369941

Archaeology and Geographical Information Systems
A European Perspective

Archaeology and Geographical Information Systems: A European Perspective

Edited by

Gary Lock and Zoran Stančič

UK	Taylor & Francis Ltd, 4 John St, London WC1N 2ET
USA	Taylor & Francis Inc., 1900 Frost Road, Suite 101, Bristol PA 19007

British Library Cataloguing in Publication Data

A catalogue record for this book is available from the British Library

ISBN 0-7484-0208-X

Library of Congress Cataloging in Publication Data are available

Cover design by Hybert Design and Type,
Maidenhead, Berkshire

Typeset by Santype International Limited, Netherhampton Road, Salisbury, Wilts SP2 8PS

Printed in Hong Kong by Graphicraft Typesetters Ltd.

Contents

Contributors

K. L. Kvamme
Department of Archaeology, Boston University, 675 Commonwealth Avenue, Boston, Massachusetts 02215, USA

D. Guillot
Ministère de la Culture et de la Francophonie, Direction du Patrimoine, Sous-direction de l'Archéologie, 4, rue d'Aboukir, 75002 Paris, France

G. Leroy
Service régional de l'Archéologie, DRAC, 5, rue Henri Daussy, 80000 Amiens, France

M. van Leusen
Field Archaeology Unit, University of Birmingham, Birmingham B15 2TT, UK

D. Arroyo-Bishop
Centre National de la Recherche Scientifique, Université de Paris 1, Pantheon—Sorbonne, GDR 880—Terrains et Théories en Archéologie, Institut d'Art et Archéologie, 3, rue Michelet, 75006 Paris, France

M. T. Lantada Zarzosa
Centre National de la Recherche Scientifique, Université de Paris 1, Pantheon—Sorbonne, GDR 880—Terrains et Théories en Archéologie, Institut d'Art et Archéologie, 3, rue Michelet, 75006 Paris, France

F. Massagrande
Department of Archaeology, University of Southampton, University Road, Highfield, Southampton SO9 5NH, UK

M. Gillings
Department of Archaeology, University of Newcastle, Newcastle upon Tyne NE1 7RU, UK

E. Jerem
Archaeological Institute of the Hungarian Academy of Sciences, Úri Utca, 49, 1250 Budapest, Hungary

F. Redö
Archaeological Institute of the Hungarian Academy of Sciences, Úri Utaca, 49, 1250 Budapest, Hungary

G. Csaki
Institute of Geodesy, Cartography and Remote Sensing, Guszev u.19, 1051 Budapest, Hungary

J. Baena
Servicio de Cartographia, Universidad Autónoma de Madrid, Cuidad Universitaria de Cantoblanco, Ctra. de Colmenar, Km.15, 28049 Madrid, Spain

C. Blasco
Servicio de Cartographia, Universidad Autónoma de Madrid, Cuidad Universitaria de Cantoblanco, Ctra. de Colmenar, Km.15, 28049 Madrid, Spain

V. Recuero
Servicio de Cartographia, Universidad Autónoma de Madrid, Cuidad Universitaria de Cantoblanco, Ctra. de Colmenar, Km.15, 28049 Madrid, Spain

M. Kuna
Archeologický Ústav AV CR, Letenská 4, 118 01 Praha 1, Czech Republic

D. Adelsbergerova
Archeologicky Ustav AV CR, letenska 4, 118 01 Praha 1, Czech Republic

E. Neustupný
Archeologicky Ustav AV CR, letenska 4, 118 01 Praha 1, Czech Republic

M. Nunez
University of Oulu, PO Box 111, SF-90571 Oulu, Finland

A. Vikkula
Department of Archaeology, University of Helsinki, PO Box 13, SF-00014 Helsinki, Finland

T. Kirkinen
Department of Archaeology, University of Helsinki, PO Box 13, SF-00014 Helsinki, Finland

M. Wansleeben
Archeologisch Centrum, Rijksuniversiteit Leiden, Institut voor Prehistorie, Reuvensplaats 4, Postbus 9515, 2300 RA Leiden, The Netherlands

L. B. M. Verhart
Archeologisch Centrum, Rijksuniversiteit Leiden, Institut voor Prehistorie, Reuvensplaats 4, Postbus 9515, 2300 RA Leiden, The Netherlands

D. Wheatley
Department of Archaeology, University of Southampton, University Road, Highfield, Southampton SO9 5NH, UK

P. Verhagen
Stichting RAAP, Universiteit van Amsterdam, Plantage Muidergracht 14, 1018 TV Amsterdam, The Netherlands

J. McGlade
International Ecotechnology Research Centre, Cranfield University, UK

R. Risch
Universidad Autónoma de Barcelona, Barcelona, Spain

S. Gili
Universidad Autónoma de Barcelona, Barcelona, Spain

Z. Stančič
Scientific Research Centre of the Slovene Academy of Sciences and Arts, Gosposka 13, 61000 Ljubljana, Slovenia

H. Watson
127 Oxford Gardens, Stafford, UK

M. Forte
Via Belojannis 1, 42100 Reggio Emilia, Italia

N. Smith
Department of Classics, College of the Holy Cross, Worcester, MA 01610, USA

J. S. Boaz
IAKN/Department of Archaeology, University of Oslo, Frederiksgt. 3, N-0164 Oslo, Norway

E. Uleberg
IAKN/Department of Archaeology, University of Oslo, Frederiksgt. 3, N-0164 Oslo, Norway

K. T. Biró
Magyar Nemzeti Muzeum, Mutargyvedelmi es Informacios Reszleg, 1450 Bp.pf.124, Konyves Kalman krt. 40, Hungary

I. Sz. Fejes
Magyar Nemzeti Muzeum, Mutargyvedelmi es Informacios Reszleg, PO Box 124, 1450 Budapest, Hungary

V. Gaffney
Field Archaeology Unit, University of Birmingham, Edgbaston, Birmingham B15 2TT, UK

S. Biswell
Field Archaeology Unit, University of Birmingham, Edgbaston, Birmingham B15 2TT, UK

L. Cropper
Field Archaeology Unit, University of Birmingham, Edgbaston, Birmingham B15 2TT, UK

J. Evans
Field Archaeology Unit, University of Birmingham, Edgbaston, Birmingham B15 2TT, UK

P. Leach
Field Archaeology Unit, University of Birmingham, Edgbaston, Birmingham B15 2TT, UK

M. Meffert
Institut voor Pre- and Protohistorische Archeologie, Albert Egges van Giffen, Universiteit van Amsterdam, Nieuwe Prinsengracht 130, 1018 VZ Amsterdam, The Netherlands

R. Wiemer
ARCHIS/R.O.B., Kerkstraat 1, 3811 CV Amersfoort, The Netherlands

S. Stead
Flat 3, 15 Northbone Road, London SW4 7DR, UK

P. Miller
York Archaeological Assessment, Department of Archaeology, University of York, Micklegate House, 88 Micklegate, York YO1 1JZ, UK

J. B. Claxton
Senior Common Room, Earnshaw Hall, Sheffield University, Sheffield S10 3EG, UK

G. Lock
Institute of Archaeology, University of Oxford, 36 Beaumont St., Oxford OX1 2PG, UK

T. Harris
Department of Geology and Geography, West Virginia University, 425 White Hall, Morgantown, WV 26506, USA

Preface

This book is based on a conference held in Ravello, south of Naples, Italy, at the beginning of October 1993. It was sponsored by the European University Centre for Cultural Heritage and held in their conference centre, the Villa Rufolo. We like to think that the inspirational affect of such dramatic 13th century surroundings was evident in the quality of the discussion at the conference and it is not difficult to see why several Popes, members of Royalty and more recently the composer Wagner have all chosen to live and work there. The conference will be remembered, at least by us as its organizers, as a rich mixture of beautiful places and scenery, excellent company, food, and wine, and stimulating discussion. It is hoped that this book goes some way towards thanking all those who took part. We especially acknowledge the debt owed to the European Centre and to Eugenia Apicella and her staff for their encouragement and help during the planning process and for their warm hospitality during the conference itself.

It is worth recounting why we felt the need for a European conference on the use of GIS in archaeology. This was first voiced, surprisingly enough, in a biker's bar in downtown Santa Barbara, California, in February of 1992. The occasion was the conference entitled 'The Anthropology of Human Behavior through Geographic Information and Analysis' organized by Herb Maschner and Mark Aldenderfer. The programme comprised mainly North American contributions with a small number of offerings from Europe. As the Bud flowed it became very apparent to us that there were differences between North America (more specifically, the USA) and Europe other than the quality of the beer. Part of those differences is to do with the types of evidence for past landscapes that are available for study in the two continents. This, combined with different approaches to archaeology, and the theory and practice of landscape archaeology specifically, was evident in the way GIS were being used and presented at the Santa Barbara conference. Coming, as it did, just a few months after the publication of *Interpreting Space: GIS and Archaeology* (Allen *et al.*, 1990) with the almost total North American bias of that landmark volume, the conference suggested to us that there was a lot of interesting work being done in Europe which was emanating from a fundamentally different set of archaeological problems.

The differences we perceived between North American and European archaeological uses of GIS are rooted in both the practical and the underlying theoretical aspects of applying the technology. Of course, this is a generalization and there will always be individual exceptions, but the lasting impression of both the Santa Barbara conference and *Interpreting Space* is an emphasis on the technology and on large-scale data integration and management. Perhaps this is related to the characteristics of the archaeological resource. Certainly over most of Europe the archaeological record of the last six millennia provides a dense and complex palimpsest both horizontally and vertically when compared to large areas of North America. This has resulted in an approach to regional landscape analysis which is concerned firstly with relationships between sites in terms of territories or spatial units representing social, political and economic interactions, and secondly with human perception of landscape especially the symbolic and ritual. With the adoption of GIS technology, this underlying theory has manifested itself in the use of Theissen polygons, cost surfaces and viewshed analysis. There is obviously an element of technological determinism here as these procedures are commonly available within GIS, although it is not the procedures themselves that are important but the underlying archaeological approaches and questions determining their use. One of the challenges facing the future use of GIS in archaeology is how the technology can escape from its roots in scientism and work with theoretical models at a more humanistic level, for example the sense of place created by the hermeneutics of Thomas (1993) and Tilley (1993). While these theoretical approaches could be said to represent a particularly European attempt to assimilate social theory within GIS technology, similar advances, although based on the different social theory involving cognition, are also beginning to emerge in North America (Zubrow, 1994).

Obviously, we are not suggesting rigid differences here because theoretical/analytical landscape archaeology is practised in North America and cultural resource management (CRM) is important in Europe, as several papers in this volume indicate. What is interesting, however, is that during this very first European conference on GIS in archaeology the most intense and heated discussion centred on the role of archaeological theory in CRM and in GIS applications generally. In particular, the concept of predictive modelling, which has been a central tenet in the development of GIS applications in North American archaeology, was seen as being inherently environmentally deterministic and a fundamentally different approach to that outlined above. This brings into question the whole purpose of CRM and whether or not there is, or indeed needs to be, explicit archaeological theory supporting it. We feel this is an important debate which takes the emphasis away from the technology of GIS and places it back within the central concerns of archaeology, certainly European archaeology. The final chapter of this book has been written specifically to address this debate and is based on many of the main themes of the Ravello discussions (thanks to Vince and Martijn for taking this risk).

Once the idea of a European conference was conceived in Santa Barbara, thanks to the European University Centre, gestation was not too prolonged or painful. The whole process was given additional momentum by Taylor & Francis, who agreed to publish this book based on the conference before it took place. We agreed with their suggestion that the book is not strictly the proceedings of the conference but rather a collection of papers which attempt to represent the state of GIS applications in European archaeology in late 1993. The spirit of Ravello survives in both the spread of countries represented and in the mix of theoretical issues and practical applications presented. We also acknowledge the importance of *Interpreting Space* and our intentions from the beginning were to complement and build on the information in that volume rather than to repeat any of it. Consequently, there are no background chapters on hardware, software or principles of GIS and we decided not to structure the book into rigid sections. However, the papers are ordered approximately into CRM applications, landscape archaeology, intra-site applications and, finally, explicitly theoretical concerns (all following a much appreciated invited first chapter from Ken Kvamme to show that we are not really anti-New World). Reading the papers will show that the boundaries between these categories are, to say the least, fuzzy although not completely meaningless.

References

Allen, K. M. S., Green, S. W. and Zubrov, E. B. W. (Eds.), 1990, *Interpreting Space: GIS and Archaeology*, London: Taylor & Francis.

Bender, B. (Ed.), 1993, *Landscape: Politics and Perspectives*, Oxford: Berg.

Thomas, J., 1993, The politics of vision and the archaeologies of landscape, in B. Bender (Ed.), pp. 19–48.

Tilley, C., 1993, Art, architecture, landscape (Neolithic Sweden), in B. Bender (Ed.), pp. 49–84.

Zubrow, E. B. W., 1994, Knowledge representation and archaeology: a cognitive example using GIS, in C. Renfrew and E. B. W. Zubrow (Eds.), *The Ancient Mind: elements of cognitive archaeology*, Cambridge: Cambridge University Press, New Directions in Archaeology, pp. 107–18.

Acknowledgements

Besides the European University Centre, we would like to thank everyone at Taylor & Francis involved in the preparation and production of this book. We are grateful for their advice and guidance. We would also like to thank Znanstveni Institut Filozofske Fakultete, University of Ljubljana, for financial assistance to help in the editing of the papers. All of the authors were extremely prompt and understanding in responding to the very strict deadlines imposed upon them and we are grateful. Mainly because of page length considerations we have been forced to take some editing liberties; any resulting mistakes or misunderstandings are readily accepted as our own.

Gary Lock, Oxford
Zoran Stančič, Ljubljana

April 1994

1

A view from across the water: the North American experience in archaeological GIS

K. L. Kvamme

1.1 Introduction

Applications of geographic information systems (GIS) in archaeology may be nearly a decade old, or perhaps a decade and a half, depending on how one defines GIS. Actual use of the term in archaeology begins to appear with regularity around 1983–85 (Hasenstab, 1983a; Kvamme, 1984; Martin and Garrett, 1985; c.f. Pomerantz, 1981), but there are strong antecedents in related technologies that go back to the late 1970s on both sides of the Atlantic.

The earliest literature mostly derives from specialized applications of computer graphics that employed GIS types of operations. Of particular relevance here are studies that utilized spatial databases which produced mapped output. Many pioneering papers also performed various types of manipulations which are clear indicators of GIS, such as weighted layer combinations and the generation of new spatial information. In the following, I focus primarily on North American examples, but also cite parallel developments in Europe.

1.2 The beginnings of GIS in archaeology

1.2.1 Archaeological surface models

The earliest applications akin to GIS clearly lie in computer graphics and statistics. Trend-surface analysis, borrowed from geology, has surprisingly numerous applications from 1975 through the early 1980s (Larson, 1975; Feder, 1979; Hietala and Larson, 1979; Bove, 1981). Typically, polynomial surfaces of various orders were fitted to artifact floors in an attempt to model the distributions of artifacts or other archaeological categories (e.g. lithics or bones). Following naturally from this, some archaeologists explored other types of surface-generating models including weighted-average spatial interpolation methods (Redman and Watson, 1970; Heitala and Larson, 1979; Jermann and Dunnell, 1979). The goal here, again, was to obtain surface generalizing models of artifact distributions to portray their overall patterns of location. This work was greatly facilitated, and in effect promoted, by the wide availability of the first successful spatial analysis and mapping software known as SYMAP (Laboratory for Computer Graphics and Spatial Analysis, 1975), in which we see ancestral forms of GIS functionality including cartographic output, interpolation and generation of surfaces, and manipulation of multiple spatial variables for a single region. It is therefore not surprising that similar applications were occurring in Europe at about the same time (Bradley, 1970; Hodder and Orton, 1976).

1.2.2 Computer mapping and regional databases

Simultaneous with these developments there was early interest in employing computer cartography to generate mapped output for regional spatial databases. One of the best examples is that of Effland (1979) who employed simple pen plotter instructions to provide archaeological site distribution maps for various time periods in the American Southwest. Other contemporary applications employed computer cartography to map artifacts, by type, on excavated occupation floors (Whallon, 1974, 1984; Copp, 1977; Clark, 1979). Here again developments were parallel on both sides of the Atlantic (Flude *et al.*, 1982; Hivernel and Hodder, 1984; Todd *et al.*, 1985), although in Europe there was also an early preoccupation with mapping the results of geophysical surveys (Scollar, 1966, 1974).

1.2.3 Digital elevation models

The application and use of digital elevation models (DEMs) of study region surfaces came quite early in the USA with the work of Scheitlin and Clark (1978), Arnold (1979), Green and Stewart (1983), Kvamme (1983) and others. By the mid-1980s, we see similar work in Britain (McKay, 1984), particularly with the efforts of Harris (1986; 1988). In most of these studies, the DEM was employed solely as a way to visualize patterns better in archaeological and other distributions within a region.

1.2.4 Computer Simulations

The computer simulation work of Zimmerman (1977) in the midwestern USA, and Chadwick (1978; 1979) in Greece, provides another source of GIS precursors. Both employed a variety of environmental data in computer form to examine the problem of human land use and occupation in a dynamic way, through time. Chadwick's work, in particular, qualifies as an early raster GIS application because he encoded multiple environmental data types in 2×2 km grid cells and manipulated these to form a weighted composite map of environmental suitability for Early and Middle Helladic Period occupations. This was quite an achievement in the late 1970s, because in terms of functionality it represents much of what we now do with GIS.

1.2.5 The Granite Reef project

The earliest example of a true, full-blown GIS application in archaeology comes from the American Southwest in the Granite Reef archaeological project of 1979–82 (Brown and Rubin, 1982). This project incorporated the services of a professional computer scientist, John Rubin, who (by 1980) had written a complete raster system for processing map data called MAPS (Rubin, 1980). Using this software the granite Reef project established distinct data layers for elevation, soils, geology, rainfall, temperature, and other surfaces over a huge area, approximately 32 000 km^2, using square grid elements of 3.4 km^2. The MAPS program allowed the full manipulation and combination of surfaces, use of map algebra techniques (Tomlin, 1991), and even the derivation of slope, aspect, and other terrain information from the elevation data. These capabilities were fully exploited by the archaeological researchers to develop models, which were weighted layer combinations, of environmental suitability for early hunters, for travel in the desert, and for prehistoric agriculture. These models were then compared against the archaeology to examine goodness-of-fit.

This project was truly remarkable because, frankly, I see little difference between it and much of what occurs in GIS-based settlement studies today (c.f. Dalla Bona, 1989; Brandt *et al.*, 1992). It is also interesting to note that nowhere in the publication is there any mention of the term 'GIS'; the time was simply too early and the term too little known. Brown and Rubin (1982: 272) refer to their operations simply as 'a computer-based cartographic analysis system'.

1.2.6 Predictive location models

The importance of predictive models of archaeological location to the growth of GIS in North American archaeology cannot be overemphasized. This phenomenon arose principally in the western United States where there exist huge tracts of federally controlled lands. Beginning in the late 1970s, various government agencies began to ask for a means to utilize patterns shown by known archaeological site locations in a region to project or predict where future sites might be located for cultural resource management and planning purposes. Sandra Scholtz–Parker (Scholtz, 1981) in Arkansas, Bob Hasenstab (1983b) in New Jersey, and myself (Kvamme, 1983) in Colorado all independently rose to the challenge and derived amazingly parallel solutions in the form of GIS. This evolution deserves comment.

The basic idea behind the methodology involves the examination of known archaeological sites or settlements in a region for statistical associations with various environmental conditions, such as ground steepness, elevation, aspect, soil, or distance-to-water preferences. Once the statistical associations are found, and multivariate discriminant functions provide an excellent and robust solution, one can go to a map at any point and, based on measurements of the relevant environmental variables, make a decision about the likelihood or even probability of archaeological site presence (Kvamme, 1990a). This procedure could be performed manually at a few locations, but what was needed for management purposes was the systematic mapping, e.g. every 50 m, of the result over large areas to produce archaeological location **decision surfaces**. In 1980, I hand-measured six variables at 256 contiguous locations spaced 50 m apart (for an 800 × 800 m region; Kvamme, 1980), and a year later Scholtz–Parker (Scholtz, 1981) did the same for 16 variables, measured at over 7000 locations at a 200 m spacing, to illustrate the potential of archaeological prediction surfaces. Scholtz–Parker, however, manually keyed these data into a statistics–graphics program (the statistical analysis system; SAS Institute, 1978) to derive computer-generated printouts of the hand-measured data and model surfaces. Full automation soon followed. By 1982, I had written computer programs to digitize fully the required map inputs, interpolate DEM, derive analytical surfaces like slope, aspect, and other terrain measures, perform distance operations, and produce predictive model surfaces printed on mylar overlays for a 150 km^2 region (Kvamme, 1983); a year later I had undertaken the same for a 1000 km^2 region (Kvamme, 1984). Hasenstab (1983b) followed independently with similar results along the Passaic River, New Jersey, using a GIS package written entirely by himself.

During the second half of the 1980s the topic of predictive locational modelling became quite popular (Kohler and Parker, 1986; Judge and Sebastian, 1988). Recent advances in Canada by Dalla Bona (1989) and others are noteworthy, and in the last few years we have seen similar work in The Netherlands (Wansleeben, 1988; Brandt *et al.*, 1992; van Leusen, 1993).

1.2.7 Summary

From the foregoing it should be obvious that if you wanted to use GIS type of methods in the early 1980s you often had to write the programs yourself. Good and reliable software was generally not available at the time. We also see at this early date a tripartite division of archaeological interests in GIS:

1. as an extender or facilitator of spatial analysis;
2. as a mapping system for regional spatial databases; and
3. as the primary tool for predictive models of archaeological location.

These categories continue to be important today.

An outcome of this initial exploratory period was a symposium, co-organized by myself and Hasenstab, at the 1985 annual meeting of the Society for American Archaeology in Denver. It was quaintly entitled 'Computer-based Geographic Informational Systems: A Tool of the Future for Solving Problems of the Past', and probably represents the world's first organized meeting on the linkage between GIS and archaeology. We had to stretch it a bit, but we managed to pull together ten papers in this session!

1.3 Growth and development

Despite the rapid and dynamic growth of the GIS industry in North America during the middle 1980s, archaeology's adoption of the technology was very slow. One explanation may lie in the relatively low number of computer archaeologists and the size of North America, which made direct interaction and communication difficult (there was no e-mail). Regular computer working groups or conferences, like the Computer Applications in Archaeology (CAA) conference which has promoted computers so well in Britain and recently Europe, have never existed. The American intellectual climate of the 1980s, which was strongly geared toward *a priori* model specification and 'deductive' reasoning, may also have been another factor. GIS tends to promote pattern-seeking in empirical data, very much an inductive process, which did not fit well in this atmosphere (this situation has fortunately changed in recent years). Finally, as we have seen, there was no readily available software; if you wanted to perform GIS kinds of operations you had to write your own code—a somewhat daunting task. Use of GIS therefore grew among isolated institutions and individuals.

During the second half of the 1980s things began to change with the availability of GIS software like ARC/INFO (Oliver and Schroeder, 1986), MOSS (Zulick, 1986), and GRASS (Westervelt *et al.*, 1986), but archaeology's access to these programs was not straightforward. Like everywhere else, archaeology was not strongly funded and most archaeologists and departments could not afford the Unix workstations required for these higher-end programs. Three developments were noteworthy at this time, however. First, certain federal agencies began to take notice of GIS as a resource management tool. Of particular importance here was the work of Dan Martin of the US Bureau of Land Management who established working archaeological databases in Colorado using MOSS (Martin and Garrett, 1985). Second, Fred Limp, then of the Arkansas Archaeological Survey, entered the GIS scene and was instrumental in the development and promotion of the GRASS program, one of the finest GIS available today. Indeed, owing largely to Limp's efforts the State of Arkansas probably represents the best example anywhere in the world of a fully automated, state-wide, GIS-driven archaeological database (Farley *et al.*, 1990). Third, a second symposium on GIS and archaeology was held at the annual meeting of the Society for American Archaeology in 1988 which was well-attended, and ultimately resulted in the publication of the book *Interpreting Space: GIS and Archaeology* (Allen *et al.*, 1990).

Interpreting Space marks a transition point into our current period of GIS awareness where most North American archaeologists have at least heard of GIS, if not used it. The ready availability of low-cost GIS software for personal computers, like IDRISI (Eastman, 1992), and the occurrence of a number of archaeological publications, conferences, and symposia on GIS, have done much to promote the topic. This parallels the situation in Europe where, for example, large numbers of GIS papers have been offered at the CAA conferences in recent years and have appeared in its published proceedings (Lock and Moffett, 1992; Andresen *et al.*, 1993). The present volume will no doubt promote even greater GIS awareness in Europe.

1.4 Data availability—data standards

The US government has made a concerted effort to provide digital GIS products at low cost for some time, in contrast to some European countries, such as England, where prices are very high (Waters, 1993). One would think that this would have promoted the use of GIS in archaeology. Yet, until recently, key digital data sets were available only for selected parts of the country. This was particularly true of high-quality digital elevation models (DEM), one of the more important data sets in many archaeological applications.

It should be noted that low-quality DEM have been available for the entire United States since the 1970s, but these were obtained from digitization of 1:250 000 scale maps and contained many errors. Nevertheless, initially, many archaeological studies did employ these data as a basis for locational analysis and modelling, but with results that were often questionable as the authors themselves admit (Marozas and Zack, 1987; Warren *et al.*, 1987). To illustrate, in a study of the effects of GIS algorithms and data sets on archaeological analyses and conclusions, I compared a computer-contoured plot of one of these DEM to 1:24 000 scale maps of the same region and found that

1. many small ridges and hills were absent;
2. minor drainages were missing;
3. large features were greatly smoothed; and

4. a major error existed in the form of an artificial, 40 m high cliff face which was erroneously placed across a well-known river valley in the digital product (Kvamme 1990b: 114).

From a locational analysis perspective, it is the types of features that were missing—the small drainages, ridges, and hilltops—that probably were most important to human settlement decisions. The solution to these data deficiencies for most archaeologists was to digitize map features in-house, and this brings up the topic of data standards.

The issue of data standards is not a trivial one. At a recent meeting of state cultural resource database managers at which 18 western states (of the USA) were represented, it was found that nearly every state had developed a unique computer database with different fields, formats, and data structures. Many of these systems appeared to be highly incompatible. As each of these states begins to adopt GIS it is clear that this variability will only be exacerbated. Unfortunately, we see a similar situation in Europe. A recent volume on European sites and monuments records (Larsen, 1992) testifies to the diversity of standards and approaches. Clearly an archaeological data standard would reduce many problems in the future, particularly in the areas of data sharing and duplication of effort, but such a goal is difficult, if not impossible, to achieve, particularly across linguistic and cultural barriers.

1.5 Regional spatial databases

In spite of the previously described problems, the bulk of archaeology's GIS future will lie in the creation of large regional or national databases of sites and monuments records for heritage management. This is a natural evolution from the present paper record formats and aspatial computer databases. Using GIS, archaeologists, managers, planners, and various government decision makers will be able to press a few keys and obtain instantaneous maps of known Palaeolithic sites, Iron Age hillforts, or Roman villas, for example. These data could, of course, be overlain on a backdrop of rivers, topography, geology, and other conditions. This capability will provide a tremendous tool to researchers as well. One excellent example already mentioned is the state-wide database of Arkansas (Farley *et al.*, 1990). In Europe, a similarly fine database is that of the ARCHIS project of The Netherlands (Roorda and Wiemer, 1992).

From my point of view, the GIS future will hopefully offer more than this in archaeology. Databases are interesting and useful, but they are an end in themselves. GIS also offers great potential in the analytical domain, an area which has barely been explored.

1.6 Problems and pitfalls

In a rather insightful remark at the recent 'GIS and the Advancement of Archaeological Method and Theory' conference (held at Southern Illinois University, Carbondale, USA, in March 1993), Stephan Shennan noted that GIS has reduced archaeological research and problem solving to the making of 'pretty pictures'. In other words, in place of the formal methods of analysis and inference that archaeologists previously employed, the goal of many GIS-based studies now seems to be merely the portrayal of results in the form of a well-designed computer graphic. Admittedly, the visualization of patterns in data is an extremely useful tool; indeed, it forms a fundamental basis of the exploratory data analysis (EDA) school of statistics (Tukey, 1977). A well-designed graphic can also more effectively communicate results. But this recent turn of events also presents a potential danger to archaeology because the human eye is easily fooled. Graphics can be constructed to highlight or emphasize certain patterns and tendencies, and down-play others, for example. We can assume that with the advanced graphics of GIS, with three-dimensional displays and phosphorescent colours numbering in the millions, this situation can be made even worse. There are already enough examples of GIS studies where one must question whether there is any substance behind the stunning visual effects.

The arguments of Lock and Harris (1992), and others, for a dual approach to archaeological spatial investigations, one that incorporates both visualization and formal statistical analysis, is germane at this point. Simple statistics cannot convey the essence of spatial pattern in the same way that an effective graphic can. At the same time, statistical tests can inform us of the existence of pattern when it is difficult or impossible to visualize, and even if we can see pattern we may wish to obtain objective measures of its strength. Both approaches to spatial analysis complement each other, they go hand-in-hand, and one should not be undertaken without the other.

Part of the problem in this debate is that most extant GIS do not offer much in the way of quantitative spatial analysis, so even with the best of intentions it is difficult to undertake. In fact, current GIS design seems to be increasingly geared toward graphical output and the making of inferences from pictures. After all, manufacturers of GIS software are market-driven and business interests, municipalities, and governmental agencies who form the bulk of the market want better graphics (and care little about spatial statistics). Consequently, the GIS industry will create software that meets market needs, of which archaeology represents a small and insignificant part.

A related problem is that archaeological research into certain topical areas might require specialized GIS procedures or algorithms that may not exist in commercial packages. For example, in the study of hunter-gatherer locational behaviour special measures of terrain form or environmental diversity might be required that are difficult or impossible to implement with the tools available in common GIS packages (Kvamme, 1992). This sort of limitation brings up the possibility that commercial GIS software might in the future actually structure or determine the nature of research, owing to limits inherent in their design and functionality.

The solution to the foregoing problems is to maintain a core of high technical expertise within the discipline. Archaeology has traditionally been well-off in this regard with a good core of programmers and even a reasonable number of archaeological software offerings (e.g. Blankholm, 1989; Wheatley, 1991; Hietala, 1993). University Departments of Archaeology should strive to maintain and promote these interests in the decades to come.

1.7 The future of GIS

It is safe to say that the software we use will ultimately improve, with increased functionality, speed, ease of use, and better user-interfaces. We may also speculate that GIS will soon become as common as word processors or spreadsheets, although most likely in the form of electronic atlases that offer a limited suite of GIS operations. One big issue of the next decade, already achieving some interest is the accuracy of spatial databases (Goodchild and Gopal, 1989), such as national data sets. We want to be certain that if we query an elevation or soil type at some locus that the result is close to reality!

Artificial intelligence (AI) is also certain to play an increased role in at least two areas. It will be used as a guide or aid to assist the user in the selection of options in spatial analysis or modelling operations, for example. At any point, the system could present the user various alternatives and warn of consequences that could arise with each. At a more fundamental level AI can act internally, behind the scenes, to select the most appropriate alogrithms when particular circumstances arise. An example here lies in the surface interpolation problem. A vast number of algorithms exist (Lan, 1983), each offering advantages and disadvantages depending on the landform context (e.g. plains-like regions vs ridges and valleys). During the interpolation process the most appropriate algorithm could be called, depending on the terrain condition sensed by the system.

Turning to archaeology, we have seen the use of GIS confined principally to spatial analysis, regional databases and predictive modelling. There are many more domains which could potentially be addressed that cry for attention. Let me hazard to mention a few.

1.7.1 Three-dimensional GIS

Nearly all GIS applications to date in archaeology have been two-dimensional (two positional coordinates for each data value). For example, most studies at the regional level have focused on the distribution of sites with respect to two-dimensional map features like terrain, soils, and vegetation. The few applications at the within-site level, noted previously, have concentrated on artifacts on occupation floors, also two-dimensional. The adoption of true, three-dimensional GIS—systems where each data value is linked with three positional coordinates (Raper, 1989)—will offer a means to better understand site structure, occupational history, and the interrelationships between stratigraphic units, for example.

1.7.2 GPS and GIS

Global positioning systems (GPS) represent a rapidly evolving technology that offers archaeologists a means to obtain accurate positional information quickly and inexpensively (Spennemann, 1992). By itself GPS will cause a revolution in the way archaeologists collect field data, but the spatial information can also be downloaded to GIS providing an excellent means for spatial data capture. Surveyor-quality, differential GPS can potentially be used to map artifacts on floors; linked with a field computer, maps could also be produced in real time, with up-to-date excavation information.

1.7.3 GIS-based simulation

The potential in this area is exciting. With GIS it is possible to introduce the actual physical characteristics of a region into computer simulations. We no longer need to make such simplifying assumptions as 'a featureless and level two-dimensional plane'. The actual landform, soils, vegetation, hydrology, and other features of a region can be incorporated for the simulation background, allowing greater realism and undoubtedly better insights.

1.7.4 Palaeo-environmental modelling

Regional applications of GIS have thus far utilized environmental data taken from modern maps in contemporary form. Often, particularly in the European theatre, there are vast differences between past and present environments. Archaeology needs to start looking at reconstructing palaeo-environmental data at the regional level, using GIS modelling tools and what is known about various geological and other processes, to retrodict into the past ancient environmental conditions and characteristics.

1.7.5 Comprehensive prospection data sets

For quite some time archaeologists have utilized computer-processed geophysical data obtained from multiple sensing devices, such as resistance, conductivity, and magnetic information, to understand site structure (Scollar *et al.*, 1990). Since GIS is designed to manage and combine multiple sources of spatial information, it is ideal for handling geophysical data sets. But why not go a step further? Incorporate into the same database aerial photography (e.g. crop marks), geochemical results, surface artifact distributions, and microtopography (a micro-elevation model of surface relief features). Such a comprehensive database, by including information from every possible dimension, would offer the most comprehensive means to assess site organization in a non-destructive way. GIS-based visualization, by allowing the simultaneous examination of multiple data sources through overlays, colour composites, or three-dimensional views, would offer great potential insight into patterns and associations which might exist, and provide an excellent site planning and management tool.

1.8 Conclusions

It is evident that there are many potential paths to follow in an exciting and fruitful future that links GIS and archaeology together. I have summarized some of the problems and issues that I see, many of which have historical roots. Foremost among these is the maintenance of a small core in the archaeological profession with expertise in computer technology, for the future of our science will continue to evolve in this direction. I think Europe is well-off here. Real computer archaeologists who can write their own code, even 'hackers', are not uncommon. With this tradition in computing, best exemplified by the largely European Commission IV of the International Union of Pre- and Protohistoric Sciences, and particularly the Computer Applications in Archaeology Conference, European archaeology is in good shape at this stage and, I expect, will take a commanding role in archaeology's GIS future.

Acknowledgements

Gary Lock provided a number of useful suggestions in the formulation of this paper, for which I am grateful. I wish to thank Gary Lock and Zoran Stančič for inviting me, out of many possible North American candidates, to contribute to this volume.

References

Allen, K. M. S., Green, S. W. and Zubrow, E. B. W. (Eds), 1990, *Interpreting Space: GIS and Archaeology*, London: Taylor & Francis.

Andresen, J., Madsen, T. and Scollar, I. (Eds), 1993, *Computing the Past: Computer Applications and Quantitative Methods in Archaeology, CAA92*, Aarhus: Aarhus University Press.

Arnold III, J. B., 1979, Archaeological applications of computer drawn contour and three-dimensional perspective plots, in Upham, S. (Ed.), *Computer Graphics in Archaeology*, pp. 1–15, Anthropological Research Papers No. 15, Tempe: Arizona State University.

Blankholm, H. P., 1989, ARCOSPACE: a package for spatial analysis of archaeological data, *Archaeological Computing Newsletter*, **19**, 3.

Bove, F. J., 1981, Trend-surface analysis and the Lowland Classic Maya collapse, *American Antiquity*, **46**, 93–112.

Bradley, R., 1970, The excavation of a Beaker settlement at Belle Tout, East Sussex, England, *Proceedings of the Prehistoric Society*, **36**, 312–79.

Brandt, R., Groenewoudt, B. J. and Kvamme, K. L., 1992, An experiment in archaeological site location: modelling in The Netherlands using GIS techniques, *World Archaeology*, **24**, 268–82.

Brown, P. E. and Rubin, B. H., 1982, Patterns of desert resource use: an integrated approach to settlement analysis, in Brown, P. E. and Stone, C. L. (Eds), *Granite Reef: A Study in Desert Archaeology*, pp. 267–305, Anthropological Research Papers No. 28, Tempe: Arizona State University.

Chadwick, A. J., 1978, A computer simulation of Mycenaean settlement, in Hodder, I. (Ed.) *Simulation Studies in Archaeology*, pp. 47–58, Cambridge: Cambridge University Press.

Chadwick, A. J., 1979, Settlement simulation, in Renfrew, C. and Cooke, D. (Eds) *Transformations: Mathematical Approaches to Culture Change*, pp. 237–55, New York: Academic Press.

Clark, G. A., 1979, Spatial association at Liencres, an early Holocene open site on the Santander coast, north-central Spain, in Upham, S. (Ed.), *Computer Graphics in Archaeology*, pp. 121–43, Anthropological Research Papers No. 15, Tempe: Arizona State University.

Copp, S. A., 1977, A quick plotting program for archaeological data, *Newsletter of Computer Archaeology*, **13**, 17–25.

Dalla Bona, L., 1989, 'Visual possibility statements: a preliminary study using geographic information systems for predictive modeling', presentation at the Canadian Archaeological Association meeting, New Brunswick.

Eastman, J. R., 1992, IDRISI (software documentation, version 4.0), Worcester, Massachusetts: Graduate School of Geography, Clark University.

Effland, R. W., 1979, Statistical distribution cartography and computer graphics, in Upham, S. (Ed.) *Computer Graphics in Archaeology*, pp. 17–29, Anthropological Research Papers No. 15, Tempe: Arizona State University.

Farley, J. A., Limp. W. F. and Lockhart, J., 1990, The archaeologist's workbench: integrating GIS, remote sensing, EDA and database management, in Allen, K. M. S., Green, S. W. and Zubrow, E. B. W. (Eds), *Interpreting Space: GIS and Archaeology*, pp. 141–64, London: Taylor & Francis.

Feder, K. L., 1979, Geographic patterning of tool types as elicited by trend surface analysis, in Upham, S. (Ed.), *Computer Graphics in Archaeology*, pp. 95–102, Anthropological Research Papers No. 15, Tempe: Arizona State University.

Flude, K., George, S. and Roskams, S., 1982, Uses of an archaeological database with particular reference to computer graphics and the writing-up process, in *Computer Applications in Archaeology 1981*, 64–75, London: Institute of Archaeology, University of London.

Goodchild, M. F. and Gopal, S., 1989, *Accuracy of Spatial Databases*, London: Taylor & Francis.

Green, D. F. and Stewart, J. B., 1983, Computer generated research aids using the forest service DTIS II system, *Advances in Computer Archaeology*, **1**, 4–25.

Harris, T., 1986, Geographic information system design for archaeological site information retrieval, in *Computer Applications in Archaeology 1986*, pp. 148–61, Birmingham: Centre for Computing and Computer Science, University of Birmingham.

Harris, T., 1988, Digital terrain modelling and three-dimensional surface graphics for landscape and site analysis in archaeology, in Ruggles, C. L. N. and Rahtz, S. P. Q. (Eds), *Computer Applications and Quantitative Methods in Archaeology 1987*, pp. 161–70, British Archaeological Reports International Series 393, Oxford: Tempus Reparatum.

Hasenstab, R. J., 1983a, 'The application of geographic information systems to the analysis of archaeological site distributions', presentation at the Society for American Archaeology meeting, Pittsburgh.

Hasenstab, R. J., 1983b, *A Preliminary Cultural Resource Sensitivity Analysis for the Proposed Flood Control Facilities Construction in the Passaic River Basin of New Jersey*, Report submitted to the Passaic River Basin Special Studies Branch, Department of the Army, New York District Army Corps of Engineers, Marietta, Georgia: Soil Systems, Inc.

Hietala, H. J., 1993, Tools for quantitative archaeology, *SAA Bulletin*, **11**, 14–15.

Hietala, H. J. and Larson Jr, P. A., 1979, SYMAP analyses in archaeology: intrasite assumptions and a comparison with TREND analysis, *Norwegian Archaeological Review*, **12**, 57–64.

Hivernel, F. and Hodder, I., 1984, Analysis of artifact distributions at Ngenyn (Kenya): depositional and postdepositional effects, in Hietala, H. J. (Ed.), *Intrasite Spatial Analysis in Archaeology*, pp. 97–115, Cambridge: Cambridge University Press.

Hodder, I. and Orton, C., 1976, *Spatial Analysis in Archaeology*, Cambridge: Cambridge University Press.

Jermann, J. V. and Dunnell, R. C., 1979, Some limitations on isopleth mapping in archaeology, in Upham, S. (Ed.), *Computer Graphics in Archaeology*, pp. 31–60, Anthropological Research Papers No. 15, Tempe: Arizona State University.

Judge, W. J. and Sebastian, L. (Eds), 1988, *Quantifying the Present and Predicting the*

Past: Theory, Method, and Application of Archaeological Predictive Modeling, US Bureau of Land Management, Department of the Interior, Washington, DC: US Government Printing Office.

Kohler, T. A. and Parker, S. C., 1986, Predictive models for archaeological resource location, in Schiffer, M. B. (Ed.), *Advances in Archaeological Method and Theory*, Vol. 9, pp. 397–452, New York: Academic Press.

Kvamme, K. L., 1980, Predictive model of site location in the Glenwood Springs Resource Area, in *A Class II Cultural Resource Inventory of the Bureau of Land Management's Glenwood Springs Resource Area*. Report submitted to US Bureau of Land Management, Grand Junction District, Colorado, Montrose, Colorado: Nickens and Associates.

Kvamme, K. L., 1983, Computer processing techniques for regional modeling of archaeological site locations, *Advances in Computer Archaeology*, **1**, 26–52.

Kvamme, K. L., 1984, Models of prehistoric site location near Piñon Canyon, Colorado, in Condie, C. J. (Ed.), Papers of the Philmont Conference on the Archaeology of Northeastern New Mexico, pp. 349–370, *Proceedings* **6**, Albuquerque: New Mexico Archaeological Council.

Kvamme, K. L., 1990a, The fundamental principles and practice of predictive modelling, in Voorrips, A. (Ed.), *Mathematics and Information Science in Archaeology: A Flexible Framework*, pp. 257–295, Studies in Modern Archaeology, Vol. 3, Bonn: Holos–Verlag.

Kvamme, K. L., 1990b, GIS algorithms and their effects on regional archaeological analysis, in Allen, K. M. S., Green, S. W. and Zubrow, E. B. W. (Eds), *Interpreting Space: GIS and Archaeology*, pp. 112–26, London: Taylor & Francis.

Kvamme, K. L., 1992, Terrain form analysis of archaeological location through geographic information systems, in Lock, G. and Moffett, J. (Eds), *Computer Applications and Quantitative Methods in Archaeology 1991*, pp. 127–36, British Archaeological Reports International Series S577, Oxford: Tempus Reparatum.

Laboratory for Computer Graphics and Spatial Analysis, 1975, *SYMAP User's Reference Manual*, Edition 5.0, Cambridge, Massachusetts: Laboratory for Computer Graphics and Spatial Analysis, Harvard University.

Lan, N. S. N., 1983, Spatial interpolation methods: a review, *The American Cartographer*, **10**, 129–49.

Larsen, C. U. (Ed.), 1992, *Sites and Monuments: National Archaeological Records*, Copenhagen: National Museum of Denmark.

Larson, P., 1975, Trend analysis in archaeology: a preliminary study of intrasite patterning, *Norwegian Archaeological Review*, **8**, 75–80.

Lock, G. and Harris, T., 1992, Visualizing spatial data: the importance of geographic information systems, in Reilly, R. and Rahtz, S. (Eds), *Archaeology and the Information Age: A Global Perspective*, pp. 81–96, London: Routledge.

Lock, G. and Moffett, J. (Eds), 1992, *Computer Applications and Quantitative Methods in Archaeology 1991*, British Archaeological Reports International Series S577, Oxford: Tempus Reparatum.

Marozas, B. A. and Zack, J. A., 1987, Geographic information systems applications to archaeological site location studies, in *GIS '87: Second International Conference, Exhibits and Workshops on Geographic Information Systems*, pp. 628–35, Falls Church, Virginia: American Society for Photogrammetry and Remote Sensing.

Martin, D. W. and Garrett, M., 1985, 'The utility of MOSS to cultural resource

management', presentation at the Society for American Archaeology meeting, Denver.
McKay, D., 1984, Landscape scaling using standard surveying data: an implementation using the BBC model 'B' micro, in *Computer Applications in Archaeology 1984*, pp. 170–7, Birmingham: Centre for Computing and Computer Science, University of Birmingham.
Oliver, S. G. and Schroeder, E. K., 1986, Archaeological applications of ARC/INFO: a geographic information system, in *Workshop Proceedings of Microcomputers in Archaeology, New Orleans*, Washington, DC: Society for American Archaeology.
Pomerantz, H. J., 1981, ACRONYM: a geographic information system for archaeology, *Southwestern Lore*, **47**, 7–11.
Raper, J., 1989, *Three-dimensional Applications in Geographic Information Systems*, London: Taylor & Francis.
Redman, C. L. and Watson, P. J., 1970, Systematic, intensive surface collection, *American Antiquity*, **35**, 279–91.
Roorda, I. M. and Wiemer, R., 1992, Towards a new archaeological information system in The Netherlands, in Lock, G. and Moffett, J. (Eds), *Computer Applications and Quantitative Methods in Archaeology 1991*, pp. 85–8, British Archaeological Reports International Series S577, Oxford: Tempus Reparatum.
Rubin, B. H., 1980, *Map Analysis and Processing System*, Tucson, Arizona: The Graphics Studio Company.
SAS Institute, 1978, *SAS Introductory Guide*, Cary, North Carolina: SAS Institute.
Scheitlin, T. E. and Clark, G. A., 1978, Three-dimensional representations of lithic categories at Liencres, *Newsletter of Computer Archaeology*, **13**, 1–13.
Scholtz, S. C., 1981, Location choice models in Sparta, in Lafferty III, R. H., Otinger, J. L., Scholtz, S. C., Limp, W. F., Watkins, B. and Jones, R. D. (Eds) *Settlement Predictions in Sparata: A Locational Analysis and Cultural Resource Assessment on the Uplands of Calhoun County, Arkansas*, pp. 207–22, Research Series No. 14, Fayetteville: Arkansas Archaeological Survey.
Scollar, I., 1966, Computer treatment of magnetic measurements from archaeological sites, *Archaeometry*, **9**, 61–71.
Scollar, I., 1974, Interactive processing of geophysical data from archaeological sites, in *Computer Applications in Archaeology 1974*, pp. 75–80, Birmingham: Centre for Computing and Computer Science, University of Birmingham.
Scollar, I., Tabbagh, A., Hesse, A. and Herzog, I., 1990, *Archaeological Prospecting and Remote Sensing*, Cambridge: Cambridge University Press.
Spennemann, D. H. R., 1992, Archaeological site location using global positioning systems, *Journal of Field Archaeology*, **19**, 271–4.
Todd, S., Colley, S. and Campling, N., 1985, Relational databases and three-dimensional graphics, in *Computer Applications in Archaeology 1985*, pp. 155, London: Institute of Archaeology, University of London.
Tomlin, C. D., 1991, Cartographic modelling, in Macquire, D. J., Goodchild, M. F. and Rhind, D. W. (eds), *Geographical Information Systems: Principles and Applications*, Vol. 1, pp. 361–74, London: Longman Scientific and Technical.
Tukey, J. W., 1977, *Exploratory Data Analysis*, Reading, Massachusetts: Addison-Wesley.
van Leusen, P. M., 1993, Cartographic modelling in a cell-based GIS, in Andresen, J., Madsen, T. and Scollar, I. (Eds), *Computing the Past: Computer Applications*

and Quantitative Methods in Archaeology, CAA92, pp. 105–23, Aarhus, Denmark: Aarhus University Press.

Wansleeben, M., 1988, Applications of geographical information systems in archaeological research, in Rahtz, S. P. Q. (Ed.), *Computer and Quantitative Methods in Archaeology*, 1988, Vol. 2, pp. 435–51, British Archaeological Reports International Series 446, Oxford: Tempus Reparatum.

Warren, R. E., Oliver, S. G., Ferguson, J. A. and Druhot, R. E., 1987, *A Predictive Model of Archaeological Site Location in the Western Shawnee National Forest*, Report submitted to Shawnee National Forest, Springfield, Illinois: Illinois State Museum.

Waters, N., 1993, The price is wrong: charging for GIS data and software, *GIS World*, **6**, 68.

Westervelt, J., Goran, W. and Shapiro, M., 1986, Developments and applications of GRASS: the geographic resources and analysis support system, in Opitz, B. K. (Ed.), *Geographic Information Systems in Government*, Vol. 2, pp. 605–24, Hampton, Virginia: A. Deepak Publishing.

Whallon, R., 1974, Spatial analysis of occupation floors II: the application of nearest neighbor analysis, *American Antiquity*, **39**, 16–34.

Whallon, R., 1984, Unconstrained clustering for the analysis of spatial distributions in archaeology, in Hietala, H. J. (Ed.), *Intrasite Spatial Analysis in Archaeology*, pp. 242–77, Cambridge: Cambridge University Press.

Wheatley, D., 1991, SyGraf: resource based teaching with graphics, in Lockyear, K. and Rahtz, S. (Eds), *Computer and Quantitative Methods in Archaeology, 1990*, pp. 9–14, British Archaeological Reports International Series 565, Oxford: Tempus Reparatum.

Zimmerman, L. J., 1977, *Prehistoric Locational Behavior: A Computer Simulation*. Report No. 10, Office of the State Archaeologist, Iowa City: University of Iowa Press.

Zulick, C. A., 1986, Application of a geographic information system to the Bureau of Land Management's resource management planning process, in Opitz, B. K. (Ed.), *Geographic Information Systems in Government*, Vol. 1, pp. 309–28, Hampton, Virginia: A. Deepak Publishing.

2

The use of GIS for archaeological resource management in France: the SCALA project, with a case study in Picardie

D. Guillot and G. Leroy

This chapter will show why and how the French Ministry of Culture decided to equip its regional offices with GIS, and then with an internally developed user interface named SCALA, especially designed to use the data in the National Archaeological Record.

2.1 Background

The Heritage Division (Direction du Patrimoine) of the French Ministry of Culture is responsible for the inventory, the protection, the conservation and for promoting knowledge of the cultural heritage, and therefore for the National Archaeological Record. Called *Carte archéologique de la France*, such a record was launched in 1975 (Dorion, 1981). At that time, an official report recommended that each regional archaeological office should hold a record of all known archaeological sites and plot them on a map (Soustelle, 1975). Computerization first commenced in 1978, and the full locational component of archaeological sites was included in the database, so as to allow the production of distribution maps (Magnan, 1989).

Initially two attempts were made at designing mapping systems: the first in the late 1970s with a home-written program, the second in the late 1980s using AutoCAD with dBaseIII. Both systems were unsatisfactory and never went beyond the experimental stage.

In 1990, in view of the threatened destruction of archaeological resources due to rapid urban and rural development and of the growing cost of archaeological surveys in advance of rescue excavation, the Ministry of Culture set the enhancement of the National Archaeological Record as one of its priorities. The government reacted favourably to its request for financial support, and granted the project FF17 million a year, for appointments for five years. A new computer application, called DRACAR, was produced. Running with Unix and Oracle on a central mainframe, DRACAR was not designed to answer specific research questions, but rather to provide a rapid tool for archaeological resource management, and to facilitate the exchange of information between the regional archaeological offices and local development agencies (Guillot, 1991, 1992). Therefore it had to be provided with mapping facilities.

2.2 Choosing the GIS technology

At this time, it was obvious that GIS technology was the most suitable answer to the needs of archaeological resource management. In order to select convenient software, we established a list of functional needs and technical requirements.

Functional needs

- Database management
- Spatial operations: buffering, overlaying, intersecting, etc.
- Statistics
- Digitizing abilities
- Use of various coordinate systems and projections—two coordinate systems are at present used in France (the 'Lambert zone' system and the 'Lambert II étendu' system) and we hope that the UTM system will soon be adopted, which will facilitate European data exchange.

Technical requirements of the software

- Running on MS DOS, which is the operating system chosen by the French Civil service
- Running on Windows, to allow a user-friendly environment
- Well serviced in France, with good training facilities

The software PC Arc/Info was chosen, meeting most of these needs and technical requirements, and being also used by a number of government agencies and local planning authorities running a GIS database, thereby facilitating the sharing of regional geographic information. Its two main drawbacks were its rather bad statistical abilities, and the lack of a Windows version, which was announced, but is not yet available. Until it is, we will use Arc/Info's macro language SML, to customize the software. Thus the SCALA (système de cartographie appliqué à l'archéologie) project was launched in 1991 and became operative during the Spring of 1993. The cost of the project was FF3.5 million, including hardware, software and some digital maps from the Institut Géographique National (IGN). Customization of the software, training and user manuals were achieved by a computer specialist of the ministry and Dominique Guillot.

2.3 Principles of SCALA

The aim was to provide a tool to be used by numerous people (about 50), most of them archaeologists, but not necessary skilled in operating sophisticated software. The system should also be capable of producing several sorts of maps using a large number of site types. The area actually covered by the project is about 550 000 km^2, i.e. the whole of France, and the number of sites likely to be mapped is about 180 000 and increasing every day.

The user has a workstation consisting of a personal computer, complete with a digitizer and a plotter. The archaeological data are imported as an ASCII file from the centralized National Archaeological Record's database managed by the DRACAR system. Of course we met with difficulties in integrating the various spatial units which define archaeological sites, such as linear features and polygonal areas. In the DRACAR system, a site can be spatially described in different ways:

1. as a point, i.e. one pair of *x* and *y* coordinates, which is usually an estimation of the site centroid;
2. as a circle, with the two coordinates of the centre and a radius length;
3. as a line, with at least two points;
4. as a polygon, giving the possibility to described the shape of any archaeological feature with a series of grid references, without any restriction in their number.

The geographical data can be either bought or digitized by the user. The National Geographic Institute has currently undertaken the digital encoding of its maps, but because of the high cost, we could only afford the administrative boundaries. The system works with multi-choice menus, based on different Arc/Info subroutines. It allows connection to the DRACAR system, for selection and capture of the archaeological data, the digitizing of geographic coverages, database management, spatial operations, and map composition.

The database can be interrogated for archaeological sites according to a combination of single or multiple attributes based on boolean logic. For instance, a basic request is which archaeological sites exist within a certain distance of a proposed development. The site can be represented either by a symbol applied regardless of its actual size, or by an area (circle or polygon) corresponding to the real coordinates input to the database. The number or the density of sites in an administrative area can be calculated and mapped.

2.4 Some problems

Having worked with the system for six months, we are now aware of a number of problems.

1. The high cost of both hardware and software due to the number of offices to be equipped led us to buy rather low quality equipment; the computer processors are 386 SX, the digitizing tablets and the plotters are only A3, which are inadequate.
2. The lack or the high prices of geographic data in digital form—geographic data are still rather rare in France—has meant we have to acquire the necessary spatial data ourselves. We only hope that the expected growth of GIS in France will lead to the future availability of vast quantities of geographic data in digital form, to profitable data exchange or at least to the reduction of the cost.
3. It has proved difficult to manage within a short time, the training of about 40 people, most of them unfamiliar with GIS technology, and it is also necessary to provide further training.
4. The completeness of the data contained in the National Archaeological Record is an important concern. To this day, the computerization of the file cards is still incomplete, and many site locations are wrong. Another problem is that some regional studies estimate that only 20 percent of the archaeological features of the oldest periods would be detected by surveys.

5. Other problems are more concerned with the spatial definition of sites, and the way Arc/Info can handle them as a vector system. For example the overlaying or the inclusion of sites described as polygonal areas is rather problematic.

It is true that because of the complexity of GIS concepts, and the lack of further training, most people will use SCALA in a very simple way. We are aware that any possibility of modelling the computerized database or establishing a predictive capability based upon multivariate relationships, for example, is still curtailed by the lack of the full archaeological record in the database, even if first results are already being obtained (see the case study below). The potential that such systems possess to handle spatial information must change the way in which a conventional archaeological database such as ours has been used, and we look forward to a fully integrated spatial database management and analysis environment which would enable DRACAR and SCALA to go beyond the present limited inventorizing stage and provide archaeology with a highly sophisticated information system.

2.5 *Picardie: a cartographic project under study*

This specific study will be the first test of the SCALA system on a rather big area and with real data, trying to use the GIS technology in a predictive way.

Picardie is a region that consists of silt laden plateaux incised by valleys, composing the North of the Bassin Parisien (Figure 2.1). For a long time, this region was spared industrial growth and kept a mainly rural nature, but now it is entering a period of change, marked by many railway, motorway, and urban developments. The archaeological potential of this region is very high. This has been demonstrated mainly by aerial surveys (Agache and Bréart, 1975), and by the important discoveries of Palaeolithic sites, made over the last century, in alluvial formations of the river Somme (Tuffreau, 1987). Therefore archaeologists need to control the region with suitable tools, and GIS is one of the tools which can be used for this goal.

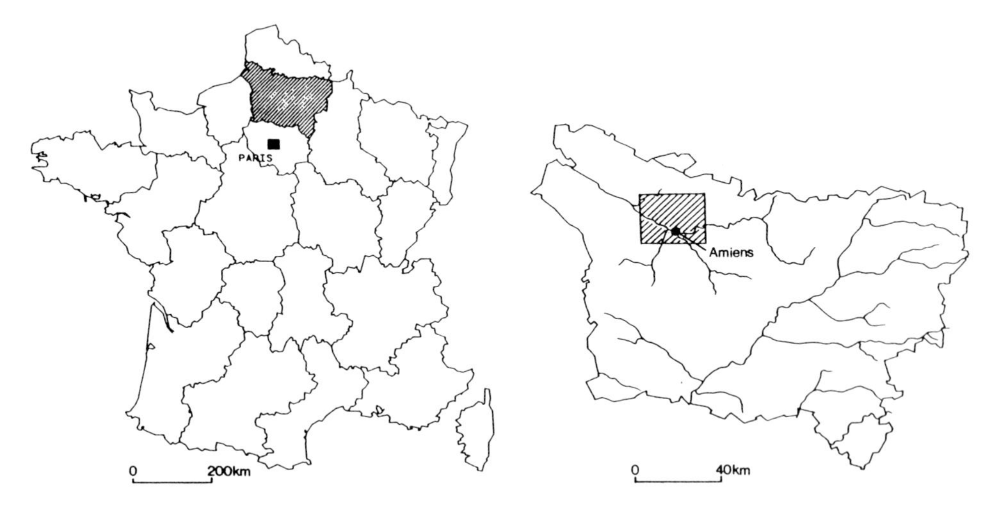

Figure 2.1 Location maps of the Picardie study area.

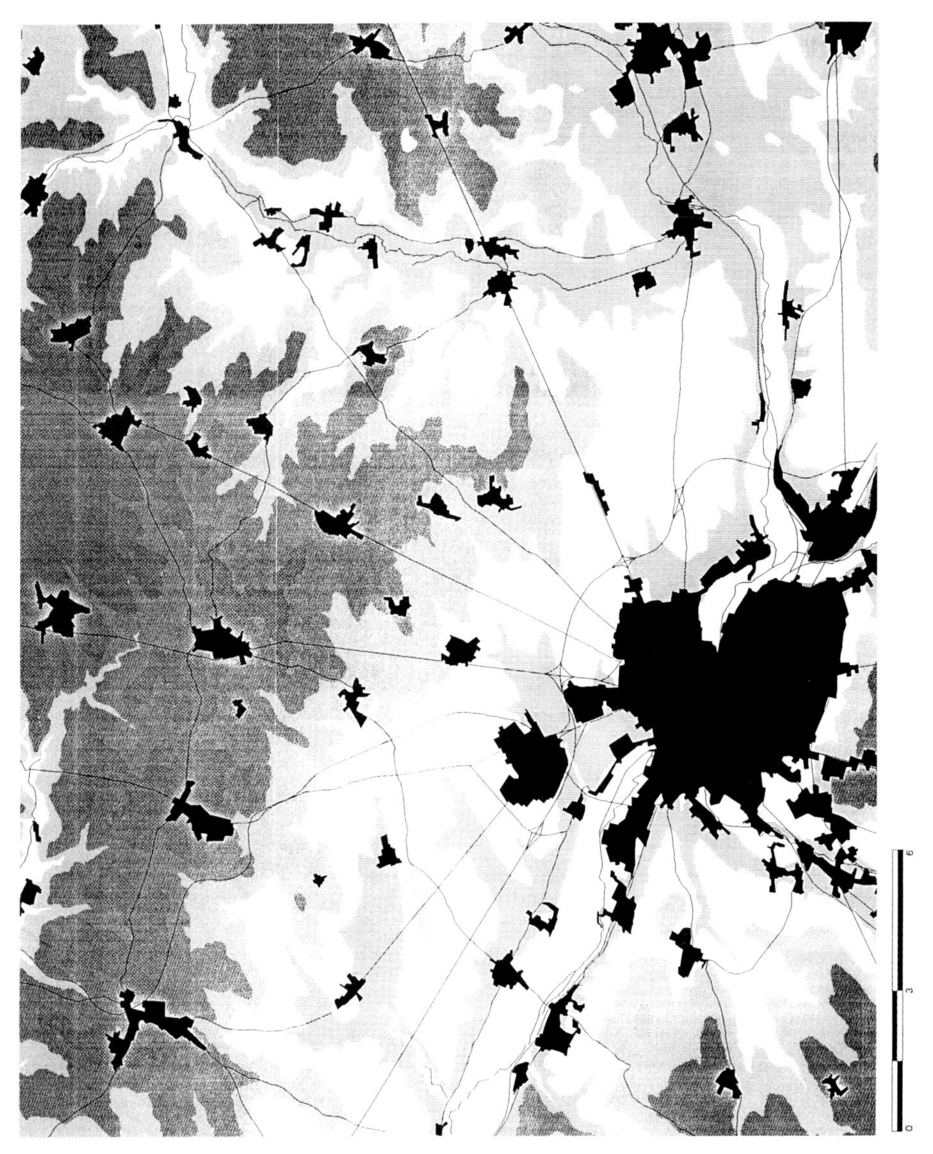

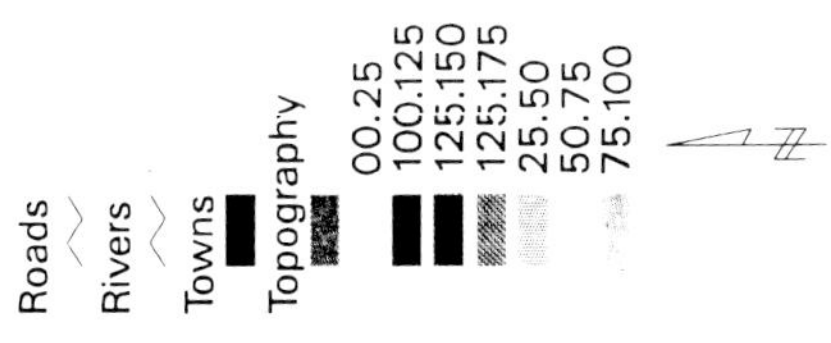

Figure 2.2 The study area as a digital map.

The regional staff of the National Archaeological Record are conducting an experimental study based on the use of GIS. The principle is to use the SCALA system provided by the Ministry of Culture, trying to take advantage of all its capacities. The aims are to implement a scientific investigation tool and to develop a mapping system for cultural resource management.

The area selected for the study is covered by an IGN 1:50000 map sheet. It represents about 524 km^2 around Amiens, the regional capital. The limitation to a relatively small area allows a rather good collection of geographical data. This area has also been well covered with surveys, and the archaeological data of more than 700 sites have all been registered in the DRACAR database. The geomorphological features are valleys, plateaux and slopes. The region is under threat from various development projects: motorways, the railway network, quarries and urban developments. The first step was the acquisition of geographical data corresponding to this area (Figure 2.2). The necessary digital coverages were: geology, hydrography, roads, topography and urban areas. Because of the high cost of digital data, most of the coverages had to be obtained by digitizing. The work was done on an A3 tablet, which is smaller than the 1:50000 maps, producing some problems due to the lack of accuracy of the data. The topographic coverage was scanned, vectorized and imported into Arc/Info using the data conversion subroutine. The method seems rather well adapted to complex coverages, and a new attempt will soon be made with the regional geological coverage.

Our needs and the general progression of geographical information in France led us to look for potential regional or national partners, to enable the sharing of the raw digital data. The Geological and Mining Research Office (Bureau de Recherches Géologiques et Minières) was the first to become interested in the project and placed some of its maps at our disposal. Other public utilities are currently considering or building a regional GIS database, and would be ready to exchange their data with the regional archaeological office.

2.6 Preliminary results

At this time, the analytical work is just beginning, although the first correlations of geographical and archaeological coverages allows us some early remarks. First, observations show how careful one must be with distribution maps, as shown in Figures 2.3(a) to 2.3(c). The protohistoric site coverage (symbols in Figure 2.3(a)) overlaid with the wind-blown silt and topographic coverages shows a clustered density of organization. The sites are gathered in an erosion zone (light grey), corresponding to a topographic depression drained by the river Somme.

The Gallo-Roman site coverage (symbols in Figure 2.3(b)) superimposed on the wind-blown silt coverage shows an inverted density compared to the former example. The highest densities are to be found in the loess-covered area, except for the Roman town of Amiens. Figure 2.3(c) shows the distribution of protohistoric and Gallo-Roman sites detected by aerial surveys, a method that is often used in this area, especially by Roger Agache (Agache and Bréart, 1975).

A first interpretation shows that a correlation could be established between settlement distribution and density and economic factors, e.g. distance to water or soil quality. This correlation, however, may be partly due to factors outside the

Figure 2.3.(a) The protohistoric sites are gathered in an erosion zone drained by the river Somme.

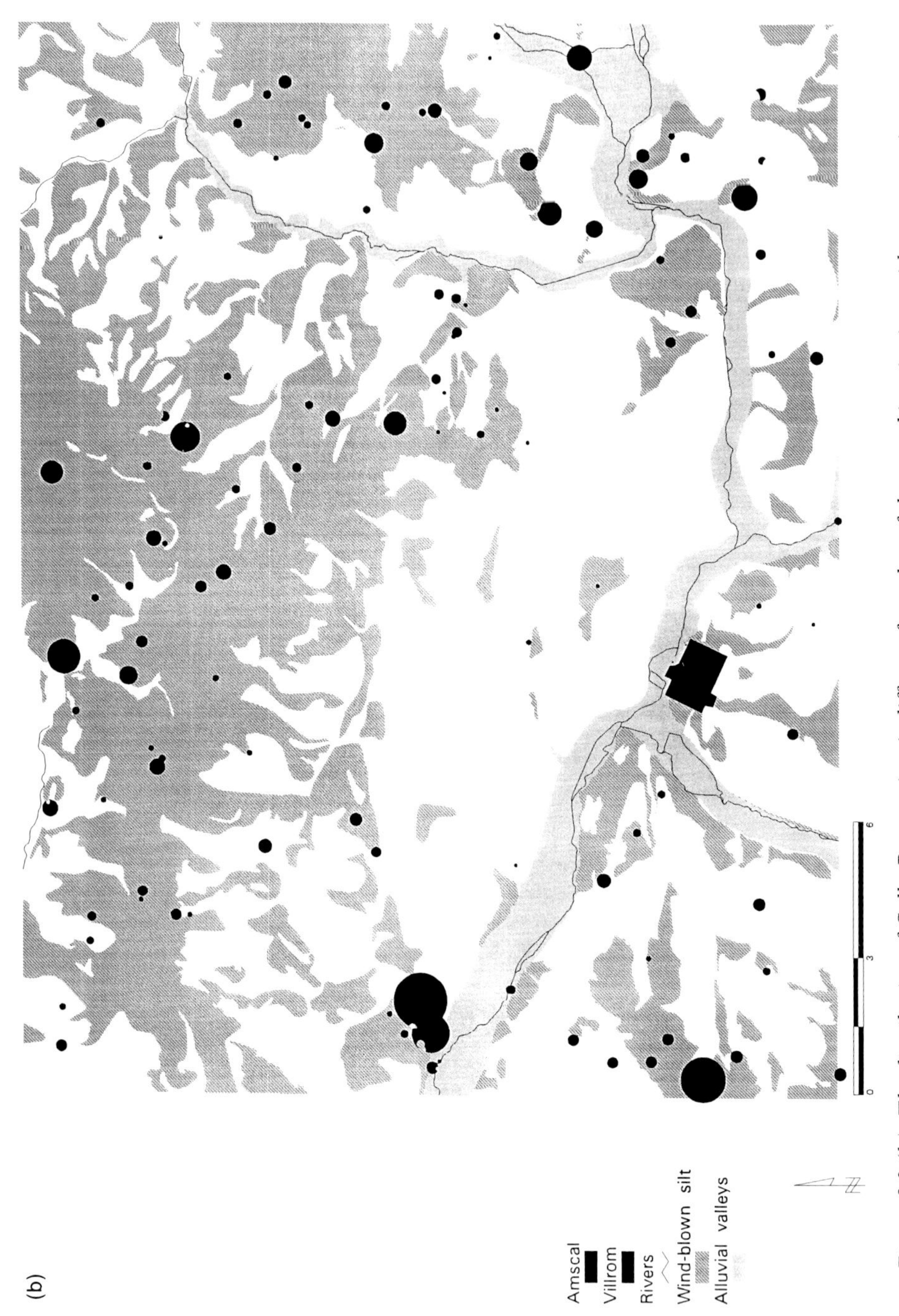

Figure 2.3.(b) The distribution of Gallo-Roman sites is different from that of the protohistoric sites, with concentrations on the loess-covered areas and around the Roman town of Amiens.

Figure 2.3.(c) Protohistoric and Gallo-Roman sites detected by aerial photography.

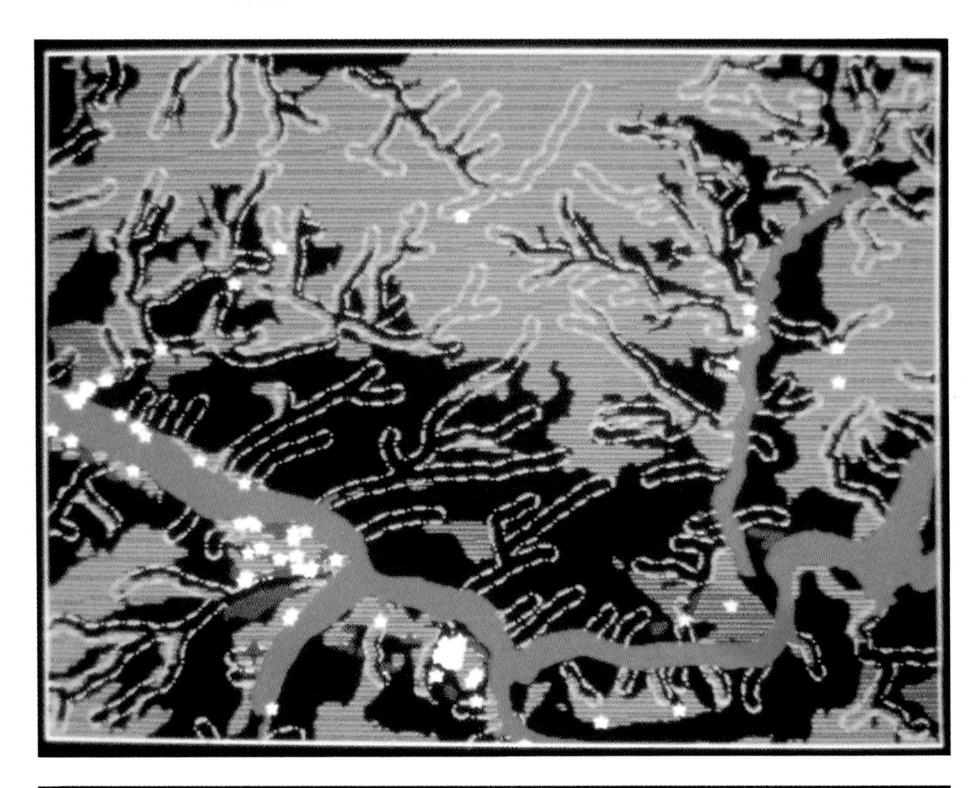

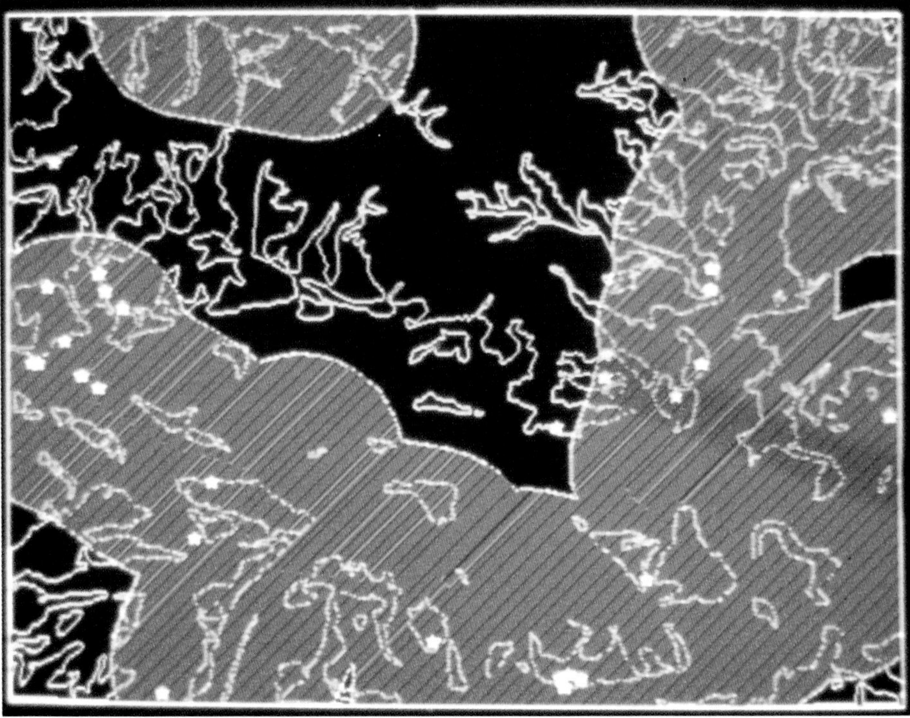

Plate 1. The Picardie study area showing Palaeolithic and Mesolithic site distributions (white), preserved alluvial banks (red), silt plateaux (green and 400 m buffers around valleys and dry valleys.

Plate 2. The distribution of Neolithic sites within the Picardie study area. The 3 km buffer zone around main water courses is shown in green.

model, such as surveying conditions. Gallo-Roman building foundations are especially visible in the loess area, and conversely the dark prehistoric pits and ditches are better revealed on chalky slopes. This first use of GIS indicates that the correlation between settlement density and landscape can be uncertain. This requires us to balance survey methods with the geographical data of the region.

Secondly, the relationship between prehistoric settlements and geographical data as shown with GIS allows for more optimism in the understanding of prehistoric site location. Palaeolithic and Mesolithic site distributions (white in Plate 1) clearly show very high densities in the preserved alluvial banks (red) and the bottom of the main dry valleys. Outside settlements are mostly located on the slopes of small valleys (green buffer, 400 m wide). Some Palaeolithic sites are situated on silt plateaux (green). The presence of Palaeolithic and Mesolithic sites is mostly due to the preservation of the Pleistocene and Holocene layers, for when the sediment cover is thick enough, sites may be preserved. Flood plains and alluvial terraces are especially sensitive areas, whereas the bottoms of main dry valleys covered with colluvium are less sensitive, followed by the secondary dry valleys and the silt plateaux. Thus we were able to provide sensitivity maps for the Palaeolithic and Mesolithic periods. Plate 2 shows the distribution of the registered Neolithic settlements associated with a buffer area of 3 km around water courses and alluvial valleys. Most sites are included in the buffer area, which fits with the observations of a research project about Neolithic settlements in the valley of the river Aisne (Plateaux, 1990). Most of the sites are also situated on plateau borders in the periphery of a wind-blown silt coverage zone. Then further investigation should allow us to produce sensitivity areas for the Neolithic period too.

Thirdly, GIS can be a tool of communication and data exchange between regional archaeological offices and developers. Figure 2.4 shows a first attempt in the study area to produce a development impact map. Data concerning the main development projects were collected and digitized, and a map was produced displaying all known sites, the sensitive areas defined above, and the boundaries of the ancient town of Amiens. The overlaying of these data should allow the monitoring

Figure 2.4 An early attempt to produce a development impact map for the study area showing archaeological sites, sensitive areas and development projects

of areas under threat from development projects, and the production of a risk map which could be provided to the main agencies responsible for regional development.

2.7 Conclusions

Using GIS we are now confronted with two types of limits, the first related to our archaeological knowledge and the second to the capabilities of the tool itself. Using GIS compels us to be more demanding about the accuracy of the archaeological data, but also about the quality of the software and hardware. In fact, at this early stage we cannot possibly anticipate every future use of GIS in managing regional archaeological resources. This preliminary study should allow us to prepare a spatial analytical method. Observations made by overlaying spatial data seem to demonstrate the short-range interest of GIS as a management and synthesis tool for the archaeological heritage. A next step should lead to a zonal partition of sensitivity for Palaeolithic, Mesolithic, and Neolithic periods, over a wider study area. This zoning will then be overlaid with the development project coverage in order to produce a map of threatened areas and also to guide future regional survey programs and methods. This operational stage will only be reached after an important investment in the acquisition of the basic geographical data, and the enhancement of both hardware and software. This experiment also attests to the importance of integrating such an archaeological project in a wider regional GIS network.

References

Agache, R. and Bréart, B., 1975, *Atlas d'Archéologie Aérienne de Picardie*, Amiens: Société des antiquaires de Picardie.

Dorion, J., 1981, Une carte archéologique de la France, *Histoire et archéologie/Les Dossiers*, **49**, 26–27.

Guillot, D., 1991, Etat d'avancement de l'informatisation de l'inventaire national des sites archéologiques, *Les Nouvelles de l'Archéologie*, **45**, 20–22.

Guillot, D., 1992, The National Archaeological Record of France: advances in computerization, in Larsen, C. (Ed.), *Sites and Monuments: National Archaeological Records*, pp. 125–32, Copenhagen: The National Museum of Denmark.

Magnan, D., 1989, L'inventaire archéologique informatisé en France, dans la perspective de l'aménagement, in Archéologie et aménagement du territoire, *Les Cahiers de l'urbanisme*, **7**, 70–78, Louvain-la-Neuve.

Plateaux, M., 1990, Approche régionale et différentes échelles d'observations pour l'étude du néolithique et du chalcolithique du Nord de la France: Exemple de la vallée de l'Aisne, Xème rencontre internationale d'archéologie et d'histoire, pp. 157–182, Antibes: APDCA.

Soustelle, J., 1975. *Rapport sur la recherche française en archéologie et anthropologie*, Paris: La Documentation Française.

Tuffreau, A., 1987, 'Paléolithique inférieur et moyen du Nord de la France (Nord Pas-de-Calais, Picardie) dans son cadre stratigraphique', unpublished Ph.D. thesis, Université des sciences et techniques de Lille Flandres-Artois.

3

GIS and archaeological resource management: a European agenda

P. Martijn van Leusen

3.1 Introduction

From its early beginnings in the USA and Canada, archaeological resource management (ARM)[1] has driven the use of, and research into, archaeological GIS (Judge and Sebastian, 1988). This is not only because the field of ARM provided early funding opportunities, but also because ARM GIS applications are very similar to established GIS applications in related fields (for instance, land-use planning and land evaluation). From these eminently practical origins, GIS have migrated to academia, where researchers are applying them for purposes other than the management of cultural resources. The approaches of both user communities overlap mainly in the area of so-called predictive modelling.

GIS can be used in ARM in two distinct ways. The first is the use of GIS as a data management and presentation toolbox, which provides archaeological resource managers with a quick method of collecting, combining and presenting the data relevant to their work. This is non-controversial and builds upon the current drive toward automating central archaeological archives at the State Service for Archaeological Investigations (ROB) in The Netherlands and other European countries (Roorda and Wiemer, 1992; Larsen, 1992). Alternatively, GIS can be used as a research instrument for the development of new and better ways of mapping and evaluating the archaeological heritage. By extrapolating from the known archaeological record, reconstructions of (paleo-) environmental variables and data about (post-)depositional processes, cartographic reconstructions of the past and/or present state of the archaeological heritage can be made.

Published Dutch applications in the area of ARM have generally been of an experimental character, with little theoretical or methodological research and no adequate review mechanism in place. Notwithstanding that, it is already clear that academic archaeologists and archaeological resource managers are on different tracks as far as the theory and methodology behind their applications is concerned (van Leusen, in press). More importantly, the future quality of GIS applications in the field of ARM is going to depend to a large extent on the results of legislative and organizational changes taking place at this very moment in The Netherlands and in other member states of the European Union.

3.2 *The Malta Convention and its implications for ARM*

3.2.1 The Malta Convention

> With increasing population and ever higher standards of living, development projects have increased in number and complexity. Examples of such projects are major public works (such as motorways, underground railways and high-speed trains, replanning of old town centres, and car parks) and physical planning schemes (e.g. reafforestation and land consolidation). The scale of such operations poses a peculiar threat to the discovery and protection of the archaeological heritage. A complex web of law and legislation is involved: there is specific legislation on archaeological materials, more general legislation on the cultural heritage, legislation on the environment, town planning, public works, building permits, etc. (Council of Europe, 1992a)

This web appears not to function properly in The Netherlands. Eerden (1993) argues that the protection of archaeological resources through the Monuments Law has proved ineffective; the legal procedures required to declare a monument take too long (about four years) and offer no guarantee of permanent protection.[2] Financial means are insufficient to acquire important archaeological terrains or to check local compliance with the law. The logical alternative to the kind of specific, after-the-fact protection offered by the Monuments Law is generic preventive protection under planning law. Currently the ROB, however, is involved in the planning process in an advisory role only. The large public sector organizations and some private sector businesses occasionally seek advice from the Service during the planning stages of a development project, but in practice this leads to the financing of preliminary research (archive retrieval, surveying) or to rescue excavation at most. Last-minute rescue excavation, often necessitated by the lack of preventive research, is extremely expensive and can seriously delay the execution of regional and municipal plans. In short, the potential of planning protection for archaeological resources has hardly been tapped.

To put a stop to the continuing destruction of the European archaeological heritage, a number of practical and legal measures, together amounting to a thorough reorganization of archaeological resource management, have been internationally advocated. These include:

1. the formation of archaeological inventories and data banks which would be communicated by archaeological resource managers to developers;
2. the creation of administrative structures capable of handling development projects involving archaeological data;
3. the adoption of legal and administrative measures necessary for archaeological data to be taken into account as a matter of course in the context of major development operations; and
4. the education of the public in the value of the archaeological heritage as a major element of European cultural identity.

An early attempt by the Council of Europe to adopt such measures (1969, Convention of London) failed. Finally, in early 1992 the European Community Ministers responsible for the cultural heritage came together at La Valetta (Malta) to sign a convention describing the way in which the cultural heritage should be protected by the member states. Figure 3.1 shows the text of the relevant Articles 5 and 6 of the Malta Convention; its two main consequences will be the regulation in

European Convention on the Protection of the Archaeological Heritage (revised).
Council of Europe, La Valetta, 16-01-92.

The member States (...),

Acknowledging that the European archaeological heritage, which provides evidence of ancient history, is seriously threatened with deterioration because of the increasing number of major planning schemes, natural risks, clandestine or unscientific excavations and insufficient public awareness;

Affirming that it is important to institute, where they do not yet exist, appropriate administrative and scientific supervision procedures, and that the need to protect the archaeological heritage should be reflected in town and country planning and cultural development policies;

(...)

Have agreed as follows:

Article 5

Each Party undertakes:

i to seek to reconcile and combine the respective requirements of archaeology and development plans by ensuring that archaeologists participate:

a in planning policies designed to ensure well-balanced strategies for the protection, conservation and enhancement of sites of archaeological interest;

b in the various stages of development schemes;

ii to ensure that archaeologists, town and regional planners systematically consult one another in order to permit:

a the modification of development plans likely to have adverse effects on the archaeological heritage;

b the allocation of sufficient time and resources for an appropriate scientific study to be made of the site and for its findings to be published;

iii to ensure that environmental impact assessments and the resulting decisions involve full consideration of archaeological sites and their settings;

iv to make provision, when elements of the archaeological heritage have been found during development work, for their conservation *in situ* when feasible;

(v ...)

Article 6

Each Party undertakes:

i to arrange for public financial support for archaeological research from national, regional and local authorities in accordance with their respective competence;

ii to increase the material resources for rescue archaeology;

a by taking suitable measures to ensure that provision is made in major public or private development schemes for covering, from public sector or private sector resources, as appropriate, the total costs of any necessary related archaeological operations;

b by making provision in the budget relating to these schemes in the same way as for the impact studies necessitated by environmental and regional planning precautions, for preliminary archaeological study and prospection, for a scientific summary record as well as for the full publication and recording of the findings.

Figure 3.1 The text of Articles 5 and 6 of the Malta Convention.

law of the relationship between development projects and the preservation of the archaeological heritage, and the provision of financial support for archaeological research. The explanatory report accompanying the Convention states that when it is implemented Article 5 'requires States to ensure that consultation [between developers and archaeologists] takes place and that time and funding is provided for an appropriate scientific study' and Article 6 'casts on those responsible for development projects the burden of funding archaeological activities necessitated by those projects'.

Full-scale implementation of the convention will start as soon as it is ratified by the required number of member states. Ratification by the Dutch government is expected in early 1995. With this international agreement to guide the national efforts to preserve the cultural heritage, what are the implications for archaeological resource management in The Netherlands and, by extension, in other EU countries?

3.2.2 Implications for archaeological resource management

The implementation of the Malta Convention entails the institution of 'appropriate national legislative and administrative measures'. Generic legislation will be instituted at the national level; administrative measures will mainly affect the executive State Service (ROB), but will also require the co-operation of academic archaeology institutes and independent ARM organizations to be successful in the area of ARM education and in the actual execution of ARM projects. In the following sections, independent ARM organizations are referred to as third parties (the ROB and academic institutes being the first and second parties to ARM in The Netherlands).

3.2.2.1 Legislative measures

The Minister responsible for the cultural heritage is currently considering several options to achieve the twin aims of integrating archaeological interests in the planning process and arranging for the necessary financial support.

To achieve the first aim, the introduction of a system of generic management measures aimed at the physical protection of the mainly invisible archaeological heritage is being prepared (Ministerie WVC, 1992). The changeover from *a posteriori* protection through the Monuments Law to *a priori* protection through the Planning Law, which is also a changeover from specific to generic ARM, puts an enormously larger burden on the executive, which must carry out the requisite mapping and evaluation for the whole land area of The Netherlands. When it is instituted national, regional and local development plans (plus a large number of mainly local waivers for unplanned development) will all have to be evaluated for their effect on the archaeological heritage, decisions about the execution and financing of various types of pre-development research will have to be made, and compliance with legal and administrative measures will have to be checked.

Options to achieve the second aim are being investigated on the Ministry's behalf by a management consultancy bureau. These options include:

1. gentlemen's agreements between the Ministry and the major soil disturbing organizations;
2. establishment of a fund fed by a fixed percentage of development costs of all building and soil-disturbing projects, from which research will be financed; and
3. legal establishment of the 'destructor's principle' (all necessary costs to be paid by those responsible for the destruction of the heritage.

The second option is most likely to be implemented.

3.2.2.2 Administrative measures

The administration of national ARM policy is the task of the State Service for Archaeological Investigations (ROB). It must implement whatever legislative measures are decided on by the Minister. Administrative measures will be needed to deal with the tasks of mapping and evaluating the archaeological heritage, administering the fund fed by development projects, controlling ARM research performed by third parties, performing whatever research is required before implementation can take place, and playing a significant role in the planning process. If the ROB is to succeed in these tasks, its involvement in ARM will have to become much deeper than it has been in the past. More manpower will be needed, along with a decentralization of ARM toward the provincial and municipal government levels (Eerden, 1993:31). Although the ROB is currently being reorganized it is not yet clear that the new organization will be able to fulfil all its tasks.

Implementation of the Malta Convention entails a complete mapping and evaluation of archaeological resources in The Netherlands. Article 6 of the Convention copies a practice established in environmental protection by insisting that research and publication be paid for by the organization that is responsible for threatening the archaeological heritage. Jumping on the environmental bandwagon, the ROB has decided to create a national archaeological structure (NAS) similar to the existing national ecological structure (NES), that will serve as a vehicle for ARM policy. The NAS will define regions that are of national archaeological importance (Klok, 1993) and will be presented to planners in the form of a generalized zoning map of the distribution and quality of archaeological resources in non-urban areas. This map will have 'children' at the regional level (archaeological policy maps, APMs) to guide regional planning in a more detailed manner.

The advantages of APMs are thought to be as follows (ROB, 1993b):

1. They are up-to-date. By using GIS to create and update an APM, it is easy to incorporate the latest relevant information and policies.
2. They are easy to use. Presenting archaeological policy in the visual form of a map is convenient and user friendly.
3. They are flexible. Future effects of land degradation processes and planning projects can rapidly be translated into a map image.
4. They are compatible. The GIS format ensures direct compatibility with physical/ecological GIS that are already used in the planning process.
5. They save money. Through their preventive working unforeseen research costs will be lower.

In summary, the national archaeological structure map and the archaeological policy maps are said to be 'instruments of preventive management, that allow the translation of scientific knowledge of the archaeological heritage into terms that are accessible to planners and policy makers' (ROB, 1993b). They 'can be used on the one hand to visualize the consequences of policies, and on the other to translate policy principles into policy measures' (ROB, 1993c).

The ROB also supervises third-party ARM research and reporting. Until recently this meant that draft versions of third-party ARM reports were sent to the ROB, which reviewed and assumed responsibility for any conclusions and recommendations that the reports contained. The ROB had no means of checking the data and methods used to arrive at those conclusions. I have argued elsewhere that the quality of GIS-supported research in such reports has been low (van Leusen, in press; see also section 3.3.2 below). Some more efficient means of controlling third-party ARM research is called for, especially since it is to be expected that in future project-based ARM research will mainly be carried out by a rising number of private companies or corporations unrelated to the ROB or the university archaeological departments.

The ROB is reorganizing to control third-party research more stringently by subjecting it to mandatory review by the Provincial Archaeologists, but it is questionable that they will be able to supervise a large number of independent organizations. If third-party ARM research is not regulated, it is likely to develop habits that we have already seen emerge in the USA and, to a lesser extent, in the UK. There will be strong emphasis on visual effects (colourful maps taking precedence over formal analysis); competitive tenders for archaeological research prior to the development of a site will force a 'minimum' approach to estimates of research costs (in terms of both time and money); economies of scale will work towards the standardization of research plans and publications. All in all a lowest common denominator, determined by the legal requirements that are set for ARM research, will become the *de facto* standard. In The Netherlands, the first signs of similar developments are that the main independent ARM organization has stopped GIS research because of lack of time and funding; it cannot pay specialist (e.g. statistician) salaries and has no separate funding for development research.

Two measures must be taken to avoid a lowest common denominator approach to ARM research. Firstly, the ROB itself should concentrate on developing and enforcing methodological standards, while leaving the actual execution of ARM research to others. It has already acted on this requirement by establishing a permanent ARM research department. The ROB thus seems set to take on the task of developing ARM theory and methodology itself, and it should ensure that third parties make use of approved research methods only. Secondly, both first- and third-party ARM research has been practised largely outside peer-reviewed channels of academic publication. Yet peer review seems the only way of providing a mechanism of independent and public discussion of issues that, in the planning process, may have important (financial) consequences; such a mechanism needs to be established, as a priority.

3.2.2.3 Education

The education of the general public in the value of the archaeological heritage, the danger it is in, and the need to preserve it, is explicitly addressed by the Malta Convention. However, nothing is said about the need to educate or re-educate archaeologists to deal adequately with the demands that ARM will place on them. Academic archaeologists in The Netherlands have not been directly involved in ARM research and have done little teaching on the subject. Thus, academically educated archaeologists tend to have little knowledge of ARM or planning science. In contrast, the ROB and third-party ARM organizations are already requiring knowledge of planning science from future job applicants. Since ARM is the major job market for academic archaeologists, students should be well prepared in this particular area of expertise.

Teaching curricula at the various university archaeology departments should reflect the changing emphasis of archaeological work to remedy these shortcomings. The Institute for Pre- and Protohistoric Archaeology (IPP) at the University of Amsterdam has started to prepare a new curriculum for its students. The relations between archaeology and planning and resource management are central to the new department of Archaeology in Society, in which the archaeology of cultural landscapes is prominently featured (Bloemers, 1993):

> Archaeology of the cultural landscape is that school in archaeology that studies the archaeological heritage as part of the cultural–historical landscape, and its relation with the natural environment, with the aim of ensuring its protection for the future. Knowledge of cultural–historical values and of the possibilities for safeguarding them through planning law is therefore of extreme importance for the future layout of the Dutch landscape. Research could be aimed at:
>
> (i) reconstructing the past use and layout of the landscape and its long-term processes;
> (ii) development of methods for the detection, monitoring and prediction of archaeological values;
> (iii) development of concepts and methods for the evaluation and preservation of archaeological values in the process of spatial planning.

With a four-year curriculum length, the new curriculum could deliver its first MAs in 1998, none too soon to participate in the implementation of the Malta Convention.

3.3 GIS and ARM: a European agenda

In the previous sections, the legal and administrative background to such GIS studies was sketched; the present section focuses on issues related to the successful application of GIS in archaeological resource management. Under the Malta Convention's Article 5, known and suspected sites can be taken into account during the planning stages of development projects. This suggests that just using GIS as a data management and presentation instrument for the central archaeological archives is not enough. Only with up-to-date surveys, inventories and maps of archaeological sites can the process of consultation work effectively. The importance of doing predictive GIS studies and of extensive surveying under this scheme is obvious.

A number of problems of theory, method, and practicability stand in the way of the immediate application of GIS to ARM tasks. The definition of 'archaeological resources' that is used, the grounds for assigning a value to these resources, and the determination of the goals of predictive studies are issues that must be decided on theoretical grounds. To serve a practical role in the planning process, archaeological policy maps (APMs) must be generalized to a certain extent and must be presented to planners in some directly useful form; the production of such maps must be streamlined in order to 'fit' within planning law procedures. Methodological issues include how to deal with gaps in the archaeological record, how to construct an acceptable evaluation scheme for archaeological resources, how to predict the presence of unknown resources, and how to use the numerical and statistical tools of GIS correctly, for map analysis and map construction. Although archaeological resource managers in other European countries may decide to manage their archaeological heritage in a manner different from the approach chosen in The Netherlands, these issues are of equal importance everywhere—which is why this chapter is entitled 'a European agenda'.

3.3.1 Theoretical issues

GIS will be used to predict and map the presence and value of archaeological resources and the effects that planned development will have on those resources. To be able to do this, it must first be clear what is meant by 'archaeological resources' and their 'value', and what is the exact aim of the predictive effort (because this determines the methodological approach to be taken).

3.3.1.1 What are archaeological resources?

Because of the way archaeological research and reporting has worked in the past, almost all our knowledge of archaeological remains is organized as 'sites'. If archaeological resource management were to proceed on the basis of current databases it would continue this bias toward the foci of human activity. If instead the 'cultural landscape' approach were taken, the stress would be on areas instead of sites, and other parts of the archaeological record might end up being protected. A decision about this issue may be made on practical instead of theoretical grounds: the fact that archaeological resources will ultimately have to be displayed in the form of a policy map forces resource managers to think of archaeological resources as areas instead of sites (ROB, 1993a).

Archaeological resource managers need a different approach to archaeological remains than do academic archaeologists. The use of the substantive resources is particularly revealing in this respect: it reflects an ahistorical view of the cultural heritage as a limited resource similar to traditional resources such as oil, gas and gravel, rather than as historical evidence of past living cultures. ARM focuses on the future state of archaeological remains that are present in our soil; academic archaeology focuses on reconstructing the past state of the archaeological heritage. The difference should be reflected in the goals of predictive modelling, since reconstructing the past means that all data formation processes should be incorporated in the model, whereas reconstructing the present avoids such difficult procedures entirely.

In other words, being exclusively concerned with the future state of the present archaeological remains may actually be advantageous: archaeological resource managers can avoid the laborious reconstruction of the palaeo-environment because they do not need a historical/causative explanation of the present distribution of the resources. The origin of most archaeological resource managers in academic archaeology has until recently prevented them from clearly recognizing and stating this difference and its important methodological consequences.

3.3.1.2 What is the value of archaeological resources?

When the prediction and mapping stage of ARM is finished, the value of the various elements of the archaeological heritage must be determined. On what basis should the evaluation proceed? One can imagine that archaeological resources may have an economic value (as tourist attractions), a scientific value (potentially telling us something about our past), a landscape value (if visibly present, reminding us of our roots and of the time-depth that is present in our surroundings), and perhaps more. Which of these should contribute toward determining the value of archaeological resources, and how much? Many problems remain to be solved before implementation of any particular evaluation scheme is possible.

For planning purposes it is not only important to know what is in the soil and how much it is worth, but also how sensitive it is to various types of destruction. This entails a second evaluation step in which the sensitivity of the resources is determined, perhaps on the basis of its depth in the soil, its position with respect to groundwater levels, the current land use and legal status of the area, etc. Because of the technical nature of this evaluation, fewer problems may be expected with its implementation.

Because the results of these evaluation steps determine the amount and type of protective measures that will be required in the planning process, the evaluation scheme has direct planning and financial consequences. The ROB must therefore be able to defend its evaluation scheme against attacks, perhaps even litigation, by interested parties. It is not clear how this can be accomplished. Will it even be possible to reach agreement on one evaluation scheme among archaeologists themselves? The discussion about evaluation schemes can only succeed if it has a sound theoretical basis. Fortunately, the build-up of expertise in this area has only recently been initiated (Groenewoudt, 1994).

3.3.2 Methodological issues

When theoretical issues have been resolved, the successful application of GIS tools to ARM depends on the development of a correct and practicable methodology. As we have already seen, methodological issues include how to predict and evaluate archaeological resources. In addition to this, archaeological resource managers will need to develop a methodology to translate their evaluation into information that can be used by planners (e.g. archaeological policy maps). They must also be able to deliver this information within a limited span of time in order to comply with the planning law procedures. In general, methodological streamlining will be necessary to ensure that prediction, evaluation, and translation into policy maps of archaeological resources is fast, robust, replicable, and transparent. The present section discusses a selection of methodological issues to illustrate this principle.

3.3.2.1 The methodology of prediction

Mapping archaeological resources requires some method of predicting the presence or absence of these resources in areas that have not been studied by some more direct means. The literature on prediction methods is quite extensive (e.g. Allen *et al.*, 1990); basically, predictive modelling of archaeological resources can proceed on the basis of induction, deduction, or a mixture of the two. Deductive modelling proceeds from *a priori* assumptions about the location of the resources; inductive modelling proceeds from properties of the known archaeological record. The main disadvantage of deductive modelling is that the *a priori* assumptions are often difficult to justify (van Leusen, in press); the main disadvantage of inductive modelling is its heavy reliance on statistical methods that are often difficult to implement. The international trend currently is in the direction of inductive modelling.

An inductive methodology of prediction would first of all have to deal with problems related to the quality of the archaeological record. One example of such problems is the biased presence or absence of archaeological data. An old saw says that 'the absence of evidence is no evidence of absence'; predictive models that disregard this are of little scientific value, which is why academic archaeologists have stressed that methods of dealing with formation processes (or, more to the point, database forming processes) must be developed (Fokkens, 1991; van Leusen, in press; ROB, 1993c). On the other hand, as was argued in the previous section, ARM need not concern itself with the whole of the archaeological record, but only with that part which is under threat of damage or destruction. Whatever the causes of the 'blanks' in the present archaeological map of The Netherlands may have been, the same causes are likely to keep those same areas blank in the future, thus 'protecting' whatever archaeological resources may be present there. This line of reasoning is perhaps difficult for a trained archaeologist to accept, but in ARM the result is what counts. Notwithstanding this, methods of predictive modelling as they are currently applied in The Netherlands are based on the often implicit assumption that the known archaeological record more or less accurately reflects the extant distribution of these resources. Confidence in the archaeological record is relatively high at the ROB (Klok, 1993):

> Although a complete and systematic inventory and evaluation of the archaeological heritage does not yet exist [1993], there is enough knowledge available to supply the data required for planning purposes and upon which to base policies.

Curiously, at the same time the Minister instituted a special financial support programme to 'fill in the remaining blanks in the archaeological map of The Netherlands' through modern non-destructive surveying techniques (Ministerie WVC, 1990).

Whatever methodology is ultimately chosen for predictive modelling, it will to a large extent be based upon the numerical and statistical analysis of rasterized maps. This presupposes that the expertise will be available to build such models, not only by correctly applying the available GIS tools, but also by critically examining the functionality of the GIS itself.

An example of what may happen if such expertise is not available, is shown by a recent report by an independent ARM research organization (Odé and Verhagen, 1992). Chi-squared analyses of the coincidence of archaeological sites per soil type, per geomorphologic unit, and per distance to macrogradient (*ibid.*, Appendices 20, 21 and 22) were produced using the standard statistical tools within the GIS GRASS. These results are purported to show the non-random distribution of archaeological site locations with respect to the physical environment within the study region. This analysis, on the basis of which planning recommendations were made that were sanctioned by the ROB, is statistically invalid because

1. the total number of sites (six) is much too small to allow a chi-squared test;
2. the test algorithm as implemented in GRASS tests against a random distribution of sites, an assumption that we already know is incorrect and which leads to futile conclusions;[3] and
3. the three environmental factors examined are strongly correlated, which means there is little sense in performing a separate chi-squared test on all three.[4]

It may be argued that such mistakes are simply the result of unfamiliarity with a GIS toolbox and spatial statistics applied in an experimental context, but these particular mistakes have been repeated in reports over a number of years (Ankum and Groenewoudt, 1990; Soonius and Ankum, 1991; Odé and Verhagen, 1992). If independent bureaux already find it impossible to line up the statistical and programming expertise needed for the professional operation of a GIS, clearly the growth in number of such bureaux will only exacerbate the problem.

3.3.2.2 The methodology of evaluation

To have any impact on the planning process, the archaeological heritage must be evaluated according to a set of rules taking into account both the intrinsic value of the resource and the amount and type of danger that it is in. To devise an evaluation scheme that is acceptable to all parties involved in the planning process, it must be clear, simple, and robust. Clarity can only be achieved if the number of factors involved in the evaluation is kept low and is based on explicitly stated assumptions. Robustness and replicability can only be achieved by applying a fixed set of evaluation rules that is independent of the particular region or archaeological data-set that is used. Speed can only be accomplished by standardization and customization of GIS procedures. Such evaluation schemes as have been devised are still in an experimental stage, are instituted at the spur of the moment and have no theoretical justification (Ankum and Groenewoudt 1990; Soonius and Ankum, 1991); it is generally recognized that the stage of operationalization will not be reached for a number of years yet.

One particularly pernicious problem is that the value of a particular resource is relative to the total of regional, national, or in some cases even international resources. The value of that resource will rise if similar resources are destroyed in the vicinity and it will drop if similar resources are discovered in the vicinity. In other words, the value of archaeological resources, like that of other resources, is dependent upon a market mechanism. The methodology of evaluation must take into account the changing scarcity value of archaeological resources, probably by using some kind of expert system approach.

3.3.2.3 Requirements of the planning process

When in Rome, do as the Romans do; cultural resource managers must learn to speak the language of planners to get their way in the planning process. They must present their assessment of the archaeological heritage to planners in a form that lends itself directly to incorporation into local, regional and national plans. In The Netherlands, the National Archaeological Structure (NAS) and APMs were chosen as the main policy vehicles that the State Service for Archaeological Investigations will use in the planning process. Such maps must be as simple as possible ('globalized' is the planning 'buzzword'). The NAS map will be large-scale and generalized; urban areas with their intricate planning structure are left out of the NAS altogether (ROB, 1993c). The regional archaeological policy maps should probably indicate no more than the type and level of protection that archaeological resource managers deem necessary for each area, since planners can't be bothered with such details as the archaeological cultures and periods that are present. Simplifying the results of a predictive study to this level requires the development of yet another set of GIS methods. Again, research into this area has only just begun and can probably only be fruitful if archaeologists are assisted by planning specialists.

The planning process also allots limited amounts of time for each stage of planning, protesting, advising and making changes. The ROB must be able to deliver its assessment of the state of archaeological resources and the amount of protection they need to the local planning agency within that time. This is only feasible if the GIS methodology is fast and robust enough to deal with the majority of development projects, and flexible and intelligent enough to allow its application to irregular and unforeseen cases. Customized GIS applications that can be operated on a decentralized level (provinces, municipalities) should take care of the majority of cases in which the archaeologist is required to play a role in the planning process—the decentralization of jobs and funding will be a major operation in itself.

Lastly, APMs are supposed to reflect the effect that proposed planning policies will have on archaeological resources. However, the planning process can be quite chaotic, with plans in different stages of development, different levels of detail, etc., being continually changed or deviated from in yet other plans (van Leusen 1990: 58–60). Reacting adequately to this shifting mass of plans will be extremely difficult, if at all possible. Archaeological resource managers at the ROB may eventually be forced to restrict themselves to a more modest role in the planning process, again perhaps by decentralizing the execution of regional and local ARM.

3.4 Summary and Conclusions

In the next few years, the whole structure of archaeological resource management in The Netherlands will undergo a shift away from the old system of incidental *a posteriori* protection of archaeological resources through the Monuments Law, toward generic preventive protection through the planning law. Funding of archaeological research (including publication) will be put on a more permanent basis than has been possible in the past. This legislative shift, accompanied by a shift in administrative procedures at the State Service for Archaeological Investigations (ROB), is part of a larger European movement to protect its cultural heritage which found its most recent expression in the Malta Convention of 1992.

Although the organization of ARM in The Netherlands may differ in some respects from that in other EU countries, it is clear that broadly the same issues are at stake there. National executive archaeological services need to find ways of efficiently playing their role in the planning process; they need to assert effective control over ARM research performed by a growing number of independent agencies; they need to take the development of appropriate GIS methods into their own hands. Control over ARM research bureaux and the development of GIS applications are tasks of the central executive service; managing the archaeological heritage at the regional and local levels should be done by decentralized means and will require a large increase in decentralized ARM staffing and funding levels.

Independent ARM research organizations could voluntarily co-operate by adhering to some set of agreements between them and the State Service for Archaeological Investigations, to ensure the quality of the research that underlies their management recommendations. Alternatively, mechanisms of control must be instituted to ensure that research (including GIS research) by independent bureaux conforms to standards to be devised jointly by the State Service and the academic archaeological community. Automation of parts of the planning process with the help of a GIS will have to overcome a host of theoretical, methodological, and practical problems, as described in this chapter. Research on most of these issues is in a preliminary stage and operationalization, especially under the stringent conditions set by planning procedures, is still a long way off. It is clear that this research cannot be performed without the active participation of academic archaeologists.

Decentralization of the administrative and financial responsibilities of the central executive, necessary to deal with the complexities of the local planning process, is not likely in The Netherlands. Nor are central staffing levels likely to increase to the extent required if ARM does not get a higher priority from society as a whole. The ROB will then be forced to limit its role in the planning process to the regional and national levels, leaving the local execution of ARM policies in essentially the same rut it has been in for the past decades.

Even if definition and execution of ARM policy is in the first instance the responsibility of the central archaeological services, even if its aims are not scientific, the academic community of archaeologists still has a responsibility to help ensure the future of the archaeological heritage in Europe. One important way it can do this is by using its research and education potential in the development of GIS applications for ARM. Hopefully, this chapter will contribute something toward that end.

Acknowledgements

This chapter presents my own personal views on the subject of GIS and ARM. I am indebted to Roel Brandt and Marco Otte (RAAP)[5] and Jan Kolen, Paul Zoetbrood and Iepie Roorda (ROB) for their help in clarifying the positions of their respective organizations, but I remain responsible for any errors in stating those positions.

References

Allen, K. M. S., Green, S. and Zubrow, E. (Eds), 1990, *Interpreting Space: GIS and Archaeology*, London: Taylor & Francis.

Ankum, L. A. and Groenewoudt, B. J., 1990, *De situering van archeologische vindplaatsen*, RAAP-rapport 42, Amsterdam: Stichting RAAP.

Bloemers, J. H. F., 1993, Transformatie en ontwikkeling, Concept-uitwerking Facultair Herstructureringsplan FRW, internal report, University of Amsterdam.

Council of Europe, 1992a, European Convention on the Protection of the Archaeological Heritage (revised). Explanatory report, 3rd European Conference of Ministers responsible for the cultural heritage, Malta, 16–17 January 1992.

Council of Europe, 1992b, Memorie van Toelichting bij het Europees Verdrag inzake de bescherming van het archeologisch erfgoed (herzien), 3rd European Conference of Minsters responsible for the cultural heritage, Malta, 16–17 January 1992.

Eerden, M., 1993, Naar een archeologische hoofdstructuur in Nederland, Bijdrage aan een actuele discussie, unpublished MA thesis, University of Amsterdam.

Fokkens, H., 1991, Verdrinkend landschap: archeologisch onderzoek van het westelijk Fries-Drents Plateau, PhD thesis, University of Groningen.

Groenewoudt, B. J., 1994, Prospectie, waardering en selectie van archeologische vindplaatsen: een beleidsgerichte verkenning van middelen en mogelijkheden, Nederlandse Archeologische Rapporten 17, Amersfoort: ROB, PhD thesis, University of Amsterdam.

Judge, W. J. and Sebastian, L., 1988, Quantifying the present and predicting the past: theory, method, and application of archaeological predictive modelling, Denver, CO: US Department of the Interior, Bureau of Land Management.

Klok, R. H. J., 1993, Structuurschema Groene Ruimte: archeologie en cultuurhistorie, Concept inspraakreactie, Amersfoort.

Larsen, C. U. (Ed.), 1992, *Sites and Monuments. National Archaeological Records*, Copenhagen: National Museum of Denmark.

Ministerie WVC, 1990, Deltaplan voor het cultuurbehoud. Onderdeel: plan van aanpak achterstanden musea, archieven, monumentenzorg, archeologie, Den Haag: SDU.

Ministerie WVC, 1992, Investeren in cultuur, Nota cultuurbeleid, 1993–1996, Den Haag: SDU.

Odé, O. and Verhagen, J. W. H. P., 1992, Aanvullende Archeologische Inventarisatie beheersobject Ugchelen-Hoenderloo-Varenna, RAAP-rapport 60, Amsterdam: Stichting RAAP.

ROB, 1993a, Natuurbeleidsplan Project 33: opzet van de archeologische onderzoekscomponent. Voorstel tot voortzetting van het pilot project, Concept dd 19/7/93, Amersfoort.

ROB, 1993b, Een archeologische beleidskaart voor de provincie Noord-Brabant, Concept dd 6/7/93, Amersfoort.

ROB 1993c, Toelichting op de Globale Archeologische Kaart van Nederland, Amersfoort.

Roorda, I. M. and Wiemer, R., 1992, Towards a new archaeological information system in the Netherlands, in Lock, G. and Moffett, J. (Eds), *Computer Applications and Quantitative Methods in Archaeology 1991*, pp. 85–88, BAR International Series 577, Oxford: Tempus Reparatum.

Soonius, C. M. and Ankum, L. A., 1991, Ede, RAAP-rapport 49, Amsterdam: Stichting RAAP.

van Leusen, P. M., 1990, De toekomst van het verleden in het Noord-Hollandse strandwallengebied, Unpublished MA thesis, University of Amsterdam.

van Leusen, P. M., 1993, Cartographic modelling in a cell-based GIS, in Andresen, J., Madsen, T. and Scollar, I. (Eds), *Computing in the Past. Computer Applications and Quantitative Methods in Archaeology 1992*, pp. 105–124, Aarhus, Denmark: Aarhus University Press.

van Leusen, P. M., in press, CRM and Academia in the Netherlands: assessing current approaches to locational modelling in archaeology, in Maschner, H. (Ed), *Geographic Information Systems and the Advancement of Archaeological Methods and Theory*, Carbondale, IL: Center for Archaeological Investigations.

Notes

[1] I have used this term instead of the more general cultural resource management (CRM), to indicate that the fields of historical geography, industrial archaeology and preservation of monuments and historic buildings are not discussed here.

[2] A court decision recently forced archaeologists to study the large and important Kops Plateau site (Nijmegen) within the next 15 years because its status as a monument for an unspecified period of time was declared illegal.

[3] When testing the location of sites against the distance to the nearest stream, for instance, it would be more useful to test against the hypothesis of a logarithmic relation than against that of a random relation.

[4] If a group of sites has a non-random distribution with respect to one of the variables (e.g. soil type), it will automatically have a similarly non-random distribution to correlated variables (e.g. geomorphology).

4

To be or not to be: will an object–space–time GIS/AIS become a scientific reality or end up an archaeological entity?

D. Arroyo-Bishop and M. T. Lantada Zarzosa

4.1 Introduction

If one contemplates using GIS for implementing an archaeological information system (AIS) for the management and conservation of our cultural heritage, certain structural necessities and obligations will soon become apparent. This chapter will focus particular attention on the analytical applications which may become prevalent, if an archaeological GIS is successfully structured and formalized. Because of the inherent correlation with the aims and preoccupations expressed at the National Archaeological Records Conference (Larsen, 1992), this chapter will begin with a partial recollection of the introduction of the paper presented in Copenhagen in the spring of 1991: 'The ArchéoDATA system: a method for structuring a European archaeological information system (AIS)' (Arroyo-Bishop and Lantada Zarzosa, 1992):

> *An introduction to current reality*
>
> Today, we are far from the self-sufficient research which has prevailed for such a long time in archaeology and which is at length being replaced with the spirit of collaboration in multi-disciplinary studies. Archaeology is trying to evolve into a science in its own right, but to attain this new status, it must now accept, as many other professions have found necessary before, that it is now essential to adopt, in this new phase of maturity, certain structuring and homogeneity in the data produced if research and administration is to be favoured and conservation efficiently carried out in the future.
>
> If archaeology is to reap some of the benefits from the information age it must prepare now. If there is to be in Europe's future an integration of archaeological research it must of course be computer based, and for computerization to work, it is an intrinsic necessity that we structure and formalize the fundamental way we record and store data. We cannot allow a myriad of databases to develop independently, each with their own themes, structures, indexes and vocabularies, as it becomes rapidly impossible for anyone effectively to research and master more than a few of them. The underlying system must be founded on basic archaeological methods and concepts, and the data registered, straitforward. The system must be logically implemented, enabling it to be easily

integrated into existing administrative and research structures. It is through the common usage of the system by surveys and excavations that its base will be built up. Information must be able to circulate widely and it must be possible to select, group and compare on a wide scale without the limitations imposed on data by the regional and national boundaries. It is no longer sufficient to decry the state of the archaeological record; it is time to establish the future framework for it to evolve.

It is imprudent, to say the least, to continue to ignore these basic facts. To face up to reality is not to consent to a grandiose scheme, but to support a scientific and administrative necessity which goes far beyond individual or group preferences and interests. Archaeology cannot expect to continue to evolve scientifically if it is stifled in its quest to dispose of the required data needed for research. It is no longer possible to continue to prepare good isolated systems; a good integrated system must now be accomplished. Some will take up the system immediately; others will need time to adapt as many possible obstacles, material and formal, do exist. This could of course take a new generation of archaeologists to put into generalized use, not to say completed, but this can only be seen as to further guarantee its longevity.

An archaeological information system is the ordered integration of archaeological and related data into a common administrative and research unit in order further to research and facilitate the management of this part of our cultural heritage. It is composed of past and present archival and bibliographic data, survey and site data, excavation reports, studies and analyses, finds and museum data, conservation information and should provide easy access to complementary information when necessary. For an AIS to be successfully established, it must be organized and structured in such a way as to encompass, from the beginning, all the different facets of archaeology. If this is not the case we can only hope to duplicate the present day situation of disparity.

Archaeology is a composite discipline and it is no longer possible to treat each field, each speciality, as a world unto itself, if one expects to be able to relate to information at all stages of research.

4.2 Back to basics

What this introduction has described are the basic necessities for using a GIS for archaeology, and the realities which oppose it. Nowhere more than in setting up and running a GIS/AIS is common data structure and data formalization so fundamental to the final operation. This is true at all scales of practice, even if the AIS component is not contemplated, data structure and formalization must be present even in the case of local or partial studies, it simply becomes ever more important as things become more complex. In the years that have passed since the ArchéoDATA project was started, some of the preoccupations expressed by field archaeologists, research workers, academics and administrators, have been addressed, and with the help of many, some have been answered and more are being considered and researched. It has been our fundamental intention, through the ArchéoDATA project, of assisting, through our contribution, to the creation of a European archaeological information system.

This chapter centres on work carried out to structure archaeological data for analysis, and in particular on the formalization developed to this end. It will not present a comprehensive overview of the ArchéoDATA System as it has been developed over the last few years. Several published articles[1] have presented different components of the system: survey, excavation, finds and inventory, as well as storage. It is important to go over these papers so as to comprehend the system's general organization, as it is fundamental to the working of the system that these aspects are also structured and formalized to enable the later phases of analysis, described here, to function correctly.

In the field of spatial analysis, much research work has been undertaken since the late 1960s, and more specifically since the mid-1980s, with the application of GIS to archaeology. (A global view is to be found in Allen *et al.*, 1990.) Here again the ArchéoDATA project has concentrated on the considerable research work that has been undertaken and has tried to synthesize the basic structure that has been necessary to accomplish it. Far from innovating, these stages focused on understanding the methodology that had been developed by others and how they had applied it to the research work they had undertaken.

This chapter presents the basic building blocks that are considered to be those on which an effective analytical structure can be based, i.e. to manage

1. geographic space
2. administrative space

and to analyse

3. the archaeological and architectural entity (AE)
4. the spatial entity (SE)
5. the temporal entity (TE)
6. the interpretative group (IG)

In our work, particular attention has been given to intra-site analysis; nevertheless all the notions presented, can be adapted and applied to any scale of inter-site analysis.

4.2.1 Basic inter-site building blocks

The two general methods that GIS use for managing space is either to relate to some form of geographic projection—the universal degree, minute, second coordinates, or some form of metric projection, such as Lambert or UTM—or by delimiting space in some artificial way, such as by administrative limits, boundaries, zones, etc.

4.2.2 Geographic space

The grid numbering is based directly on the universal transversal Mercator (UTM) projection of latitude and longitude, adopted by the Council of Europe as the future geographical coordinate system for Europe.[2] Each grid block comprises 100 squares numbered from 0 to 99 and related in three stages to the hundreds, tens and units of the UTM coordinates. These correspond directly with the metric system's own structure of kilometre, hectare, and metre. The resulting number locates the exact unit directly on the surface of the earth. When the excavation grid is set up, the number of each square is determined directly by its x and y coordinates. Since no two places can have the same absolute coordinates, each set of excavation data is unique and can be related to other data recorded in the same way.

4.2.3 Administrative space

To manage administrative space, common national and international denominators have been found, as it should be possible for the archaeologist to process data at all spatial scales. The code derived is based on a 14-digit number structured in three significant parts: the first three numbers identify the country, and are based on the code originally developed for international telecommunications, the following seven numbers are the postal union code, that localize the site on a national level; and the last four numbers position the site and survey data. For example, the Roman/ Medieval cemetery 'Cementério del Carmen' (site 17 in the township), found to be part of the deserted village of 'Malcantes', today the Dehesa de Macintos, which belongs administratively to the town of Villoldo (131 in the province), in the province of Palencia (34 in the country), in Spain (34 in the world) is coded as 34 0034131 0017

It is not only administrators who will find themselves at home with being able to manage their data according to administrative limits and boundaries, but also researchers who will adapt them to their work through the use of spatial entities, as will be presented below. For research purposes, most administrative boundaries are either very similar, or are easily adaptable, to those of historic periods, thus reflecting periods of historical space, occupation, culture, influence, etc.

4.2.4 Using space

To manage administrative space, both national and international, by referencing data to a grid it becomes possible to directly use raster-based GIS software. Combined with a structure that permits the recording of point, line and polygon data, it is also applicable to vector-based systems. These bring to the archaeologist the geographic configurations necessary for basic spatial analysis.

4.3 Current intra-site data management methodology

Methodologically speaking, the basic recording units for field archaeology were defined in the first part of this century and subsequently refined in the latter part. The problem has been that this traditional organization of archaeological recording into only stratigraphic unit, feature and structure within a zone has become inadequate, compared to the data relating potential that GIS and impending computer applications promise.

finds → stratigraphic unit → feature → structure

A study was undertaken to see which were the limiting elements and what variants could be introduced to extend the breadth of interpretation. The aim was to develop a complementary set of building blocks that could be formalized and permanently configured into the recording process, to be used whenever necessary in the interpretative and analytical stages. It was stressed that these components were to be embedded within ArchéoDATA's general structure, to be used extensively, and not as an analytical afterthought.

4.4 Intra-site building blocks

4.4.1 Archaeological and architectural entities

The archaeological entities (AEs) assign a precise archaeological identity to a series of related stratigraphic units, they formalize a 'feature' that has been interpreted as a specific archaeological manifestation. If a 'feature' is identified as being a silo, it is given its real world name in the form of a three-letter abbreviation[3] (SIL) and numbered uniquely (SIL 1, SIL 2, ...). No longer should we reason at this stage of interpretation in the basic structural units necessary at the initial stages of recording, but instead in the building blocks created by people at the time: walls, floors, silos, etc. (Figure 4.1). By identifying them clearly it becomes possible to select, group and compare readily, to quantify the contents of a series of pits to determine their origin, use and chronology, to determine the components of the walls of a structure, or present clearly the exact components of a specific phase of occupation, so that:

$$\text{finds} \rightarrow \text{stratigraphic unit} \rightarrow \text{feature} = \text{archaeological entity}$$

To be able to use more fully the possibilities of the relational model, two types of archaeological entities are available to the archaeologist. The first, as we have seen with SIL, is the basic component used to identify an elemental archaeological manifestation, the second is the composite AE which brings together several AEs and SUs into a structured archaeological entity: several walls (WALs), floors (FLRs), etc., make up a house (HSE). The use of archaeological entities is extended to sequence and phase diagrams, rendering them easier to create and verify. The architectural entity exists for recording and analysing standing structures and varies only in name from its archaeological homologue. In practice, it has been found that it is better to use two names to characterize each specific area. Would an architectural study consider a refuse pit as being an architectural structure, for example?

4.4.2 Spatial entities

The basic necessities for the spatial management and interpretation of data in archaeology are, as we have seen, similar to those of practically any other discipline: the basic geographic spatial units of x, y and z and the distinctive units related to the field being studied. In the case of archaeology, these units of interpretation were determined to be those which make up the natural environment, human modification of this environment, and lastly, the spaces created in and around places of human activity. It is from these elementary precepts that we have tried to formalize the components necessary to manage the data to be spatially studied.

The concept of the spatial entity (SE) is set forward as the spatial unit for managing assemblages of spatially related data so as to determine and compare usage of a given space. The space is characterized by material remains present and

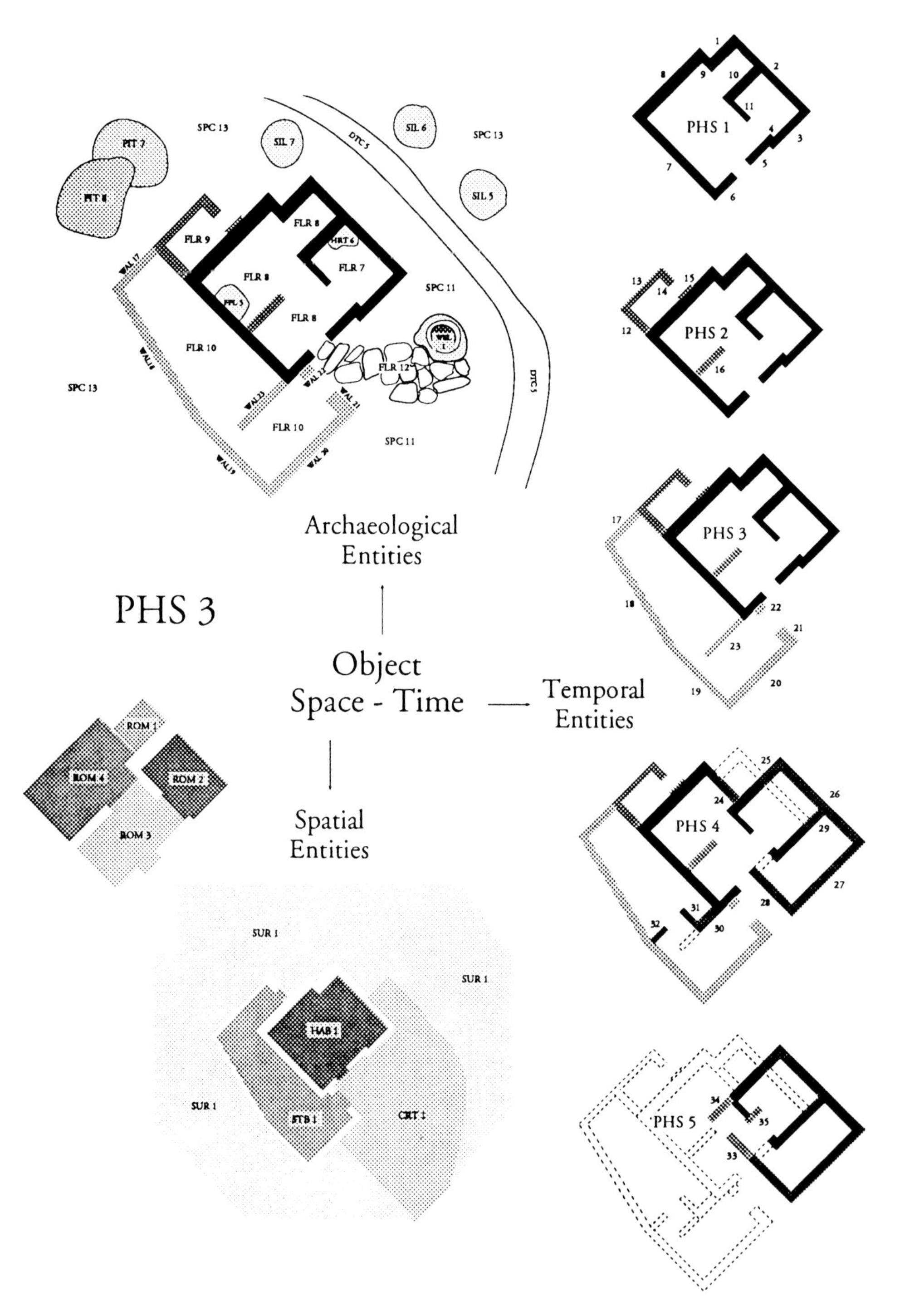

Figure 4.1 Archaeological entities, spatial entities and temporal entities in an object–space–time archaeological information system.

through them it is possible to contrast different spatial elements that make up a structure. Spatial entities can be configured to cover practically any space. Some are logical and their configuration is dictated by the physical shape of a room, a corridor or a courtyard (Figure 4.1), while others, the surroundings of a house, for example, can stay indeterminate as to their use, until later stages of interpretation when they might be characterized and then designated as areas of storage or disposal. The interior areas of a longhouse can be chemically analysed, its sherds quantified and spaces of occupation determined and structurally compared to other houses. Other spaces can be designated as work areas, where specific actions have taken place. This gives:

finds → stratigraphic units → archaeological entities = spatial entities

Archaeological entities can belong to one or more SEs: a wall separating two rooms can belong to two distinct SEs, but the coatings on each side of the wall will only belong to one or the other of the SEs. Even though they are essentially a means for spatially structuring intra-site data, SEs are not necessarily limited to this scale of analysis, and they can be adapted and defined by the archaeologist to cover larger units of space depending on research needs. As with the AEs, two types of spatial entities are available to the archaeologist. The first, as we can see in Figure 4.1, ROM 1, ROM 2, etc., are the basic components used to identify elemental spaces; the second, HAB, is the composite SE which brings together several SEs into a structured spatial entity.

4.4.3 Temporal entities

If archaeology is in essence a scientific study of human chronological/historical evolution through the study of material remains and of the traces left through time, it is fundamental that a spatiotemporal component be structured to analyse the data in this dimension. Archaeology is an object–space–time relationship that needs to manage all three of these factors in order to arrive at valid conclusions. In order to advance analysis, and in particular to model evolution effectively, it is necessary to access spatiotemporal data efficiently, that is, to process the dimensions of both space and time in relation to other data. At present, there is no GIS software to process effectively the temporal dimension, but there is a fundamental necessity to foresee these developments by structuring this essential dimension into general recording and analytical structure of data. A small segment of GIS professionals and programmers have been working these last few years on defining the scope of time and the way it should interact in information systems.

In contemporary archaeology, these changes have been organized into sequences and phases, but as they were not structurally related to space, they could only infer this dimension. This logical organization of the passage of time has worked very well for as long as it was done mentally because the human brain bridges the gaps in the information structure. Because of the indirect manner in which space is addressed, associated data cannot be managed directly, and as such, cannot be used to structure an automated analytical system. The basic precepts, developed and perfected through years of use in field archaeology, are quite satisfactory; what is needed is to formalize them as part of the analytical system that manages the temporal dimension.

To illustrate this, it could be said that the internal and external composition and organization of a building will, in most cases, change during its existence. We can, through the use of archaeological entities, identify objects such as walls and floors, and with spatial entities, we can also identify spaces such as rooms and corridors. But to assemble chronologically contemporary data, we use temporal entities, which have the inherent capacity to integrate one or more AEs and/or SEs, into chronologically coherent manifestations of human presence at a given moment (Figure 4.1).

There has been an enhancement of the basic traditional methods at the site, inter-site, period, etc., levels of analysis. Structured temporal entities can be used to manage any level of chronology and to associate related data to them; for example, MED 1, PRD 6, etc., as shown in the following sequences:

FNDs → SUs → AEs → SEs → SEQ 1

FNDs → SUs → AEs → SEs → SEQ 2 =

SEQ 1 + SEQ 2 = Structural chronology

FNDs → SUs → AEs → SEs → SEQ 3

FNDS + SUs + AEs + SEs → PHS 1

FNDS + SUs + AEs + SEs → PHS 2

PH 1 + PH 2 = Intra-site chronology

FNDs + SUs + AEs + SEs → PHS 3

Site 1 + Site 2 + Site 3 + Site 4 → MED 1

Site 2 + Site 4 + Site 5 + Site 6 → MED 2

MED 1 + MED 2 = Regional chronology

One cannot underestimate the possibilities that will be offered to archaeology by a future generation of GIS software that can handle the temporal dimension, but their use in archaeology will depend on archaeologists having suitably structured data ready for input.

4.4.4 Interpretative grouping

An interpretative group (IG) permits diverse data to be momentarily assembled for study. The use of an IG permits the selection and integration of homogeneous and heterogeneous data, isolating subject matter clearly for a specific analysis or to build a particular synthesis or model. The first use of IGs is to globalize the final stage of a hierarchically structured system; the second has the potential for grouping data independently of previous types of structuring. Throughout the ArchéoDATA project development process, the obligation to utilize uniformly structured environments has led to the efficient flow of information within the system. But, as it is only possible to structure what is known, it became apparent that to carry out certain stages of research efficiently, it was frequently necessary to assemble data in a heterogeneous and piecemeal manner. It was, therefore, necessary to develop a procedure for grouping data and data structures in an independent manner, and at the same time, to benefit from previous structuring.

To illustrate this, a farm may consist of only one building, although usually there are also barns, stables, enclosures, etc., all of them structures, as well as fields, grasslands, etc. Together these make up a farm (FRM), or in other words, an analytical unit (IG) if we were to study rural life at a given moment. An interpretational grouping can also include other IGs as is the case where several farms (FRM 1, FRM 2, FRM 3, ...) and a church (CRH 1) make up a village (VLG 1), to be studied with other villages (VLG 2, VLG 4, VLG 5, ...), which make up a geographical feature such as a valley (VAL 3), a natural or user-defined region (REG 1), etc.

4.4.5 From intra-site to inter-site analysis

The relationships that we model here in an excavation can be readily extended to data from other sources. This form of organization is particularly well suited to setting up data for inter-site analysis using survey data, cultural evolution, land use, etc. It is through the flexibility offered by this approach, that it is possible for the archaeologist to intuitively structure a study in an innovative manner, selecting data which are considered relevant and, using the methods which are appropriate, still benefit from the inherent basic data structure of the system itself. To give an example:

$$\text{FND} \rightarrow \text{SU} \rightarrow \text{AE} \rightarrow \text{STR} \rightarrow \text{SE} = \text{IG} = \text{FRM}$$

$$\text{FRM} + \text{FRM} + \text{CHR} = \text{VLG}$$

$$\text{VLG} + \text{VLG} = \text{VAL}$$

$$\text{REG}$$

4.4.6 A new structure for prehistoric excavation recording and interpretation

The ArchéoDATA project's main preoccupation has been to adapt the best current archaeological methodology into a well-structured system. This has led on occasion to the possibility of innovating and advancing this same methodology. These analytical entities can be perceived in the procedures that have been developed so that traditional three-dimensional prehistoric recording techniques can now be structurally interpreted, and above all analysed, along the same lines that have been developed over the past decades for the archaeology of historic periods. When archaeological recording methods were being examined, it became clear that there were only two basic techniques of recording excavation work in order to reconstruct the past: first, by the stratigraphic unit; and secondly, by the three-dimensional recording of each unit of information. After investigation, it became apparent that it should be possible to structure further the analytical phases of prehistoric excavations and align them with those used in historic periods. With the aid of archaeological and spatial entities, it has been possible to structure prehistoric interpretation explicitly along the lines of what was developed for recording by stratigraphic unit. In the method developed, only the first stage of recording, where each object is uniquely recorded by its x, y and z coordinates, is conserved due to the inherent nature of the excavation work that imposes this method of identification, but all further interpretation is structured along the lines of the stratigraphic method.

4.5 The finality: the integration of data into the GIS structure

It is essential for the generalization of AEs, SEs, TEs and IGs, that they be imbedded within the basic archaeological recording structure, as normal and routine elements of excavation, post-excavation and survey work. It is hoped that the formalization of this series of building blocks, will, through the inherent structuring possible, contribute to the efficient usage of GIS in intra-site and inter-site data usage and further their integration into everyday archaeological recording, interpretation and analysis.

4.6 Conclusion

Will an object–space–time GIS/AIS become a scientific reality? Yes, it is inevitable, because there are scientific and administrative imperatives which will have to be answered sooner or later, with all the accountability that this will entail for the future. Archaeology, whether material or intellectual, is part of our cultural heritage and as such it is destined to be conserved as well as possible, for as long as possible. But there is no doubt that substantial loss to the archaeological record, past and present, is occurring and will continue to occur for many years to come.

We may not admit it today, but future generations of archaeologists, will with all reason, wonder how the archaeological community at the end of the twentieth century, could still not take concerted action to assure the perpetuity of the archaeological record.

References

Allen, K., Green, S. and Zubrow, E. (Eds), 1990, *Interpreting Space: GIS and archaeology*, London: Taylor & Francis.

Arroyo-Bishop, D. and Lantada Zarzosa, M. T., 1992, The ArchéoDATA System: A Method for Structuring a European Archaeological Information System (AIS), in Larsen 1992: 133–156.

Larsen, C. U., (Ed.), 1992, *Sites and Monuments: National Archaeological Records*, Copenhagen: National Museum of Denmark.

Notes

[1] A book covering the ArchéoDATA project in detail is available from Editions Errance (Paris), 1995.

[2] The disparity between the different European cartographic systems, together with the availability of good NATO UTM maps, has resulted in the suggested standardization by the end of this century. Nevertheless, due to known limitations of the UTM coordinates to display and manage extended geographic space, we also use the traditional longitude and latitude coordinates for our analysis.

[3] Examples of the abbreviations used are as follows.

- For excavation work: stratigraphic unit, FiNDs, etc.
- In naming archaeological and architectural entities: SILo, WALl, FLooR, or structured archaeological and architectural entities, HouSE, STaBle, etc.
- In naming spatial entities: ROoM, CORridor, or structured spatial entities, HABitat, CouRTyard, etc.
- In naming temporal entities: PHaSe, SEQuence, or structured temporal entities, ROMan, VACcean, etc.
- In naming interpretative groups: VilLaGe, VALley, REGion, etc.

5

Using GIS with non-systematic survey data: the Mediterranean evidence

F. Massagrande

5.1 Introduction

A very large amount of archaeological data is collected every year, all over the world, during non-systematic field surveys. As these surveys are normally carried out over a number of seasons, the local archaeological units often possess huge quantities of such data, which is traditionally painstakingly transcribed onto paper records by enthusiastic local archaeologists. The spatial distribution of this type of data often suffers from biases introduced by the random nature of the collection, that is, an unplanned survey carried out wherever the archaeologists feel there might be something interesting. The data are also usually supplemented by records of occasional finds by local people who might take in 'that old bone that the dog dug up' just to see what the local Indiana Jones has to say about it. The fate of these data tends to be the same, regardless of the intrinsic qualities of the data or the greater or smaller degree of commitment of the archaeologists: the data lie abandoned or forgotten in the archives of the archaeological units.

This state of affairs does not depend on the laziness of archaeologists or arcane political reasons, but rather on the volume and nature of the data. The total number of archaeological sites which have been recorded during a non-systematic field survey that lasted several years can amount to several hundred, making the analysis of the set of data as a whole very difficult. The data from these surveys are also very rarely quantified, with information being recorded only about the presence or absence of certain types of artifacts (e.g. pottery, glass, coins) and of architectural remains (walls, pieces of columns, cisterns). Often, if specific diagnostic pottery is missing, only the location of the supposed site is recorded, with no mention of what revealed its presence on the ground. Alternatively, sites can be recorded just as scatters of tiles, bricks or pottery, without trying to identify the pottery types. When the pottery is actually identified, it is often to different degrees according to who carries out the survey, or depending upon the experience of individual surveyors. As a consequence, sites in a certain section of a survey area can yield terra sigillata chiara A, C or D, while sites in other sections of the same survey area can be recorded simply as containing terra sigillata chiara (with no sub-groups specified), or generically terra sigillata pottery.

This chapter is concerned with trying to develop a methodology for gathering information from non-systematic field survey data that up to now has usually been regarded as bulky and largely useless. The sample survey data that are being used to develop this method will also be briefly described.

5.2 The project

The Institute of Archaeology of University College, London and the Department of Archaeology of Southampton University are co-operating in a project which aims to develop a methodology to use GIS techniques with non-complete data from non-systematic field surveys. Though the project is mainly focused upon Roman data from the Mediterranean basin, ideally the developed methodology should be universally applicable to non-systematically collected data, regardless of the area and period of origin and the surveying technique used. The specific field of archaeological interest in the development of the GIS method for non-systematic survey data is the study of the relationship between town and country in the Mediterranean region during Roman times (from the Late Republic to the early Middle Ages). From an archaeological point of view, a study of this sort is interesting because, though such a relationship has already been investigated for a number of places in the past, the regions covered were limited in extent. Computer techniques allow the storage and manipulation of much larger sets of data, while the application of the same analytical technique to the archaeological data sets from a number of sample areas from different parts of the Mediterranean region will eventually allow a comparison to be made between these areas. It will then be possible to assess the degree of similarity and regional variation in different regions of the Mediterranean which fell under Roman control. It will also be possible to make inferences about similar or differential development of these areas though time.

It was decided to use low-cost GIS software, as the economic factor always has a major role in deciding which equipment to use in archaeological research. The IDRISI[1] GIS package was chosen, as it does not have special requirements for disk space or memory. IDRISI was supplemented by a number of other packages for the data entry, storage and manipulation.[2]

5.2.1 The database structure

One of the aims of the project was to devise a structure for the database of non-quantified data which could be used by anybody else interested in carrying out a similar study. It is generally a good idea to become familiar with the data before any storage or analytical strategy is planned, so that different options can be considered and the best one chosen. Because of the non-quantitative nature of the data, the only information we had about the archaeological assemblage from a site consists of whether or not a certain type of material was present. Consequently, the database was structured in such a way that information about each site could be retrieved according to its content and/or period of occupation. This involved creating a site database structure with one separate presence/absence entry for each pottery type and object type that we might expect to occur at the site, rather than having a simple descriptive character string field to accommodate everything. The relevant archaeological materials were identified in advance, to avoid having to waste more time transforming the database structure, after the data had been input, to 'make space' for materials which had not been taken into consideration when designing the structure.

5.2.2 The sample of non-systematic survey areas

A number of sample survey areas in the Mediterranean region were selected to develop and test the method, according to a number of criteria:

- that the non-systematic survey had been carried out for enough seasons to cover a large enough region;
- that the data recorded in the surveys were available to the public either as published material or from the archives of local archaeological units;
- that enough information was recorded about each site (i.e. site contents, not just site location); and
- that the survey areas, taken as a whole, should offer a good sample of the different geographical, geological and environmental conditions that are found within the Mediterranean basin.

Of all the surveys in the region of the Mediterranean basin that fulfilled these criteria, four were chosen: Veii, Maresme, Tarragona and Seville.

Veii: the area around the Etruscan and then Roman town of Veii, north of Rome (Italy). The source of information on Veii is a report of a survey published in 1968 in the Papers of the British School at Rome (Kahane *et al.*, 1968). As this information is currently being revised and improved, it was judged better to treat the data with some caution and thus to use the area for testing rather than drawing conclusions. The background data were digitized from maps of the region. The region of southern Etruria, where Veii is situated, is a volcanic area, characterized by round lakes and fertile soil.

Maresme: the region of the Maresme, north-east of Barcelona (Spain). The information concerning the Maresme region was stored in the archives of the Archaeological Service of Catalonia (*Servei de Arqueologia, Generalitat de Catalunya*) in the form of a card catalogue, each card containing information about one of the sites. The sites were grouped according to which urban centre was nearest. The background data were obtained from Spanish army maps of the region. The sites of the Maresme are along the Mediterranean coast. The area is low lying .

Tarragona: the area around modern Tarragona, to the south-west of Barcelona (Spain). The site information for the Tarragona area was kept at the Archaeological Service of Catalonia in Barcelona and stored in the same way as the data for the Maresme. The background data were digitized from maps. Tarragona is located, like the Maresme, along the Mediterranean coast of Spain. The elevation is slightly higher and the coastline more rugged than in the Maresme, but the valley of the river Francolí divides the region into two parts, which differ in geology and land forms.

Seville: the region of Seville in the Guadalquivir valley (Spain). The site data for the province of Seville were collected during surveys carried out in the Guadalquivir Valley, in south-west Spain, until 1986. The more recent data collected in the 1989 survey were not yet available. For the Seville area, maps of the different soil types rather than those of the geology are available (De La Rosa and Moreira, 1987), therefore a slightly different type of analysis will have to be carried out.

Part of the site data was kept at the Direcció General de Bienes Culturales in Seville. The data were stored in a card catalogue and covered the whole of the province of Seville. Other site data for the Guadalquivir valley were obtained from the four books published by Ponsich (1974, 1979, 1986, 1992) containing information he collected during a number of survey seasons in the area. The Guadalquivir valley is (nowadays) one of the most important agricultural areas of Spain, with very fertile soils.

5.3 *What is a site?*

It is important, at this point, to define exactly what is meant by the word 'site', as there is no unique definition of the term, at least in an archaeological context. The problem of the definition of 'site' in archaeological terms has been discussed by a number of authors, and the whole issue can be summarized with a quotation from Schofield (1991: 4): 'The term most widely used in describing surface distributions, therefore, appears to mean something different to the majority of people responsible for their interpretation.'

Because the data used in the project have been collected by a number of different people and different regions have been covered, a standard definition of what a 'site' is must be decided. For the purposes of this study and throughout the rest of this chapter, from now on the term 'site' will be used to refer to any location where archaeological material has been found, whatever its nature, and that has been recorded as an individual entry in the card catalogue of an archaeological unit or in a published source.

5.3.1 The problems with this approach

Because of the very nature of non-systematically collected information, there are a number of problems with this sort of large-scale approach to survey data. It is important to be aware of these problems before any analysis is attempted on the data sets, especially when the different data sets are going to be compared.

As the data have been collected in separate surveys carried out by a number of different people, the data have been recorded in a variety of different ways. The dating and interpretation of the function of each site is bound to be biased by the personal preference of whoever carried out the survey. While copying the data from the card catalogue, it was also clear that different people who surveyed the same area interpreted the surface data in different ways. As the name of the surveyor who recorded each site was included in the site record, it was quite evident, even without using statistics, that different people had different ideas about how to interpret material scatters in the same region. It was noticed that some surveyors would almost always identify sites as villas, while others would never find them, even when surveying the same area; in the same way, some areas, surveyed by the same person(s), would have no medieval pottery at all, but it would appear in a good number of the sites surveyed by other person(s).

Apart from inter-survey variation, intra-survey variation is quite evident and can be quantified. A good example of this is the number of 'generic' (my definition) sites to which no function has been assigned by the surveyors; in the Tarragona area the number is nine out of 218, while in the Maresme the number is 97 out of 198 (i.e. almost half of the recorded sites). In the Veii area 181 'generic' sites are known, while 252 (out of 594) sites are just defined as 'pottery scatters' with no interpretation attempted. In the province of Seville the number of 'generic' sites is 705, while those classified as 'rural sites' are 669 (out of 2166). Moreover, the sites in the Tarragona region are often dated to a very specific time of the Roman period, such as the Republic or Early Empire, while the sites of the Maresme and the Seville province are often just classified as Roman or Medieval. There does not seem to have been any attempt at all by the surveyors of the Veii area to date their sites.

5.3.2 A possible solution to diverse collection and recording

The proposed solution to these problems is to try to create a standard classification of the sites in one area based on the study of the contents of sites excavated in the region. The material found in excavations is much more indicative of the real nature of the local assemblage than just surface data, which always tend to be biased. It has been pointed out by Hodder and Orton (1976: 105) that field-walking always seems to yield a large amount of fine wares because these tend to attract attention more than other wares. Younger material is also more likely to be on the surface than older material, therefore introducing another type of bias in survey assemblages. The study of stratified material from local excavations would allow a broad chronological sequence for the area to be built and could also reveal clues as to specify the function of a site.

Because the data are not quantified, it is easy to fall into the trap of classifying each site with fine pottery as a villa. As we have no information about the amount of each class of material from each location, we would not know whether we are dealing with a single isolated sherd or a deposit of pots. It is important, therefore, to consider the presence of other indicative elements. If there is no single definition of site, the archaeological literature certainly abounds with definitions of what villas were. Before any attempt is made to divide sites into categories, it is necessary to define the categories themselves. Any country site which can be shown to have been a high status one will therefore be labelled a villa, while a country site without evidence of high status will be called a farm. It is not always easy to draw a line between what was a villa and what was a farm, especially when no information about the extent of a site is known. Equally, there can be no unique definition of what a site should yield to be called a villa, as items which were considered a luxury in one place were probably common where they were produced and their presence there does not necessarily imply high status for the site where they were found. This is especially true for pottery types. A particular fine ware can be found in low-status sites in the region where it was reproduced, but, if it was exported to a different area, it is likely to be only found in high status sites in the importing region. Other elements, such as the variation of the pattern of supply of certain types of pottery to certain regions also have a big impact on the resulting archaeological assemblage (Marsh, 1981).

When deciding on the criteria to be used to classify sites in a region, it is vital to remember that not all the criteria used for another area necessarily hold. A careful examination of the archaeological assemblages of each of the regions of interest is therefore required to make the whole study meaningful. The sole presence of fine pottery should not be taken as an indicator of the presence of a villa, but if this is found together with other elements such as mosaic floors, evidence of large structures (e.g. cisterns), or architectural elements (e.g. columns), the site can be safely classified as a high status one. Of course, villas and farms will not be the only categories to be applied to sites.

From a purely statistical point of view, the larger the number of excavated sites used to build up local categories the better, but there are some limitations that must be taken into consideration. The two main limitations are the time that can be spent in dealing with this part of the method and the availability (and easy accessibility) of detailed excavation reports for the regions under study. It will be necessary to work out the most suitable number of excavated sites to be used for this purpose, keeping in mind the limitations mentioned above and that the statistical error will be smaller the more data from excavations are used. When the work on the excavated data is finished, it will probably be possible to date the sites into rather broad date spans, i.e. Republic, Early Empire, Middle Empire, Late Empire, Early Middle Ages and Later Middle Ages. After the diagnostic elements of each site category (and/or period) have been established, it is very easy to write a simple program to run on the site database and thus to determine the site type according to the elements contained in each site. This information can then be directly exported to the Idrisi package for use in the GIS study. Of course this method is not perfect and it is more likely to underestimate the status of a site than to overestimate it, which goes against the common trend in archaeological studies. The status of the sites is likely to be underestimated because the classification relies on what materials found during surface survey have been recorded on paper. The classification program will require a certain number of distinct elements to be present at the same time before a site is classified as a high status one. Also, the lack of information about the extension of sites and the amounts of materials found in them influences the classification. A large site with lots of fine pottery but no other distinctive elements is obviously different from a small site with little fine pottery and no other distinctive elements, but, because the information on site size is not available to us, the two sites would be classified in the same category.

At present, the standard classification and dating of sites has been worked out for the Maresme survey area, based on excavation data published by Marta Prevosti Monclús (1981), who also used information gathered in older surveys and excavations. The excavated sites used to build up the classification are: La Miranda, Can Sentromá and Ca l'Alemany (this last site was excavated by Prevosti Monclús herself). The problem presented by this approach is that usually only the 'rich' sites (i.e. villas for the Roman period) are considered worthy of excavation, and thus it is difficult to gather information about other types of sites.

5.4 Other problems encountered with this approach

Other problems linked to non-systematic data cannot be solved so easily. A good example is illustrated by the distribution map of the Maresme (Figure 5.1). It is clear

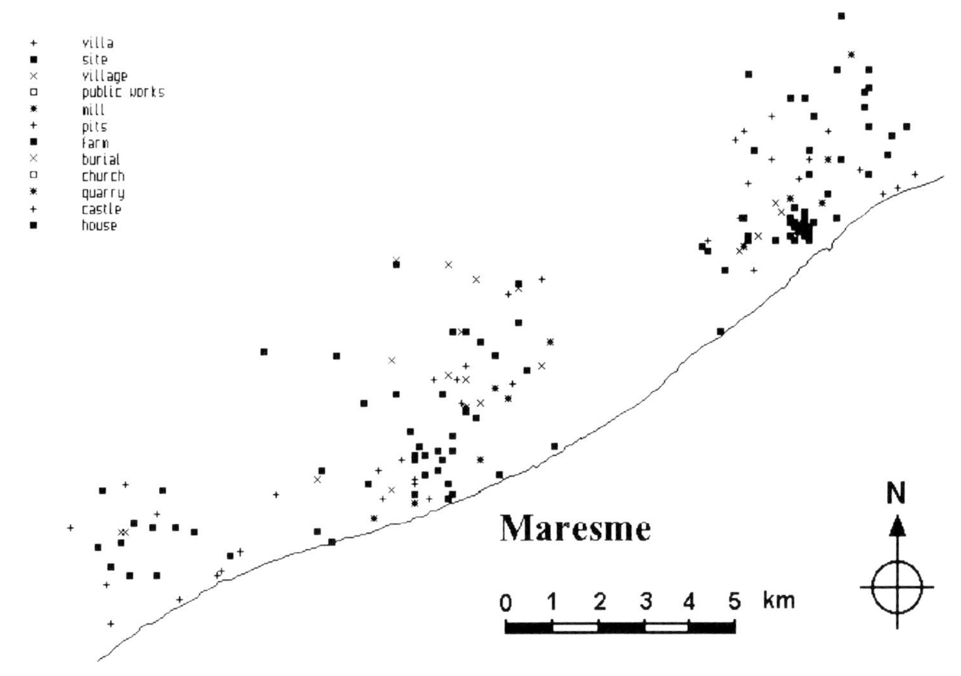

Figure 5.1. The site distribution of the Maresme region.

that there are two main concentrations of sites, one centred around the Roman town of Iluro (modern Matarò) and the other to the south-west of this, with a remarkable lack of archaeological sites in between. The reason for the existence of this pattern seems to be the fact that the archaeological units in charge of surveying the two halves of the area kept strictly to what they saw as being their own area. This sort of pattern is obviously artificially created by recovery biases and it is quite important to keep this point in mind when doing spatial analysis on these data.

Another example is the distribution in the Guadalquivir valley (Figure 5.2). To the south of the town of Seville, there are major concentrations of sites around the two towns of Lebrija and El Coronil, while the area north of Seville has been uniformly surveyed. The pattern of recovery of Roman sites is also quite clearly influenced by the position of modern artificial features. A good example of this is the pattern around the modern town of Carmona, where the sites follow the main roads, especially the C432 motorway and the major rivers. A cost surface was created from the DEM of the valley of the Guadalquivir, having modern towns and modern major roads as starting points for the cost distance buffers. Figure 5.3 shows the frequency of sites plotted against distance from the modern features: the largest concentration of archaeological sites lies within 5 km from any major feature, with the curve dropping sharply afterwards, with the increase of distance (the sites of 0 distance account for sites found in towns).

Such a distinctive patterning is not detectable by the naked eye in the pictures of the other survey areas. There can be two reasons for this: either a patterning induced by modern features and communication routes is not present in the Maresme and in the region of Tarragona, or this pattern is actually made more discernible in the Seville picture by the scale. The image covering the survey of the

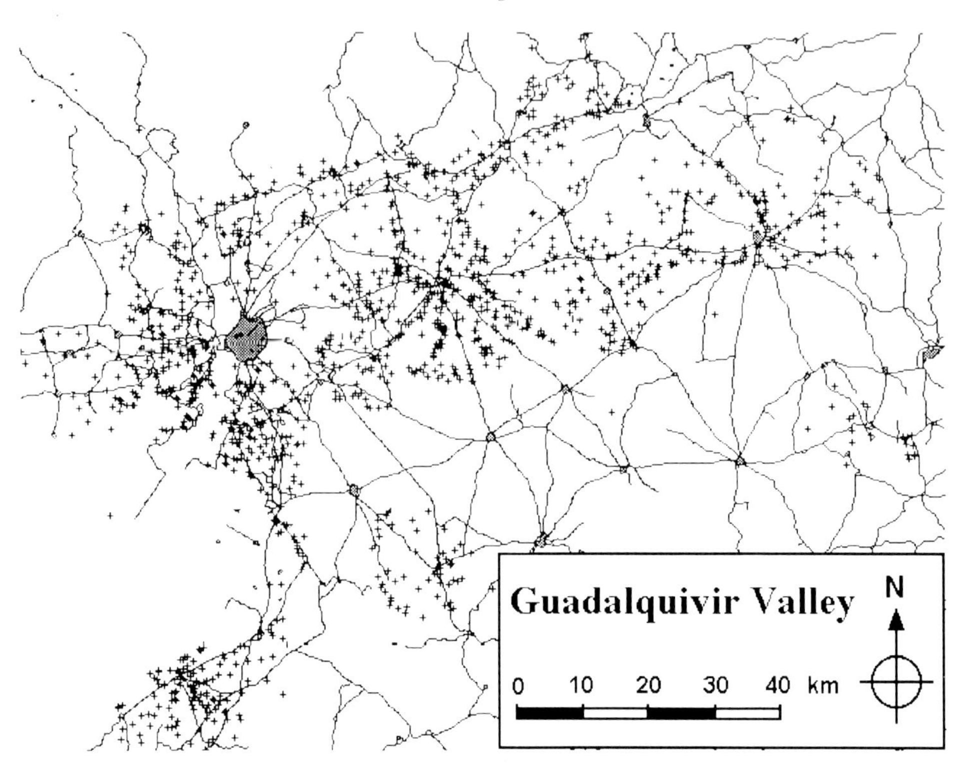

Figure 5.2. The site distribution and the modern towns and road network in the valley of the Guadalquivir. The large town is Seville.

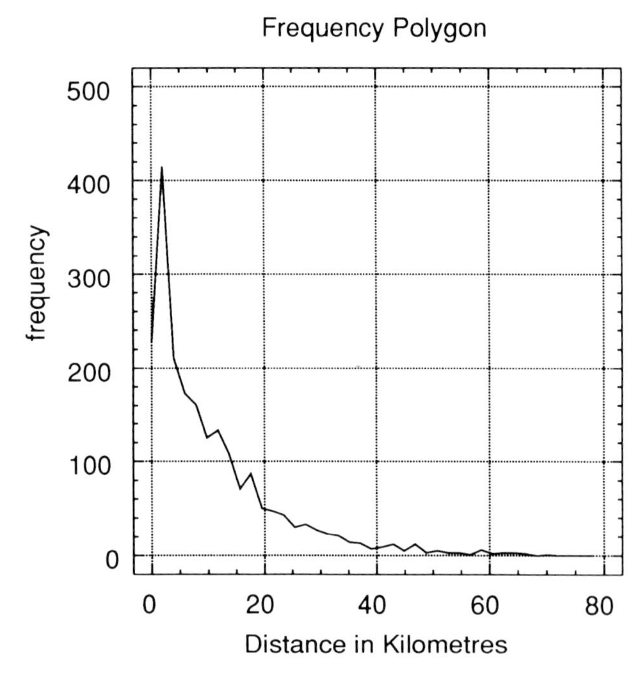

Figure 5.3. The site frequency plotted against the distance from the nearest modern town or road.

province of Seville represents an area on the ground of 143 (east) × 108 (north) km, the area of the Tarraco survey is 34 (east) × 23 (north) km, the Maresme survey area is 20 (east) × 15 (north) km and the Veii survey image covers an area on the ground of 11 (east) × 18 (north) km. As the survey of the valley of the Guadalquivir covers a much larger area, more absolute spatial information is present and large-scale patterns, which could appear confused in small scale images, are more easily detected. If such patterns exist in the Tarragona and Maresme regions, but are hidden by the scale of the images, it should still be possible to prove their presence statistically.

5.4.1 Computer problems

Apart from the problems posed by the nature of the data and the influence of human subjectivity, other practical problems were encountered in setting up the system and inputting the data in the right format. One example is the data from the province of Seville, which spans two UTM zones so that the line representing the border between the two UTM zones cuts across the city of Seville making it difficult to exclude one of the halves from the analysis. As IDRISI (and most other GIS) requires that all the elements are referenced to the same coordinate system, it is impossible to use these data directly. The site coordinates contained in the database are obviously referenced to the grid covering the area in which each site occurs and suffer from the same problem.

The most straightforward solution is to create two different GIS databases, each containing information about the background and the sites on one side of the section border line, but this would then make it impossible to use the two sets of data together for a single spatial analysis. After evaluating a number of possible alternatives, it was decided that the best solution was to write a program to perform a mathematical transformation on the database coordinate data. This program uses standard geometrical transformation formulae to ensure that the whole of the data set uses the same coordinate grid. After the transformation, the data were transferred into IDRISI using a standard export program.

A similar problem was presented by other data for the province of Seville published by Ponsich (1974, 1979, 1986, 1992). The sites reported by Ponsich are all referenced on the Lambert rather than the UTM grid. Again, the direct use of these data in the GIS database was not possible and the program was modified to perform the transformation of the site coordinate data from the Lambert to the UTM grid.

Other pieces of software had to be produced to compensate for some of IDRISI's limitations. One problem with IDRISI is that it will not let the user automatically extract the value of background variables (i.e. elevation, geology type, soil type, slope, etc.) in the cell where a site occurs. Such an operation can be performed in IDRISI using the extract module, but for large sets of data this is extremely awkward and it involves a large amount of manual work. Since one of the aims of the project is to adapt IDRISI to the needs of archaeological research, a program was written in Turbo Pascal to extract the background values in the form of an ASCII file which can be easily exported to statistical packages such as Statgraphics and MV-ARCH or other types of software. Other software was produced to make AutoCAD (v. 11 and higher) fully compatible with the IDRISI vector file format and to link the site image in IDRISI directly to the site database in dBaseIII+.

5.5 The last stage of the project

At the time of writing (early 1994), the project is not yet completed. Though all the data for the four survey regions has been almost completely input into digital format, very little analysis has yet been carried out.

Once the data entry has been completed, the analysis will involve sampling the data, trying to work out the elements that influenced site location, i.e. distance from towns, soil type (as inferred from the geology types), site types, etc. All distances will be calculated from cost surfaces rather than basic distance buffers to give a better estimation of the accessibility of any point from the others. The friction surface to be used to create the cost surface will include both geographical elements (terrain, geology type, etc.) and artificial elements (known Roman roads, Roman bridges); a different weight will be given to each element in the cost surface depending on assumptions that will be made about their relative importance. The analysis will be done for all the large-span periods to which the sites can be dated, and the variation of the settlement pattern in time will be studied. Inferences will be made about the possible reasons for the variation in such a pattern and hypotheses about it will be tested with the aid of spatial statistics. After this has been done for all the survey areas, the results for each will be compared to check for similar or differential development of regional settlement patterns during the Roman period.

References

De La Rosa, D. and Moreira, R., 1987, *Evaluación ecológica de recursos naturales de Andalucía*, Seville: Agencia de Medio Ambiente–Junta de Andalucía.

Hodder, I. R. and Orton, C., 1976, *Spatial analysis in archaeology*, Cambridge: Cambridge University Press.

Kahane, A., Murray Threipland, L. and Ward-Perkins, J. 1968, *The Ager Veientanus north and east of Veii.* Papers of the British School at Rome, **36**.

Marsh, G., 1981, London's samian supply and its relationship to the development of the Gallic samian industry, in Anderson, A. C. and Anderson, A. S. (Eds), *Roman pottery research in Britain and north-west Europe*, BAR Int. Ser: 123 (i) 173–238.

Ponsich, M., 1974, *Implantation rurale antique sur le bas Guadalquivir*, Vol. I, Paris: Publications de la Casa de Velazquez, Série Archéologie.

Ponsich, M., 1979, *Implantation rurale antique sur le bas Guadalquivir*, Vol. II, Paris: Publications de la Casa de Velazquez, Série Archélogie.

Ponsich, M., 1986, *Implantation rurale antique sur le bas Guadalquivir*, Vol. III, Paris: Publications de la Casa de Velazquez, Série Archéologie.

Ponsich, M., 1992, *Implantation rurale antique sur le bas Guadalquivir*, Vol, IV, Paris: Publications de la Casa de Velazquez, Série Archéologie.

Prevosti Monclús, M., 1981, *Cronologia i poblament a l'area rural de Baetulo*, Monografies Badalonines No. 3, Badalona: Museu de Badalona.

Schofield, A. J. (Ed.), 1991, *Interpreting artefact scatters. Contributions to ploughzone archaeology*, Oxbow Monograph 4, Oxford: Oxbow Books.

Notes

[1] IDRISI is rather inexpensive and will run on a standard PC without special requirements for hard disk space or memory (though the more recent versions require a 486 processor). While there are no standard raster GIS functions that IDRISI cannot perform, the quality of the display and hardcopy output are rather limited, though the display capabilities have been improved in the 4.1 version of the package. No vector GIS functions are supported, apart from simple plotting (with some bugs!).

[2] AutoCAD for Windows v.12 was used for inputting the map data, dBase III+ to keep the site databases, Statgraphics 6.0 and MV–ARCH to supplement the statistical modules of IDRISI, and finally Turbo Pascal 6.0 was used to produce a series of programs to extract information from IDRISI image files in a format suitable for export to the statistical packages, and to solve some problems with the implementation of some of the data.

6

Flood dynamics and settlement in the Tisza valley of north-east Hungary: GIS and the Upper Tisza project

M. Gillings

6.1 Introduction: the Upper Tisza project

Since 1991, the Anglo–Hungarian Upper Tisza Project (UTP) has been involved in detailed research in north-east Hungary. The principal aims of this interdisciplinary research project are to define and explain changes in environment, subsistence, settlement, land use and social structure over the last 10 000 years (Chapman and Laszlovsky, 1992). The project is a co-operation between the University of Newcastle upon Tyne, Eotvos Lorand University Budapest and the Magyar Tudományái Akadémiai Institute of Archaeology.[1]

Since 1991 fieldwork has been targeted upon three distinct ecological zones:

1. the main Tisza valley (1991 fieldwork season);
2. the low lying plain of the Bodrogkoz with alluvial areas and associated river terraces (1992 fieldwork season); and
3. the piedmont zone of the southern Zemplen hills (1993 fieldwork season).

To study the three areas, a variety of analyses are being undertaken. These range from enhanced soil mapping and detailed core analysis to programmes of geophysical survey and lithic characterization. One important area of research has been the identification of archaeological features through a programme of systematic field surveys. The aim has been to identify and map ploughzone concentrations of archaeological material.

The role of the GIS within the project structure has been to provide the analytical framework for the organization, management and articulation of the varied data sets. The initial application of the GIS has concentrated on one facet of the research problematique: the investigation and analysis of changes in human settlement and exploitation strategies through time. The first stage of this study concentrated upon data generated during the 1991 field season. Research effort has been directed towards the examination of the patterning of observed artefact concentrations within the landscape.

To undertake such a study and attempt to examine the organization of activity foci in the context of the surrounding landscape requires the following fundamental assumption to hold true. The modern observable landscape form, and its related dynamics must be closely comparable to those in operation during antiquity. In the

Tisza area, large-scale changes in the topography and hydrology of the flood plain over the last 200 years have resulted in significant alterations. To be able to examine and model past human behaviour these alterations have to be undone and past landscape forms accessed.

This chapter attempts to illustrate how the GIS is being utilized to facilitate a more detailed examination of past social dynamics through the retro-modelling of past land-forms, in particular the dynamics of an active flood plain.

6.1.1 The nature of the archaeological data

The 1991 research season concentrated upon a 432 km^2 landscape block centred upon the town of Polgar (Figure 6.1). This effectively targeted two distinctive landscape zones, the main Tisza river valley, with its characteristic low-lying flood-plain, and the adjoining terraces of loess, often capped by sand deposits. A total of 15 per cent of this block was covered by a systematic transect survey, yielding a total of 148 sites.[2]

Once each site had been identified, its area was determined and the position recorded on a 1:10000 field map. A sample collection of diagnostic lithic and ceramic material was then taken to provide dating evidence. On the basis of this, the primary archaeological data set can be defined as a series of point coordinate locations. These points were related to presence/absence summary data for 26 archae-

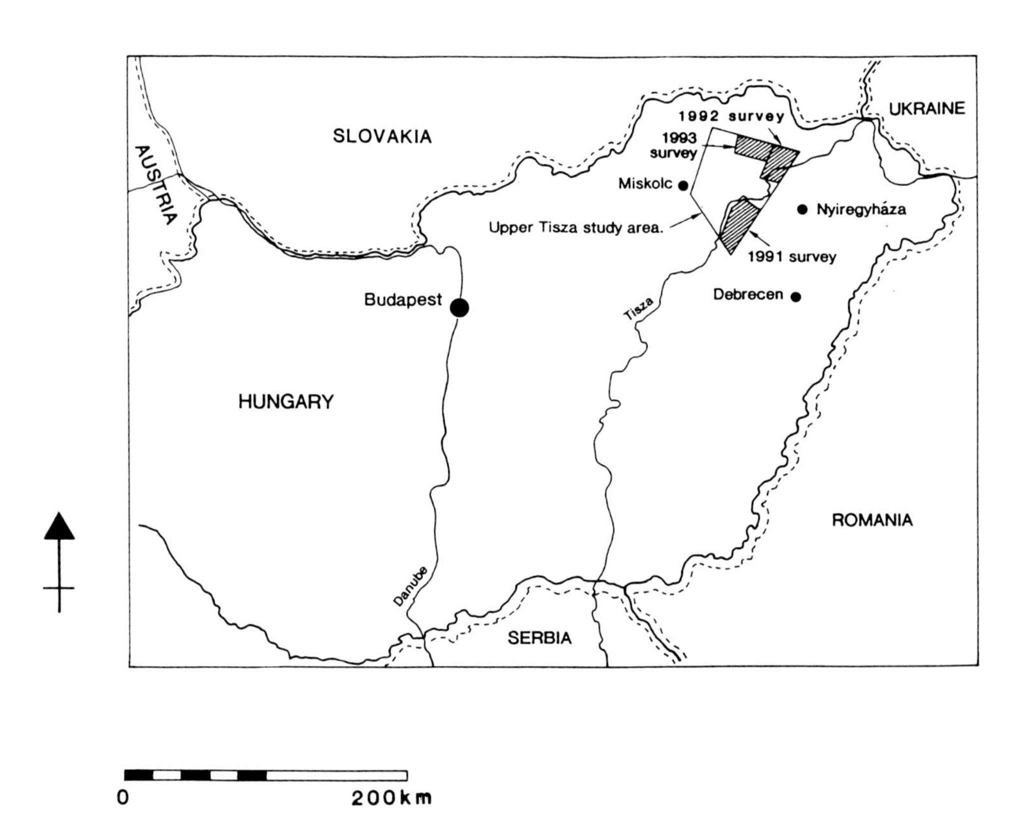

Figure 6.1 Location of the study area.

ological phases, spanning the period between the early–middle Neolithic (5200 CAL BC) and the present day.

The relevance of the calculated site area as a useful analytical parameter was thrown into doubt by subsequent analysis of the artefact assemblages. It was revealed that, of the 148 sites discovered, 137 could be dated to more than a single period. To utilize the area in any study of diachronic change presupposes some degree of stability between utilization phases. It is clear that a variety of scenarios could have shaped the final recorded area. For example, a number of small, adjacent, horizontally stratified occupation events could result in a misleadingly large artefact spread. The same is true of the vertical stratification of large and small occupations. In each case, as soon as the site distribution is broken down by phase, the value of the recorded area becomes questionable. This limitation is currently being addressed through systematic, gridded intra-site collection of a range of multiperiod sites.

6.1.2 The character of the flood-plain landscape

Before any analysis of the site distribution data could be undertaken, the fundamental nature of the Tisza flood plain had to be taken into consideration. Far from being static, the landscape had been subject to a complex set of nested cyclical dynamics, long-term processes and unpredictable 'wildcard' events throughout the Holocene period. All of these factors had an important influence upon its evolving form.

The first and most obvious of the dynamics comprised an annual, seasonal flood event occurring through the months of April and May. This predictable phenomenon operated within a secondary cycle of flood events, the green floods, appearing in June. These secondary floods occurred during particularly wet years, operating on an approximate seven-year cycle. The effects of these floods upon the landscape of the region were enormous. Historical sources suggest that the cycle of floods resulted in the inundation of 30–50 per cent of the available land surface. These areas could remain under water for up to one-third of the year (Loczy, 1988: 167–74).

The principal wildcard occurred as a result of extreme flooding of the river Danube. This rare event had the effect of causing the Tisza to back up from the confluence of the two great rivers causing catastrophic flooding. Between 1730 and 1888, eleven devastating floods are attested and historical records for the period 1846–1857 describe how one-third of the Tisza flood plain could spend almost an entire year under water, with a further third inundated during the annual flood period.

The response to these extreme and catastrophic events was to initiate extensive river control works through the construction of levees and canals. The result has been effective control of the dynamic flood cycles since the mid-nineteenth century.

In conjunction with the cycles of flooding, longer-term processes were in action. Of these, the most important was the gradual meandering and relocation of the course of the river Tisza across the flood plain. This is evidenced in the landscape today by the large number of distinct relic river channels and substantial ox-bows.

Operating within this overall framework of change would have been many small-scale events associated with human exploitation strategies. Although more temporally and spatially restricted, clearance and drainage activities could have had a marked effect upon the dynamics and resultant shape of a given flood event (Ward, 1990: 272).

What is clear is that the modern, static landscape cannot be used as an analytical template for the study of past settlement and exploitation strategies. Throughout antiquity, large portions of the available land surface could have spent part of the year dry and part flooded. This would have had a crucial impact upon settlement location, optimal agricultural regimes and movement through space. What is also apparent is that between flooding events, large areas of land would have spent part of the year in a semi-waterlogged intermediate state as the flood waters drained away. This land may have proved too wet to build on, cultivate or walk across and yet too dry to fish or raft. It is obvious is that any attempt to study the spatial organization of activity foci must incorporate and accommodate the dynamic complexity of the study area.

6.2 The retro-modelling of landscape dynamics

Attempts to assess the effects of the cyclical flood events upon the earlier phases of landscape exploitation were hampered by the marked absence of quantitative, and readily accessible data sources. The extensive control measures prevented the use of direct observations and the available historical records tended to be largely descriptive. In the case of pre-control maps, factors such as flood extent could at best be inferred from negative evidence, such as an absence of settlement. In response to this, the first stage of the GIS program constituted an attempt to simulate a typical, predictable seasonal flood event. To undertake such a procedure first required the generation of an analytical, topographic surface.

The topography of the study zone was digitized from 1 : 10 000 base maps as a set of contour lines, taken at elevation intervals of 0.5 m. This interval was selected as best encompassing the subtle topographic variations characteristic of the flood-plain environment. This took the form of key dune and terrace systems, typically represented by elevation changes of less than 3 m, spread over hundreds of square kilometres. Once digitized, the contour layer had to be transformed into an analytical surface—a digital elevation model (DEM). The quality and robustness of the DEM effectively determines the quality of subsequent simulations based upon it; therefore it was crucial to ensure the preservation of the critical micro-topographical variations inherent in the land form, and encoded within the contour data. This criterion was achieved by the employment of a triangulated irregular network (TIN) surfacing technique prior to generation of the DEM. In practice, the digitized contour lines were sampled along their lengths to yield a set of irregularly spaced points of known elevation. The points were then linked by a network of triangles. The relevant slope and aspect data for the inter-point areas were encoded within the facets of the linking triangles. The derived TIN surface was then converted into a raster grid format to yield a continuous elevation surface, the DEM, upon which subsequent analyses could take place.

At this point, the spatial resolution of the analytical area had to be decided, as this was to be encoded within the resolution of the raster grid.[3] This question was resolved inductively by consulting the survey records to determine the minimum size of observed artefact spread. This was found to be typically 20 m on the shortest

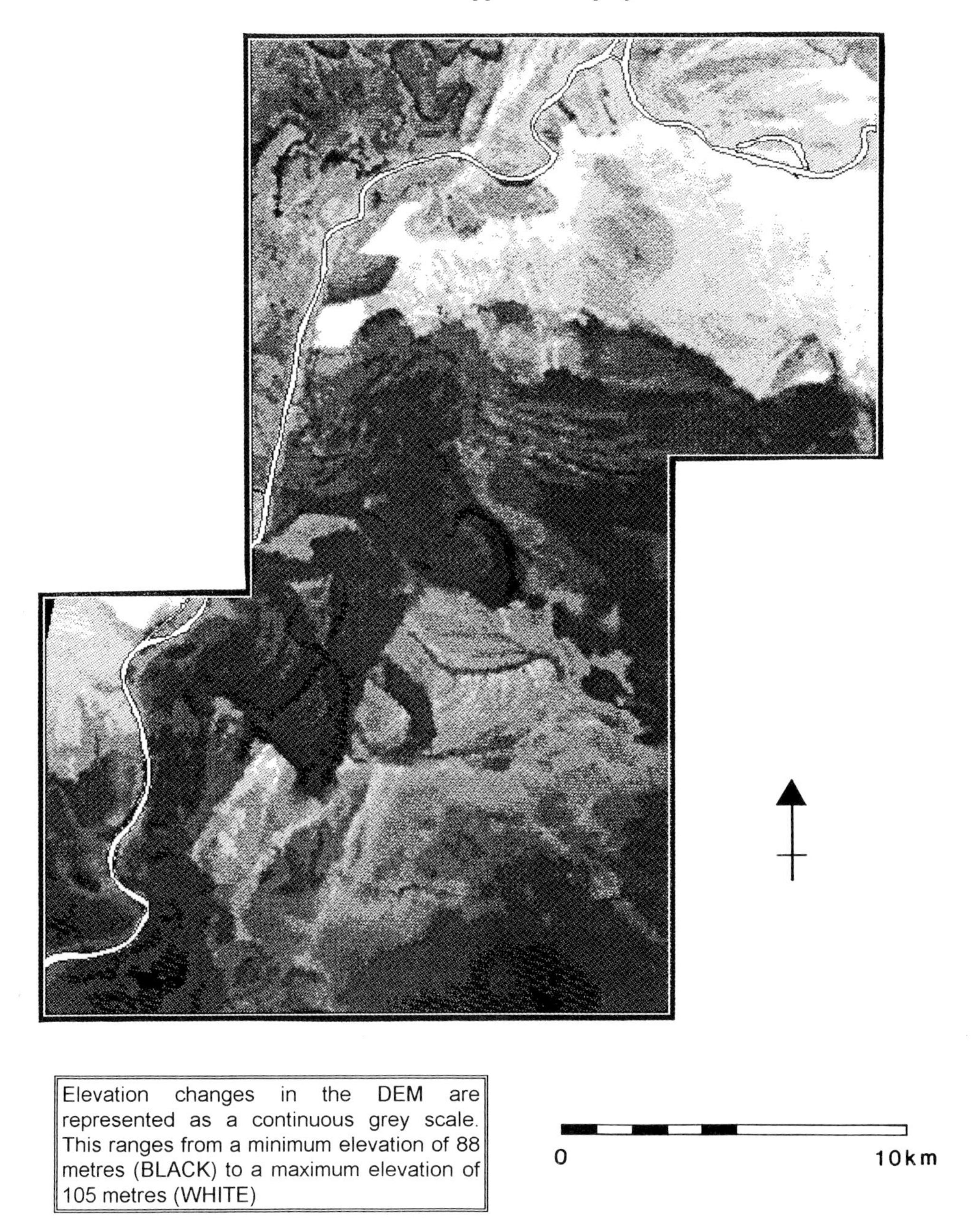

Figure 6.2 The digital elevation model (DEM).

axis. While any grid size coarser than 20 m could lead to site clumping and a loss of information, any finer grid resolution would add nothing to the quality of the information. On this basis, the cell size was set to 20 m square. It should be noted that the grid cell resolution was retained at all stages of the analysis.

The resultant DEM is illustrated in Figure 6.2. The effectiveness of the surface in summarizing the micro-topographic detail is evidenced in the complex network of relic channels and ox-bows that can be clearly defined. For reference purposes the modern course of the river Tisza is indicated in white.

6.2.1 The flood simulation

Historical sources indicated that a typical flooding event would have resulted in inundation up to an elevation of 91.5 m above sea level (Shiel, 1992: 25–7). The first stage in modelling such an event was to identify what can be termed the 'optimum' flood zone. To delineate this area the DEM was interrogated with the aim of identifying all areas of the landscape falling below the critical threshold. To simulate a flood, water was then introduced into these areas.

In this form, the simulation was far too simplistic. One important limitation proved to be that the optimum zone did not respect the concept of blocking terrain. This resulted in a large number of small lakes isolated from the river and the main bulk of the flood zone. As the encroaching flood waters would not have had direct access to these areas, it is questionable whether they would have suffered inundation, though the rise in water table and soil percolation effects may have resulted in appreciable water accumulation. The next stage in the process was to refine the optimum model. To achieve this, the GIS was used to launch a flood event from the course of the Tisza. In practice, the river course was identified as the primary source of intrusive water, and the flood-zone edge was permitted to spread out from the river, crossing the surface of the DEM until blocked by intervening terrain. The result was a much more robust and inductive simulation (Plate 1).

Various questions can be raised concerning the integrity of the model, based as it was upon modern topography and the present-day course of the Tisza. Factors such as river mobility, the extensive re-modelling of the nineteenth century and the cumulative effects of successive deposition cycles were not explicitly incorporated into the simulation process. They were, however, carefully considered during the simulation design.

Recent research into river mobility has suggested that the Tisza adopted its present course at the boundary of the Pleistocene and Holocene periods (Loczy, 1988: 169). On this basis, it can be assumed that no significant large-scale river channel migration will have taken place during the period under study. In the case of the flood control works, the main impact was upon the original hydrological network. Stream channels were heavily restructured and canalized, with construction of major levees directly adjacent to the river. The overall topography of the landscape was little altered, though the effect upon the spatial form of the existing drainage network was enormous. The question of deposition effects is more problematic. Detailed mapping has identified substantial remnant Pleistocene surfaces within the flood plain, none of which are significantly covered by blankets of alluvium (Rónai, 1987). This suggests almost negligible sedimentation over the bulk of the flood plain, away from the immediate environs of the river itself. Shiel (1992) has interpreted this phenomenon as reflecting flood water moving slow enough to deposit silt and sand close to the river and yet fast enough to prevent carried clay particles from settling out. Despite this, the results of the preliminary soil coring programme have identified fluvial deposits 6 m deep within the flood-plain, away from obvious depression features such as ox-bows and relic channels. Cores taken at depression sites showed sedimentary deposits exceeding 7 m in depth. This suggests

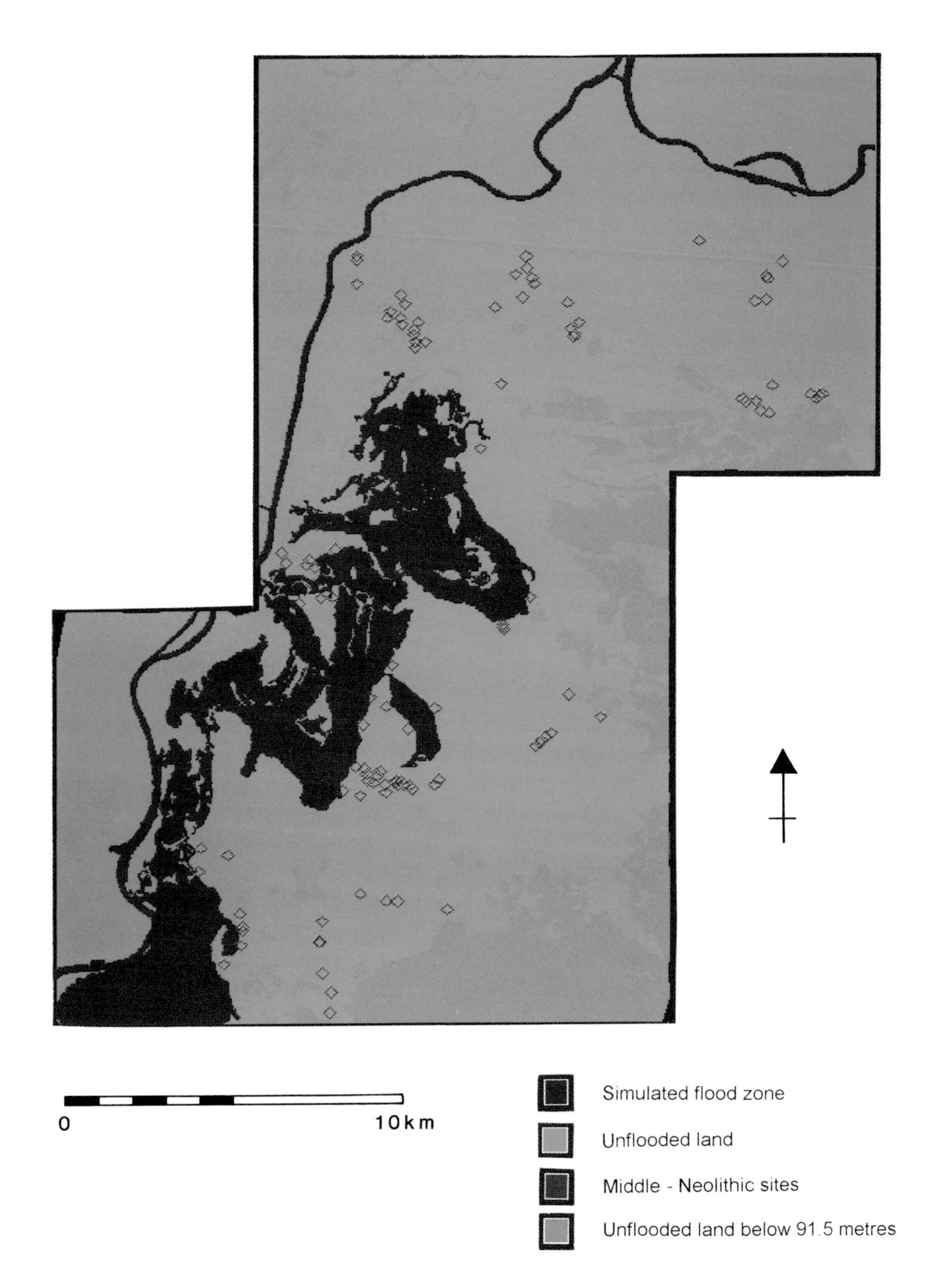

Plate 1. Flood simulation and the distribution of Middle Neolithic sites in the Tisza valley, Hungary.

that a blanket of sedimentary material exists across the flood plain below 93 m above sea level. To assess and compensate for possible alterations in the basic landform requires refinement of the original simulation. The mechanics of this refinement are discussed below.

One clear limitation of the model concerns the question of spatial context, in particular the influence of topographical trends beyond the boundaries of the defined study area. As the study block did not represent a bounded spatial entity, such as an island, the accessibility of certain portions of the flood plain to the encroaching waters could be dependent upon topographical trends falling beyond its boundaries. In an attempt to compensate for any contextual problems the wider landscape form was examined and an additional 48 km^2 digitized to ensure that the principal topographic trends were encompassed. The problem of boundedness is of particular importance in procedures such as drainage basin delineation, and has been acknowledged at all stages of the simulation process.

What is clear is that the simulation constitutes an idealized end product, a very general first-stage analytical model based upon very simple dynamic assumptions. To address more detailed problems such as the influence of localized rainfall events and the conservation of water volume with increasing surface area, more research has to be undertaken. With the simulation model in place, efforts can be targeted to test it. Refinements will best be achieved through a programme of enhanced modelling, fully integrated with fieldwork programmes. This process is already underway with a programme of soil coring being organized to place a transect through the postulated dry and flooded zones. As the data from this and the detailed soil mapping become available, the basic nature of the flood-plain soils and presence, depth and textural composition of alluvial deposits will be established. This will supply an important check upon the integrity of the modelled flood boundary and will yield important data concerning drainage and surface flow rates. Returning to the question of alluvial modification to the landscape, on the basis of textural analysis and estimates of the transportation potential for the flood water, the GIS will be used to delineate deposition zones at increasing distance from the river course. These data will then be used to simulate the progressive build-up of alluvial deposits over time. This information will be fed back into the first stage of the modelling process, enabling the DEM to be restructured, to reflect the removal of chronological blocks of alluvium prior to flood-zone delineation.

In the context of advanced modelling, several shortcomings have already been addressed in an attempt to enhance the existing simulation. One of the most important of these is that factors such as the rate and form of the flood have not been taken into consideration. Quantification of these parameters required a detailed understanding of the flood water routes in and out of the plain. This in turn required careful examination of the associated hydrological network, in particular the stream branches connecting the river course to the bulk of the flood plain.

6.2.2 Hydrological modelling

As discussed above, the flood control works of the nineteenth century completely remodelled the pre-existing hydrological network. The modern system comprises regular canals and angular canalized streams. To examine the routes in and out of the flood zone, the pre-control network had to be re-established. This was also crucial if variables such as proximity to water courses were to be considered as factors in any analysis of site location during antiquity. For the purposes of subsequent archaeological analyses, two crucial components of the overall hydrological system had to be isolated—the stream network and the drainage catchments, or basins. Drainage basins are of particular importance. They can be defined as discrete areas draining water to a common outlet. For hydrological purposes, the basin comprises one of the most important analytical study units: it represents the discrete area within which hydrological processes are integrated.

Derivation of the required data from the existing DEM required the application of advanced hydrological modelling algorithms, which served to simulate the movement of fluid across the continuous analytical surface. The first stage in the process involved the smoothing of the DEM to remove small-scale imperfections, such as isolated sinks and peaks, that could distort the simulated passage of water. The DEM was then analysed and new raster data layers were extracted. The first of these layers encoded the direction of flow across the surface for each component grid cell. The second data layer summarized the accumulated volume of water at each of the points.

The directional data layer was utilized in the identification of drainage basins. Here the problem of spatial context once again became apparent. Before delineation of the drainage basins in operation within the region, the principal water outlet points from the study area had to be identified. As the study area was not a discrete spatial entity, these exit points had to be identified manually. This would involve a high degree of subjective interpretation that could markedly influence the final form of the model. To overcome this potential source of error, a preliminary analysis was undertaken, using the DEM to identify points along the study zone boundary where simulated water flows were most likely to exit. The data generated by this analysis were incorporated along with the flow direction information into the basin generation algorithms. The result was a raster-based area map defining all of the basins and fractions thereof contained within the study block. The resultant map is illustrated in Figure 6.3.

The next stage in the modelling procedure centred upon the generation of the pre-control stream network. To achieve this, the accumulation data layer was analysed to identify the routes of maximum accumulated flow through the landscape. In practice, a threshold was applied to the flow accumulation data in the form of a 50-cell catchment. All grid cells exceeding the threshold were positively coded and written to a new raster grid. The result was a branching map of linked channel courses covering the survey area. In this form, the data were undifferentiated. There was no indication of the relative magnitudes of each of the integrated stream segments. To quantify the network, the component streams were ordered using the Strahler method. Here the end branches of the network, corresponding to the smallest ephemeral tributaries, were ordered 1. When two first-order channels joined, the resultant stream channel was ordered 2. The calculated order increased as progressively higher ordered streams merged (Ward, 1990: 316). The final stage in the process was to convert the resultant data grid into a vector form, comprising nodes linked by ordered stream segments.

With the drainage basin and stream data in place, the original flood simulation could be contextualized and refined, with the identification of the bounding area and principal flood water routes in and out of the plain. Vectors describing the course of the Tisza were combined with the ordered stream data and a network

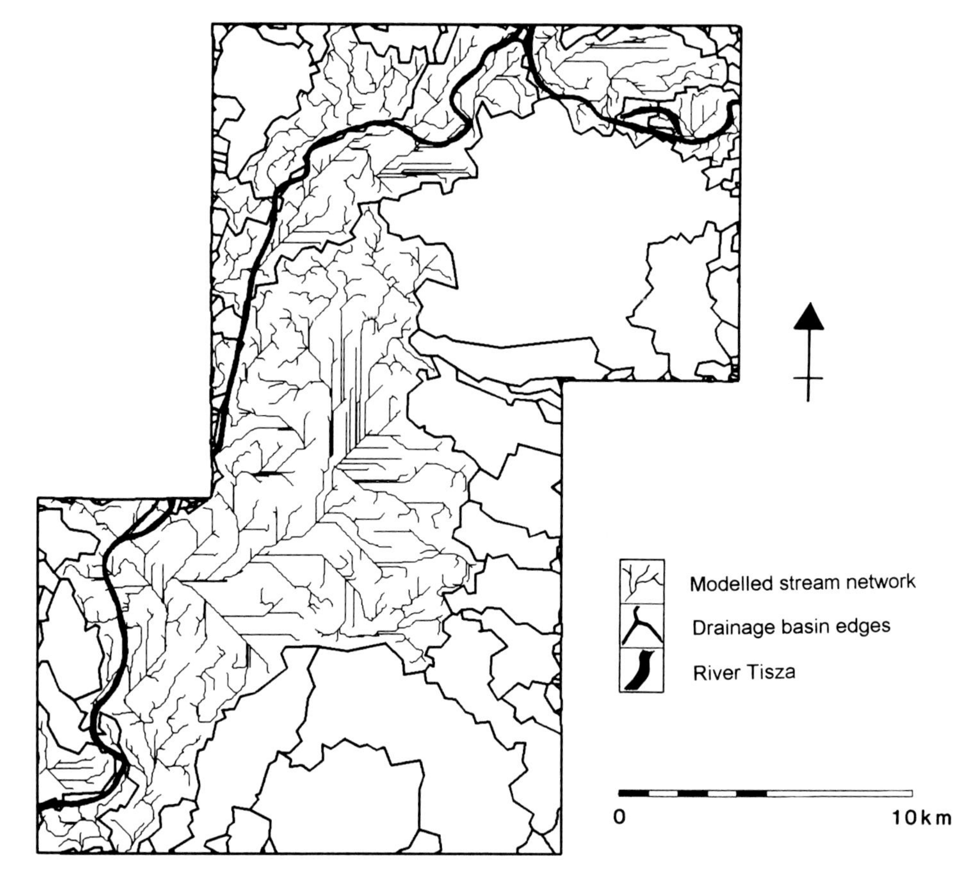

Figure 6.3 The modelled drainage basins and stream network.

analysis was performed. This served to identify and isolate all stream segments and nodes through which a path could be traced directly to the line of the river itself, in effect stream branches draining directly into the Tisza. The networked stream data are shown in Figure 6.3.

6.2.3 Summary

Through the application of the GIS a retro-modelling exercise was undertaken in an attempt to simulate a typical annual flood event. To realize this simulation, a variety of integral hydrological analyses has been performed. The principal drainage basins for the study area have been defined and the pre-control stream network has been identified, with isolation of tributaries flowing directly into the river Tisza.

As discussed, the model is a long way from completion. At present the 'before flood' and 'after flood' stages have been identified and the key flood routes detailed. It still remains to characterize the transition stages between the wet and dry states. These refinements will be derived from carefully targeted fieldwork to test aspects of the model, along with the integration of detailed soil coring and mapping data as they are processed. Utilizing these data will enable factors such as drainage potential, evaporation rates, localized rainfall effects and the impact of clearance activities to be studied. The results will be fed back in to the simulation process, enabling the identification and modelling of both temporal stages in the dynamic and the effects of wildcards.

6.2.4 The results of the simulation

The combined results of the flood simulation are shown in Figure 6.4. The maximum flood extent can be seen in conjunction with the contributing stream network and drainage basins. Study of the drainage catchment boundaries shows how clearly the integrated facets of the overall model are contained within the major study zone basin.

In terms of the impact of the flood event upon the landscape as a whole, the simulation resulted in the inundation of 15 per cent of the available land surface. This is considerably less than the 30–50 per cent level suggested by earlier researchers. The most striking feature of the flood-zone generated was the way in which the observable landscape was partitioned by the intrusive waters. Instead of a simple wet block–dry block contrast, the flood zone adopted a shape largely dictated by the complex of large ox-bows and relic channels that characterize the

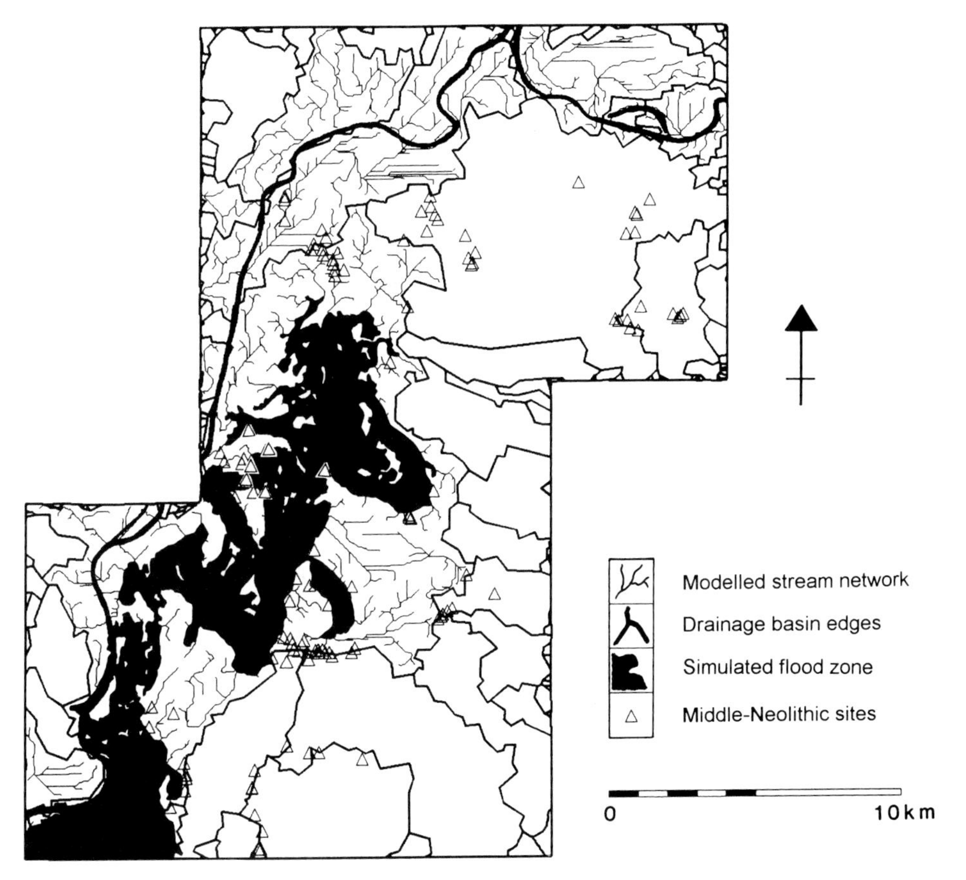

Figure 6.4 Combined flood simulation, modelled hydrology and the distribution of Middle Neolithic sites.

region. This resulted in the creation of a number of discrete, dry islands truncated from the bulk of dry land by the flood waters.

Analysis of the low-lying ox-bow and relic channel areas with respect to the simulation enabled two important conclusions to be drawn. The first was the fact that, during the flood season, these natural lakes would have been fully subsumed into the flood zone. The second concerned the relation of the features to the modelled stream channels. A number of stream branches could be identified draining into the natural lakes. This would imply a degree of permanence, with a guaranteed flow of water entering the ox-bow and channel beds. This would make them a very valuable economic resource throughout the year, in terms of economic factors such as reeds, fish and aquatic wildfowl. What is clear is that the seasonal flood dynamic gives rise to a complex restructuring of the exploitable and observable social and environmental space.

6.3 Case study

During the historical period, the flood cycles had an important influence upon the preferred settlement and economic regimes. Town and village sites were confined to the high terrace zones adjacent to the flood plain, with livestock rearing the dominant economic activity (Loczy, 1988: 172–3). In such a dynamic and potentially catastrophic landscape system, some degree of general environmental determinism is inevitable. Key settlement foci are unlikely to be sited in areas where repeated flood events are the norm. To reiterate, the original impetus for the flood simulation programme was an attempt to identify the major hydrological variables at work in the landscape. Once characterized, these parameters could then be incorporated into subsequent landscape analyses to enable more reliable interpretation of patterns of past exploitation.

The analysis of the phased site data will be illustrated in the form of a case study, looking at factors influencing the location of sites from a single archaeological period, the Middle Neolithic (4900–4600 CAL BC) or Tiszadob group, one of the regional groups of the Alföld Linear Pottery complex (Kurucz, 1989). This period was chosen as it represented a phase of a very intensive exploitation of the available landscape resources (Chapman, 1992: 14). Of the total of 148 sites discovered by the UTP, 112 contained material datable to this period. The aim was to determine whether the dynamic hydrological cycles had an influence upon the utilization of space and resources by the population. By choosing a period of high density exploitation, it was hoped that any trends in settlement patterning would be more clearly visible.

6.3.1 The approach

In an attempt to explore the relevance of the flood dynamics to settlement and exploitation strategies during the Middle Neolithic, the observed site distribution was plotted with respect to the modelled hydrology. The site distribution map forms part of Plate 1.

A cursory visual examination of the data suggests a number of trends. The most obvious of these is that, in contrast to the recent historical picture, sites are not confined exclusively to the higher elevation loess terraces, adjacent to the flood plain. An appreciable number of sites are located within the confines of the plain itself. Within this flood-plain distribution, there appears to be a clear clustering of sites around both the flood-zone fringe and along the drainage basin edges. To investigate these possible trends, the site distribution was studied in the context of five variables:

1. the active stream network,
2. the drainage basin edges,
3. the principal ox-bows and relic river channels,
4. the extent of the flood zone, and
5. elevation above sea level.

The principal mode of investigation was by the establishment and testing of null hypotheses through the chi-squared test of statistical significance (Shennan, 1990: 65–74). In this initial stage, the nature of the archaeological data-set, dated points within the landscape, does not support more detailed modes of multivariate and exploratory analysis. Only when more detailed archaeological studies are undertaken to determine factors such as phased site area and functional artefact composition will the resultant data complexity permit more structured study.

6.3.2 The analytical results

The landscape of the study area was partitioned into the two principal geomorphological regions: the flood plain and its associated loess terraces. The number of sites occurring on each of the land-forms was determined through a simple spatial database query. A null hypothesis was then established, reflecting an area-proportionate distribution of sites across the regions.

Although the loess terraces represented only 19.2 per cent of the total area of the study zone, 38.4 per cent (43) of the sites were grouped within it, a result which rejects the null hypothesis at both the 0.05 and 0.01 significance levels. This suggests, rather unsurprisingly, that the sites were concentrated upon the higher portions of the landscape, insulated by elevation from the effects of seasonal floods.

Within each of the regions the distribution of sites with respect to the identified variables, the stream networks, drainage basins, natural lakes, extent of the flood zone and elevation, was then investigated.

In the flood plain, the principal flood routing channels took the form of stream branches networked directly to the river Tisza. As such, the utilization potential of such streams as sources of fresh water or energy may have been tempered by their rapid susceptibility to the influx of flood water. The terrace region would have been characterized by a much more stable stream network.

Investigation of the location of sites with respect to these systems required processing of the network of linear vectors comprising the individual streams. To ensure that only the more robust and permanent water courses were considered, only those with an order of two or greater were included in further study. A selection procedure was initiated to trim away the far extremities of the network, in effect removing the most ephemeral tributaries.

To address the question of proximity, buffer zones were established around the stream courses at distances of 100 m and 200 m and the frequency of sites falling within the two proximity zones was determined. A null hypothesis was established reflecting an even distribution, with no clear clustering of sites with respect to the stream network. The significance of the observed site frequencies was then tested.

In the case of the terrace zone, the null hypothesis was not falsified. There appeared to be no significant patterning of sites with proximity to the stream channels. This could be contrasted with the result for the flood plain. Here, there was a significant clustering of sites away from the stream network. This is particularly striking in the case of the 200 m buffer where only 26.3 per cent of the total flood-plain area lay outside the buffered catchment and yet contained 50 per cent (34) of the flood-plain sites. In this case, the null hypothesis was rejected at both the 0.05 and 0.01 significance levels.

As discussed earlier, the drainage basin can be viewed as the basic spatial unit in hydrological analyses. In the context of risk management, and the siting of features in a flood-prone region, it is the drainage basin edges that are of prime importance. These represent the lowest-risk areas with respect to flooding. To examine the relation of the basin interfaces with the site distribution, the drainage basin area map was converted to a vector form, and the resultant basin boundary lines were buffered to generate 100 m and 200 m proximity zones. The frequency of sites falling inside each of the zones relative to the remainder of the landscape was then determined.

Once again, on the terrace zone, the null hypothesis was not rejected, there was no significant clustering of sites with respect to basin edge proximity. This can be contrasted with the result in the flood plain. Here the null hypothesis was clearly rejected, with 36.2 per cent (25) of the sites present in the flood plain located within 100 m of the basin edge, a corridor representing only 13.2 per cent of the available land surface.

To investigate the distribution of flood-plain sites further, a proximity analysis was undertaken with respect to the maximum extent of the 91.5 m flood simulation. The results showed a significant cluster of sites, at both the 0.05 and 0.01 levels, located within a 100 m buffer zone around the edge of the flooded area. This included a number of sites concentrated upon the truncated flood islands and a group of three sites submerged within the flood zone itself.

6.3.3 Summary of the results

The analyses revealed several interesting trends. Whilst the site distribution is not confined exclusively to the higher loess terraces, there is a significant clustering of sites towards this landscape block, suggesting a preference for the higher regions insulated by elevation from the flood waters. Analysis of the spatial distribution of sites with respect to the principal hydrological features identified clear and contrasting trends between the terrace and flood-plain environments. On the terraces, factors such as proximity to water courses and drainage basin boundaries do not appear to be important criteria in the location of sites. In the flood plain, the opposite is true, with a significant clustering of sites away from the stream network, directly adjacent to the basin edges. This appears to reflect a low-risk strategy with the aim of minimizing the impact of flood events. There does, however, appear to be a contradiction within the flood plain itself. Alongside the low-risk sites are a significant group located directly adjacent to the flood-zone edge. As the present flood simulation represents a 'typical' as opposed to 'extreme' flood event, placing sites in this area would appear to represent a markedly high-risk locational strategy.

6.3.4 Site location, elevation and the natural lakes

There appears to be a strong level of patterning within the flood plain, with one group of sites concentrated around the basin edges and a second, high-risk group, located at the interface between the flood water and the dry land. To assess this supposed risk factor, the DEM was queried to establish the elevations of the total site population. Of the 112 sites discovered, only 4.4 per cent (5) fell at an elevation less than 92 m above sea level. A total of 78 per cent (97) were sited at elevations greater than 1 m above the typical flood level. The implication here is that, although horizontally adjacent to the flood zone, the vast majority of sites were displaced vertically. This can be most clearly evidenced by the observable cluster of sites upon very localized high-ground features such as the flood plain islands. A further test can be undertaken by assessing the effects of a more extreme flood event. This was achieved by re-running the simulation with the maximum inundation threshold increased to 92 m above sea level. The result is a dramatic increase in the extent of the inundated flood zone. The 15 per cent total surface area of the initial simulation is expanded to 46 per cent. Despite this marked increase in area, the number of sites falling within the enlarged flood zone increased by only two. This suggests strongly that the original interpretation of a high-risk strategy was flawed. Although sites were clustered adjacent to the flood zone edge, the risk of inundation was minimized by the maximum exploitation of available high ground.

This observed patterning prompts an immediate question: why were sites located so close to the flood plain edge? One obvious interpretation involves the large ox-bows and relic channel segments that shape much of the observable flood plain. As discussed earlier, these areas would have been an important economic resource for the occupants of the region. An alternative notion is that the observed site patterning with respect to the flood zone is a secondary phenomenon, the result of the location of sites relative to the low-lying lake areas and the importance of these features in shaping the final form of the given flood event.

To test this hypothesis, the proximity of flood plain sites was analysed with respect to the principal ox-bow and relic channel features. Once again, 100 m and 200 m buffer zones were established around the feature perimeters and the site frequencies calculated. The result of the analysis identified a significant clustering of sites, at both the 0.05 and 0.01 levels, within 100 m of the flooded channel shores. This serves to support the interpretation of site location favouring the natural lakes.

One group of sites still appeared to reflect a high-risk locational strategy, namely the three sites located within the main body of the flood zone itself. To investigate this phenomenon further, a close examination of the respective micro-locations was undertaken. One of the sites (Polgar 61) was positioned at the immediate flood water interface. It was located on the edge of a small flood-zone island, at an elevation of 91.4 m, very close to the critical inundation threshold of 91.5 m. Given the inherent 20 m spatial resolution of the analyses it is difficult to reliably infer that this siting was the result of a deliberate, high-risk locational strategy. In the case of the remaining sites, Tiszagyulahaza 3 and Ujtikos 6, such an interpretation can be more reliably tendered. Both of these features were located at an elevation of 90.8 m above sea level and each was situated within the main body of an ox-bow or relic channel. As discussed in section 6.2, these ox-bow and relic channel features were instrumental in shaping the final form of flood events and would have been particularly susceptible to the rapid influx of initial flood waters. In the case of these sites, it appears that high-risk areas of the landscape were being actively exploited.

6.3.5 The Middle Neolithic response to flood dynamics

Examination of the distribution of Middle Neolithic sites revealed clear signs of spatial patterning with respect to both the simulated flood zone, and the modelled facets of the hydrological system. If risk is defined as the possibility of inundation as a direct consequence of the seasonal flood dynamic, three general locational site classes can be identified:

Class 1: Low-risk exploitation of the terrace environment, with a significant concentration of sites on the higher loess terraces.
Class 2: Low-risk exploitation of the flood plain, with siting along basin interface zones and localized high-ground features.
Class 3: High-risk exploitation of the flood zone itself.

The obvious interpretation is of a structured pattern of landscape exploitation, with site security balanced against economic benefits. In this context, a risk-management strategy can be envisaged, with communities cultivating a main crop on the flood plain, and an insurance crop on the higher loess terraces. Evidence to support such an interpretation has been found during recent excavations at the Late Neolithic tell of Polgar–Csoszhalom. Here archaeo–botanical investigations have identified weed species indicative of cereal and pulse cultivation on both wet and dry soils (Fairbairn, 1993: 71–5).

Of particular interest are the sites located upon the flood zone islands. During the flood season, these areas would have been completely isolated, truncated by water from the terrace region and main flood zone edge. Given the limited area of a number of these islands, often of the order of 3–4 km^2, permanent year round occupation seems unlikely. In the case of the sites discovered within the flood zone itself, permanent occupation would have been impossible. These factors combine to suggest a further level of complexity in the perceived flood response regime, with limited transhumance between the identified locational classes linked to the annual flood dynamic. The possibility of seasonal exploitation and the complex flood-water partitioning of the available social space implies the presence of optimal communication pathways between the seasonal nodes. An important research priority will be to establish and test these routes through the network capabilities of the GIS.

6.4 *Discussion and conclusions*

It is clear that the flood cycles had a marked impact upon the settlement and exploitation of the region during the Middle Neolithic. The GIS-based simulation and retro-modelling programme has enabled the identification of a number of spatial trends. Further testing of these interpretations will require more directed GIS analyses, fully integrated into an enhanced programme of fieldwork. Crucial to this process will be the careful re-examination of a number of targeted site data sets supported by limited excavation, along with the assimilation of detailed environmental data. As the range of social and environmental variables becomes wider, more detailed exploratory statistics will be employed to refine, enhance and evaluate both the interpretations and the simulation models upon which they are based.

Though the hydrological cycles have a clear influence upon the organization of the landscape in the Middle Neolithic, it cannot be assumed that they, or the environment as a whole, constitute the only factors influencing exploitation. There are still important questions to be answered concerning the micro-organization of the landscape. For example, why did communities chose to exploit flood islands when equally good areas, defined in environmental terms, could be found on the mainland edge? The solution undoubtedly lies with a combination of social and environmental factors. As data are enhanced and models refined, the ultimate aim of the UTP GIS programme will be to address these very questions.

In conclusion, the value of GIS in the context of the Upper Tisza Project cannot be over-stressed. Through simulation and retro-modelling procedures, facets of past land-form systems have been characterized, enabling the introduction of more robust archaeological analyses.

References

Chapman, J. C., 1992, Settlement patterns in the Polgar area, in 'The Upper Tisza Consortium. 1992. Upper Tisza Project Report to University Research Committee 1991–92', unpublished report, University of Newcastle upon Tyne.

Chapman, J. C. and Laszlovsky, J., 1992, The Upper Tisza Project 1991: report on the first season, *Archaeological reports 1991 (Durham and Newcastle upon Tyne)*, pp. 10–13, Durham: University of Durham.

Fairbairn, A., 1993. Archaeo-botanical investigations, in 'The Upper Tisza Consortium, The Upper Tisza Project Report on the Second Season (1992)', unpublished report, University of Newcastle upon Tyne.

Kurucz, K., 1989, *A nyíri Mezöseg neolitikuma*, Nyíregháza: Jósa András Múzeum Kiadvanyai, **28**.

Loczy, D., 1988, Cultural landscape histories in Hungary—two case studies, in Birks, H. H., Birks, H. J. B., Kaland, P. E. and Moe, D. (Eds), *The Cultural Landscape: Past, Present and Future*, pp. 165–76, London: Cambridge University Press.

Rónai, A., 1987, *Hajdúnánás*, Budapest: Magyar Állami Földtani Intézet.

Shennan, S., 1990, *Quantifying Archaeology*, Edinburgh: Edinburgh University Press.

Shiel, R. S., 1992, Introduction to the landscape, in 'The Upper Tisza consortium. 1992. Upper Tisza Project Report to University Research Committee 1991–92', unpublished report, University of Newcastle Upon Tyne.

Ward, R. C., 1990, *Principles of Hydrology*, London; New York: McGraw-Hill.

Notes

[1] The material in this chapter has been presented on behalf of the Upper Tisza Consortium. The UTP is grateful for financial support from the following bodies: the National Geographic Society, the British Academy, the Society of Antiquaries of London, the Prehistoric Society and the University of Newcastle upon Tyne.

[2] At this point a definition is in order. In the context of this chapter, the term 'site' is used to represent a plough soil concentration feature indicative of some past human activity. Where interpretation permits, form–functional refinements are offered, such as 'settlement' or 'industrial'.

[3] The decision is of fundamental importance, as too coarse a resolution could result in the merging of significant spatial features, whereas too fine a resolution may lead to the accumulation of spurious detail.

7

Data recording and GIS applications in landscape and intra-site analysis: case studies in progress at the Archaeological Institute of the Hungarian Academy of Sciences

Gy. Csáki, E. Jerem and F. Redö

7.1 Introduction

Archaeological applications of GIS are a natural consequence of this system's increasing popularity: archaeological attributes associated with geographical loci may include historical data as well. This presents the possibility of tracing the distribution and movements of, as well as interactions between archaeological cultures. Trends may even be outlined for entire continents. In comparison with traditional distribution maps, this type of macro-regional survey facilitates the evaluation of complex problems by virtue of simultaneous data processing.

During the course of evaluating smaller territorial units (micro-regions) using the same technique, GIS databases may be complemented by information obtained applying other types of scientific analyses which are used in the reconstruction of both the landscape and the immediate environment of the archaeological site itself. In this way, the possibilities of historical–archaeological interpretation are further expanded.

Intra-site documentation and analysis using GIS makes the individual and comprehensive study of settlement strata possible, and also offers an opportunity for the integration of environmental archaeological data. Thus, in addition to following spatial changes in various time horizons, histories of individual features or buildings may be reconstructed. Fragmentary architectural remains and decorative elements may be compiled or, if necessary, modelled. Should a feature-oriented approach be necessary (e.g. the identification of fine stratigraphic details) GIS can be efficiently used if appropriately gathered data are available.

Experience accumulated during the application of GIS as an analytical tool in archaeology could be most fruitfully used in two major fields in Hungary. The first of these is the ongoing national archaeological survey, known as the 'Archaeological Topography of Hungary', a high priority project. In a broader sense, expertise in using GIS is an important contribution to the continuous development of analytical procedures in archaeology.

In this chapter, the authors present three themes within an ongoing project of the research team for computer applications in archaeology at the Archaeological

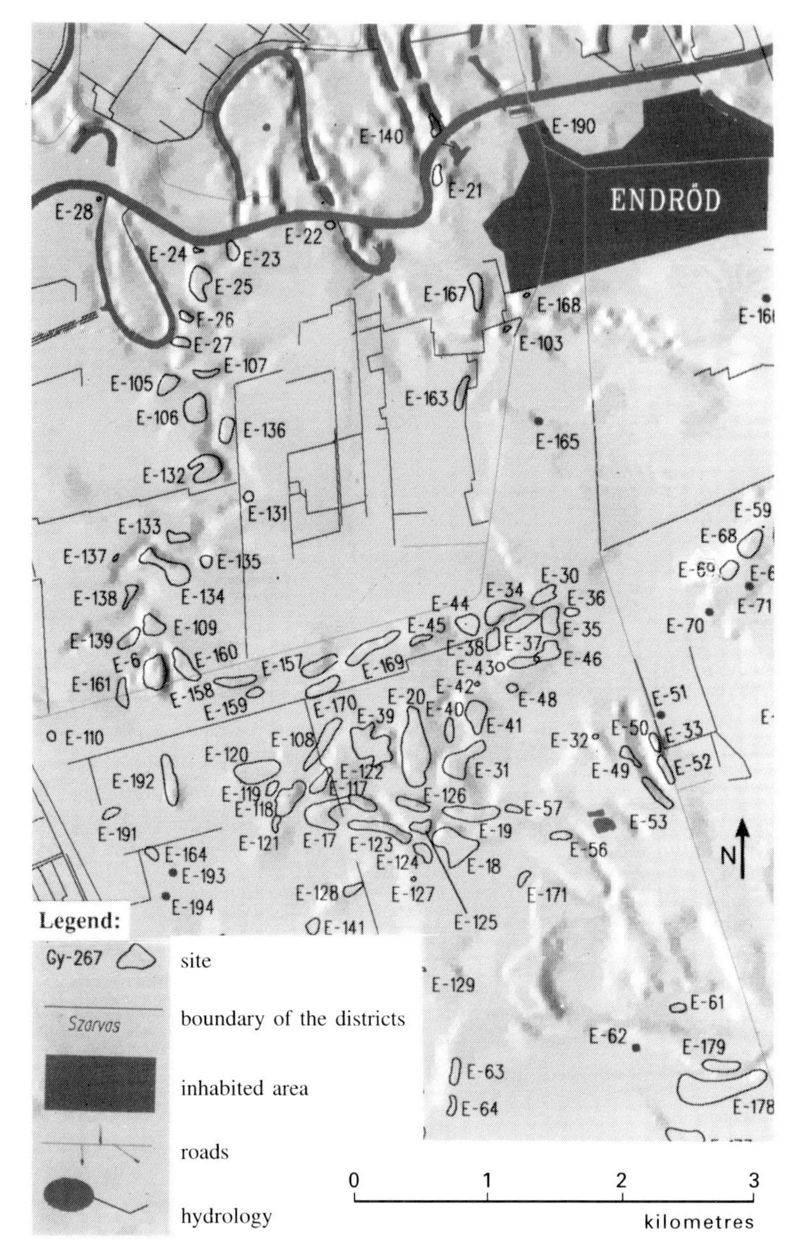

Figure 7.1 Archaeological sites covered by the micro-regional survey project in the area west-southwest of Endröd.

Institute of the Hungarian Academy of Sciences. In some manner, they all represent a contribution to landscape archaeology, integrating archaeological and environmental data. This approach facilitates the recognition of new inter-relationships as well as the comprehensive evaluation of results and our examples illustrate applications based on the previously mentioned different scales.[1]

7.2 *The micro-region project in the Great Hungarian Plain*

The first application presented here is closely associated with the settlement research programme carried out by the Archaeological Institute of the Hungarian Academy of Sciences. This project took place in southeast Hungary and included more than 200 archaeological sites scattered over an area of 42 km^2. Chronologically speaking, these sites ranged from the early Neolithic to the Middle Ages. Similar to some other regional evaluations (Crumley and Marquardt, 1989; 1990; Gaffney and Stančič, 1991) our aim was to integrate regional site survey data with site excavation data.

Ancient geomorphological conditions and palaeohydrological formations of decisive importance were studied using aerial photography, field walking, geophysical prospecting and core drilling in the Körös river region of southeast Hungary, Great Hungarian Plain (Jerem *et al.*, 1992).

The large-scale river regulation projects carried out in the last century made an enormous impact on the landscape. Nowadays, it is hardly possible to recognize the old hydrography of this territory. Most of the rivers cut their beds into the Pleistocene alluvial deposit which has been identified several times during core sampling. The actual shape of the river beds known prior to regulation may have evolved by the beginning of the Holocene. A variant of riverbed alterations is represented by periodic floodings during which, several tens of kilometres were covered by water for varying lengths of time. The more interior areas were completely uninhabited, particularly in prehistory, because of deteriorating climate. The other typical aspect is the formation of meanders whose ox-bow-like curves are one of the most quickly changing of riverbed formations. The speed and quantity of water decreased once meanders were cut through and sealed off and since the remaining water could not be drained, the meanders turned into cutoffs. With the passing of time these lakes silted and filled, thus contributing to the formation of new meaders. These events have been documented by mapping and soil core sampling.

Our three-dimensional terrain model is based on a 1:10000 digitized contour map. A strong, artificial distortion was introduced in order to express slight differences in altitude (*c.* 80–86 m above the Adriatic sea level). The image clearly shows the high density concentration of settlements along former river beds (Figure 7.1). Accumulations of archaeological sites may be observed in particular on slightly elevated loess backs which offered protection from flooding.

The area was populated with different intensity through time. A finer, differentiated reconstruction of such changes may commence once exact chronological data are available at the end of ongoing archaeological analyses. The results of palaeoecological investigations will also be included in the database. This will allow comprehensive graphic presentation and evaluation of data.

7.3 *Research in north-west Hungary*

Large-scale development projects in the Lake Fertö Basin located within the alpine foreland, as well as just initiated highway construction connecting Budapest with Vienna, made both field surveys and well-organized intensive rescue excavations necessary. Uncovering large, continuous surfaces often resulted in the more or less complete excavation of smaller, village-like settlements (Jerem *et al.*, 1984–1985;

Jerem, 1986). The results of target-oriented environmental–archaeological sampling (also included in the database) fit within the framework of comprehensive archaeological evaluation. In this case, attention is focused both on individual sites and the region as a whole, since these settlements formed part of a major network.

The digitized form of the eighteenth-century hydrological map showing the situation preceding large river regulation work (Figure 7.2) indicates that areas between the Danube river and Lake Fertö were almost constantly under water or were at least always exposed to major floods. Consequently, they are rather poor in archaeological sites. Human habitation occupied the margins of formerly uninhabited wetlands only during periodic climatic improvements such as the Late Migration Period and the Árpád Dynasty (ninth to eleventh centuries AD). The location of late Iron Age (Celtic) sites along the previous, prehistoric shoreline of Lake Fertö, east of the Alps, shows that the lake's water level was significantly higher than in modern times. With a few exceptions, all settlements and cemeteries can be found at altitudes of 100–250 m above sea level. These arable areas were located in the proximity of water, however, on terraces relatively well protected from flooding (Figure 7.3).

In 1990–1991, large scale excavations were carried out at the multi-period settlement of Ménföcsanak–Szeles in the vicinity of Györ. Over 700 features were uncovered within an area of 2 ha (Figure 7.4, and *Régészeti Füzetek*, 1992, 1993). This village site lay directly by the ancient riverbed of the Rába river between 112 m and 116 m above sea level (Figure 7.2). As a result of multiple occupation, settlement features were often found in overlapping positions thus creating an extremely complicated horizontal stratigraphy (Figure 7.5).

Our digital plan is based on the 1:2500 map of this area. Since the entire documentation is available in computerized form, spatial exploratory data analysis

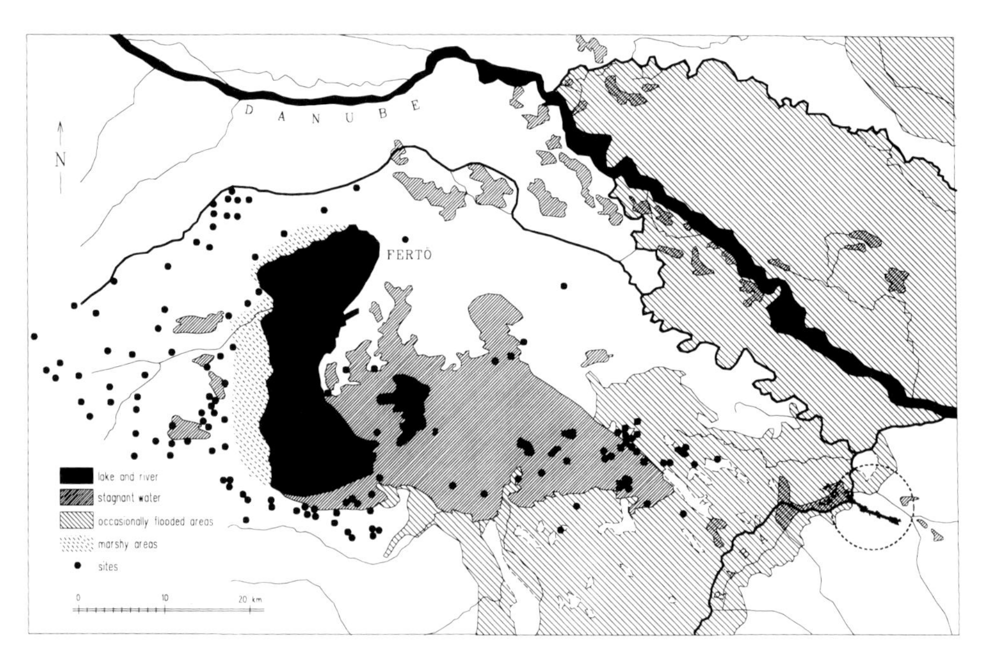

Figure 7.2 Palaeohydrological map of the Small Hungarian Plain with the location of sites in the Lake Fertö Basin and the location of the Ménföcsanak–Szeles excavations (circled).

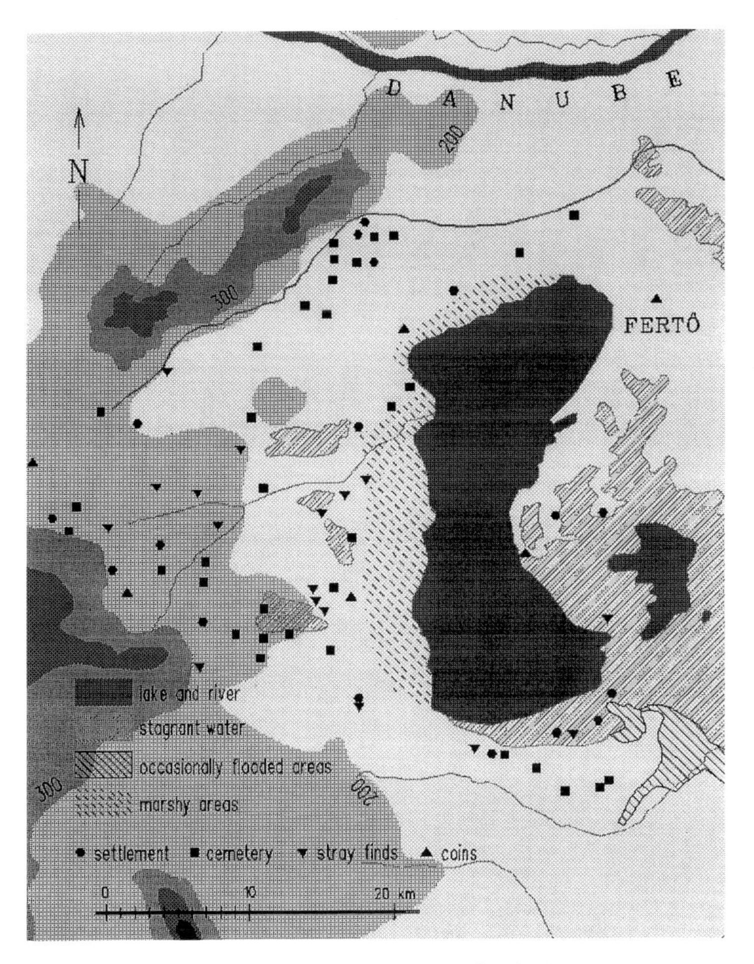

Figure 7.3 Iron Age sites around Lake Fertő.

could be carried out, for example Bronze and Iron Age houses and pits were all located on elevations with favourable geomorphological positions. At the same time, the Roman period indigenous settlement and especially the village from the period of the Árpád Dynasty lay in areas near the water taking advantage of the more favourable climatic conditions (Figure 7.4). Using this method, it is possible to study individual features within their context. Their profiles and various sections can also be analysed.

A variety of techniques were applied to reconstruct the site formation processes and the past use of the former village. Remote sensing and core drilling helped us to define the geomorphological and micro-stratigraphical conditions. *In situ* soil development profiles and cultural fill were examined in order to follow the local evolution of sediments. Many samples were taken from each level for micromorphological and chemical (ICP-OES, X-ray diffraction) investigations. Water movements were decisive, occasional floods and stagnant water playing the biggest role in site modification and determining the soil conditions.

Our sampling strategy involved a significant amount of wet sieving and flotation both from cultural and off-site deposits. These methods yielded molluscan,

Figure 7.4 Plan of the Ménfőcsanak–Szeles excavations.

microvertebrate and botanical evidence. Lithic, palynological and zoological data together form an environmental database constructed following hierarchical principles (Plate 1). Based on the results of these studies, post-depositional processes as well as changing climatic and ecological conditions could be traced between the Bronze Age and the early Medieval Period.

Finally, all the excavations of an Iron Age settlement site in Sopron, the accumulation process producing a hut's fill was modelled on the basis of the find material deposited in it (Jerem and Somogyi, 1992). This database contained data on over 1200 ceramic sherds as well as the parameters of 31 vessels (Plate 2).

7.4 Excavations at San Potito di Ovindoli (AQ)

Archaeological excavations have been carried out in the Abbazia area of San Potito di Ovindoli (AQ) since 1983 with the support of the soprintendenza dell'archeologia degli Abruzzi (Gabler and Redö, 1986; 1988; 1988; 1991). In addition to small quantities of Iron Age sherds, the remains of a grandiose villa from the period of the Roman Empire were brought to light at this site. Above the Roman ruins, a late medieval church and the adjacent cemetery were excavated as well although the Roman villa is in the main focus of this study.

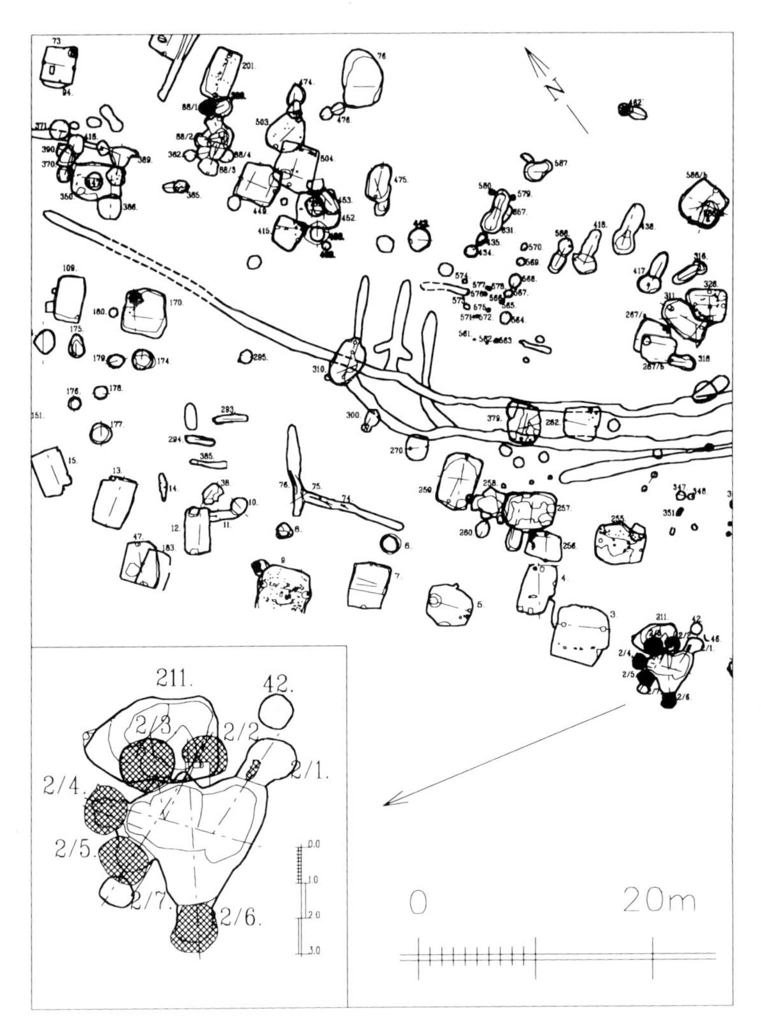

Figure 7.5 Detail from the plan of the Ménföcsanak–Szeles excavations showing features from various periods. The inset shows ovens from the Árpád Dynasty built on top of the fill of a Hallstatt pit.

The villa was built during the first half of the first century AD, but at least three major phases of construction could be identified and it was probably in use as late as the second half of the third century. During this time, the history of this area was influenced by arable land being available some 6 km south of the site, in the basin of Lake Fucino which had been drained by Emperor Claudius (Amato, 1980). Alba Fucens, the ancient colony established at the turn of the fourth and third centuries BC, is located southwest of the site along the edge of the former lake basin. Corfinium is located somewhat further away, however, still within the area of Fucino. The area under discussion here is crossed by the via Valeria, or major road, which ran in an east–west direction (Grossi, 1991).

The villa itself was located at an altitude of 1000 m above the Adriatic sea level, in a narrow valley surrounded by 1700–2000 m high mountains. The valley is open toward the south.

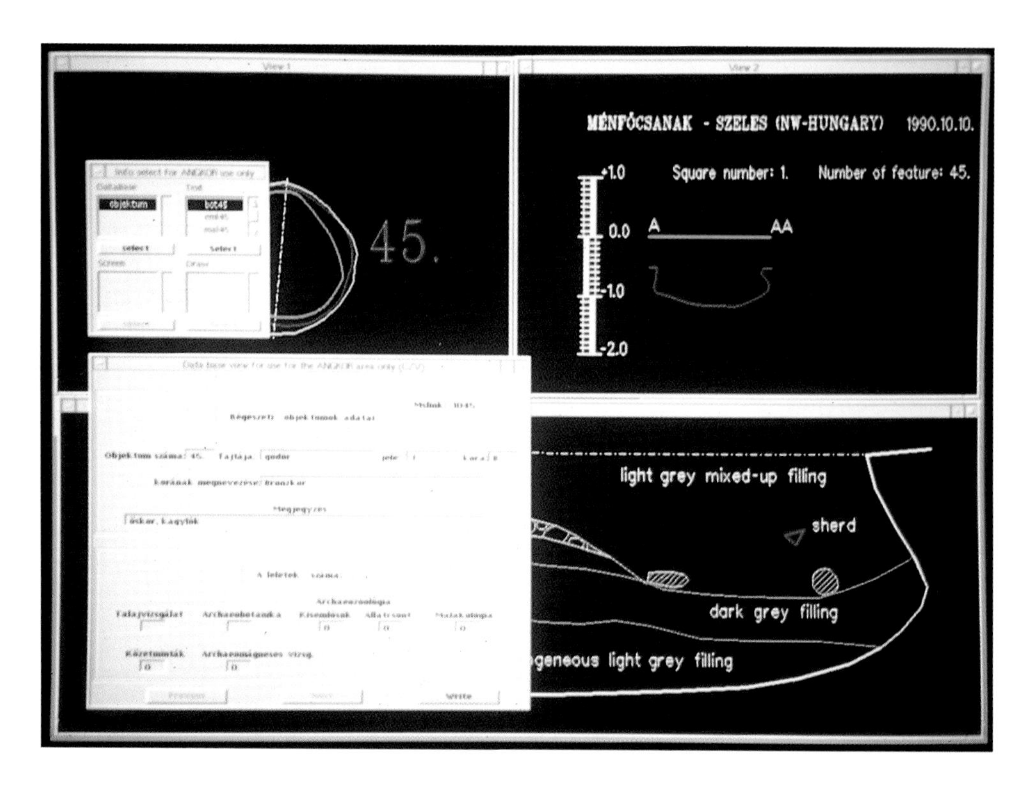

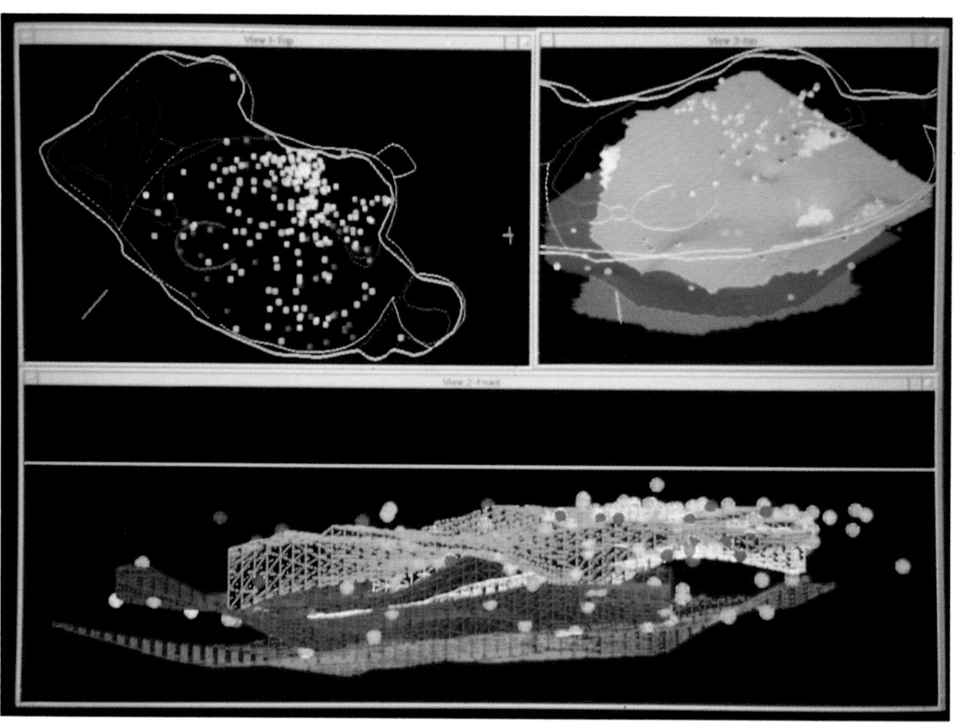

Plate 1. Bronze Age pit number 45 with its environmental and archaeological database entry, from excavations at Ménföcsanak–Szeles, Hungary.

Plate 2. The section and surface model of Feature 334 from the Iron Age site at Sopron–Krautacker, Hungary.

It was necessary to study the villa's environment independently of current environmental conditions, since modern settlements, roads, mines and other features of modern civilization may, in many ways, have distorted the original conditions. This problem was solved by creating the surface model of this area based on contour maps, lines within these maps were set to differences of 10 m in altitude. In places where lines were too concentrated, however, main levels were considered at 50 m intervals (Figure 7.6(a)). This surface model does not divide the area into squares of uniform size, but into triangles defined by altitude within the 10 m and

Figure 7.6 The site of San Potito di Ovindoli and its topographic setting: (a) a contour map; (b) the first stage of the surface model; and (c) the finished surface model.

50 m intervals respectively (Figure 7.6(b)). The increasingly steep slopes of these triangular units are marked by increasingly dark colours. Horizontal surfaces appear as practically white (Figure 7.6(c)). This method yields a good quality surface at a scale of 1 : 10 000. A variety of reference files may be integrated within this surface. These files contain antique topographic information acquired by surveys carried out in the wider or immediate neighbourhood of the excavation. They also include the documentation of feature plans showing the various periods. In the area under discussion here, both a medieval church and a cemetery lay above the Roman ruins. This makes the presentation of boundaries around real estate traditionally owned by the clergy particularly interesting (Figure 7.7).

Topography itself justified the selection of this site for the purposes of settlement. The villa was built on possibly the highest, horizontal surface from where Fucino could be both reached and seen. There is a spring on this small plateau and

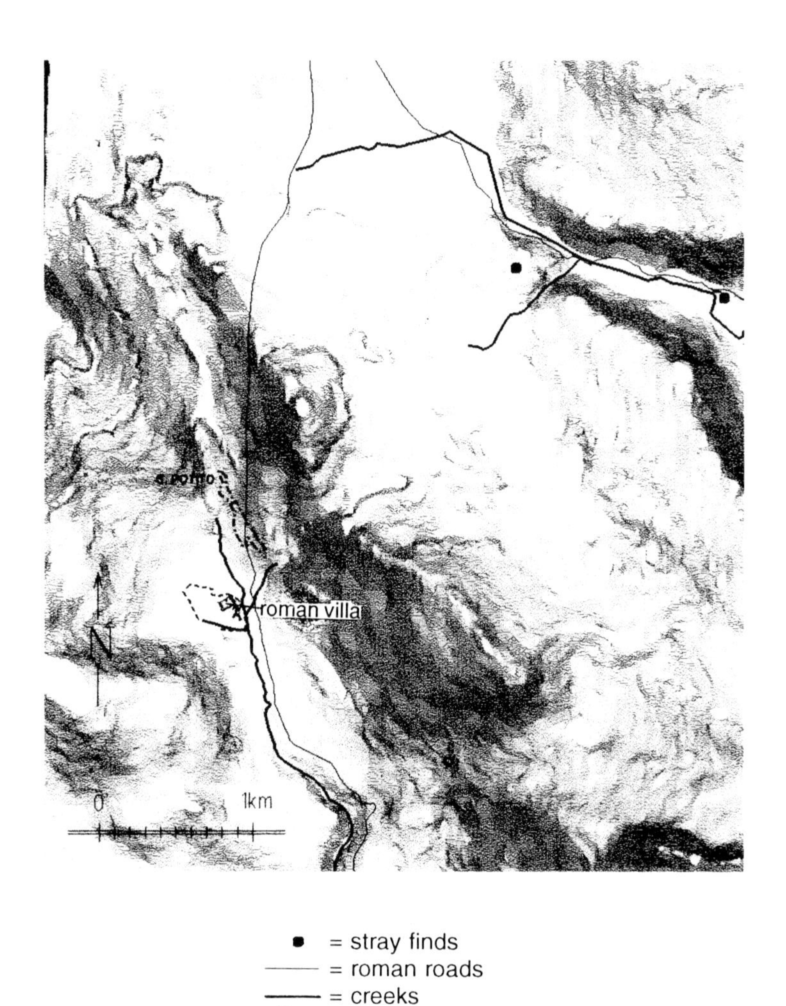

Figure 7.7 The surface model of San Potito di Ovindoli's wider environment with the topographic data.

three small creeks meet here. The spot is defended on three sides by surrounding hills, and a steep mountain path winds its way to the north from here. It connected the villa with a larger plateau located to the northeast some 350 m higher. Although the archaeological–topographic survey of this highland has not yet been completed, two smaller excavations are already known to date. They yielded Roman period ceramic sherds and a Roman road, respectively. This latter ran around the northern side of a major hill east of the villa, then followed the Fucino coastline all the way to the next settlement (Figure 7.7).

Connections with the highland must have been important for the inhabitants of this villa since the sheer size of this compound makes it unlikely that its population could have subsisted on the smaller plateau alone. The only form of agricultural income guaranteed by the topography of this area would have been alpine grazing.

Periods of the villa may be incorporated in the surface model. The only technical problem was posed by the fact that contour maps used in the construction of the surface model have a lower resolution than the site plan. In this case, separate contour maps had to be compiled. Another problem is that features from different periods often overlap each other or earlier structures were *de facto* incorporated within later ones (Figure 7.8).

The evaluation shows that the *cortile* found in the northern section of the villa represents two verifiable periods. This is shown by superimposed floor levels, walls and blocked entrances. Its south-eastern corner is adjacent to the central block of

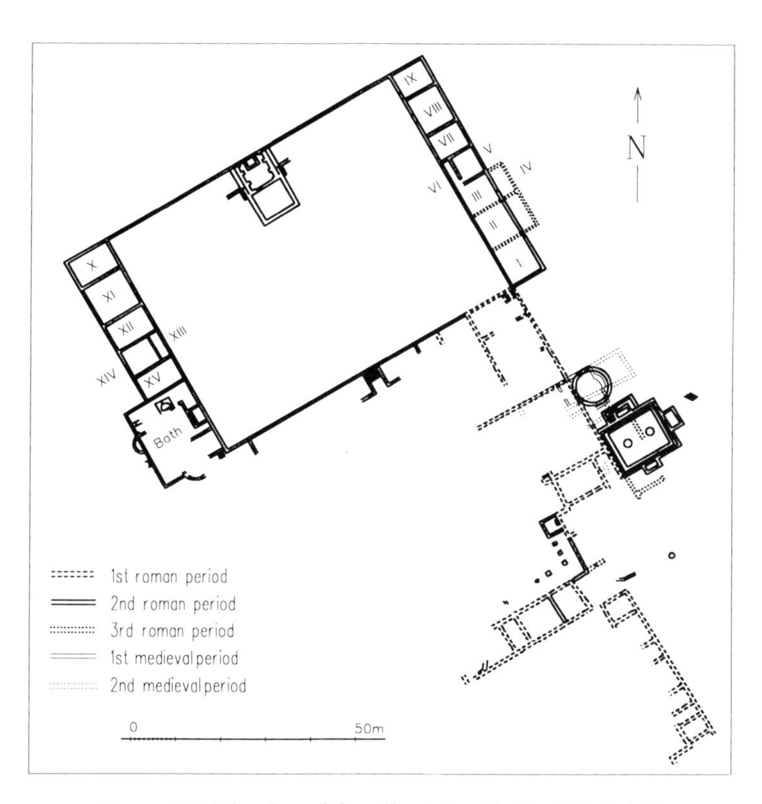

Figure 7.8 The plan of the villa at San Potito di Ovindoli.

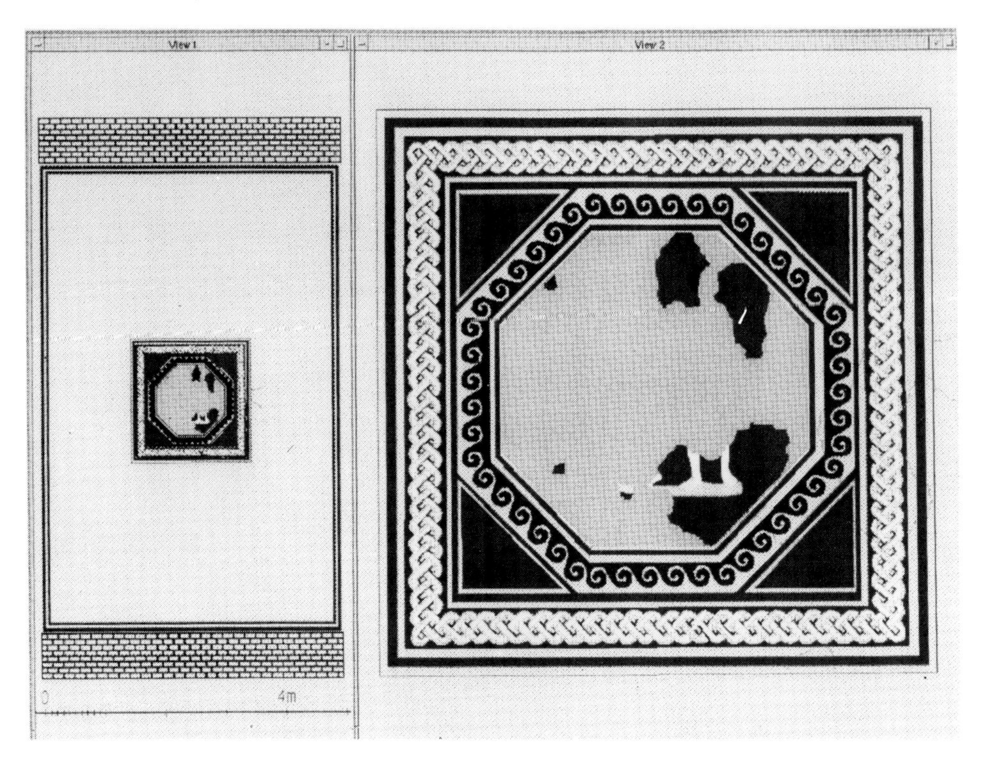

Figure 7.9 The mosaic floor and reconstructed image from room XII of the San Potito di Ovindoli villa.

the compound which was an earlier construction. This section has not yet been sufficiently investigated. It is apparent, however, that three of its periods may be distinguished, and two of these periods (the middle and last phases) are parallel with the periods identified in the *cortile*.

The documentation of individual rooms may also be activated within the database thus created, with the most significant elements being the mosaic floors. Their fragmented original state and reconstructed form may both be attached to the database of this site. The options for documentation provided by the graphics software may prove particularly useful for this purpose. With the help of these facilities, figures may be made beyond the precision levels of the human hand (as well as the eyes) and integrated within the system (Figures 7.9 and 7.10).

7.5 Conclusions

Studies of the spatial distribution of sites representing various periods can reveal, on the one hand, the environmental conditions in ancient times and, on the other hand, the dynamics of the formation of settlement networks. The proportion between populated and uninhabited areas may be appraised, prehistoric and present-day access routes may be pinpointed and changes in commercial and other road networks can be reconstructed.

Figure 7.10 The reconstruction of the fragmentary mosaic from room V of the San Potito di Ovindoli villa.

Characteristics of the geographical setting of the multiphase, riverside rural settlements, such as changes in the altitude above sea level, adaptation to changing hydrology and occasional floods as well as considerations of soil quality (as determined by altering climatic conditions) can be traced with GIS.

Stratigraphic data as well as post-depositional effects may thus be integrated with the archaeological evaluation of the find material. This study has showed the relationship of each internal unit to the others, as illustrated by the three-dimensional modelling of a hut on the basis of over 1200 finds.

The documentation of the San Potito Roman period villa, its relationships with the most immediate as well as wider environments, and the location of this compound were analysed within the context of a comprehensive, integrated GIS setting. This system also included the modelling of both entire features and some of their details. The methodology is applied in such a way that this system is open and therefore future research results can be added as well.

Acknowledgements

Grateful thanks are due to Annamária Csáki who digitized our maps and contributed figures to this study. The help of László Bartosiewicz and Ildikó Perjés in the compilation of this chapter must also be acknowledged.

References

Amato, D., 1980, *Il primo prosciugamento del Fucino*, Avezzano.

Crumley, C. L. and Marquardt, W. H. (Eds), 1989, *Regional Dynamics. Burgundian Landscape in Historical Perspective*, London: Academic Press.

Crumley, C. L. and Marquardt, W. H., 1990, Landscape: a unifying concept in regional analysis. in, Allen, K. M. S., Green, S. W. and Zubrow, E. B. W. (Eds), 1990, *Interpreting Space: GIS and Archaeology*, Taylor & Francis, London, 73–9.

Gabler, D. and Redö, F., 1986, Gli scavi della villa romana a San Potito di Ovindoli (AQ) 1983–1984 *Acta Arch. Hung.*, **38**, 41–87.

Gabler, D. and Redö, F., 1988, Gli scavi a San Potito di Ovindoli 1985–1986. Seconda relazione preliminare, *Specimina nova dissertationum ex Instituto Historico Universitatis Quinque-ecclesiensis*, 69–94.

Gabler, D. and Redö, F., 1991, Gli scavi della villa romana di San Potito di Ovindoli, in *Il Fucino e le aree limitrofe nell'antichita. Atti del convegno di archeologia—Avezzano 1989*, Roma, 478–500.

Gaffney, V. and Stančič, Z., 1991, *GIS approaches to regional analysis: A case study of the islands of Hvar*, Znanstveni inštitut Filozofske fakultete; University of Ljubljana, Yugoslavia.

Grossi, G., 1991, Topografia antica della Marsica (Aequi-Marsi e Volsci): quindici anni di ricerche, 1974–89, in: *Il Fucino e le aree limitrofe nell'antichita. Atti del convegno di archeologia—Avezzano 1989*, Roma.

Jerem, E., 1986, Bemerkungen zur Siedlungsgeschichte der Späthallstatt und Frühlatenezeit im Ostalpenraum (Veränderungen in der Siedlungsstruktur: archäologische und paläoökologische Aspekte) in: Hallstatt Kolloquium Veszprém *Mitt. Arch. Inst.*, **3**, 107–118; 363–65.

Jerem, E. and Somogyi, P., 1992, Zur statistischen Auswertung von Keramik aus Siedlungsobjekten, *Acta Arch. Hung.*, **44**, 161–92.

Jerem, E., Kiss, Zs., Pattantyús A. M. and Varga, A., 1992, The combined use of archaeometric methods preceding the excavation of archaeological sites, in, *Cultural and Landscape Changes in South-East Hungary. I: Preliminary Reports on the Gyomaendröd Project*. Budapest: Archaeolingua **1**, 61–98.

Jerem, E., Facsar, G., Kordos, L., Krolopp, E. and Vörös, I., 1984–85, A Sopron–Krautackeran feltárt vaskori telep régészeti és környezetrekonstrukciós vizsgálata (The archaeological and environmental investigation of the Iron Age settlement discovered at Sopron–Krautacker) *Archaeológiai Értesít*, **111**, 141–69; **112**, 3–24.

Régészeti Füzetek, 1992, (Excavation reports) Ser. 1 **44**, 11–13; 1993, **45**, 12–14.

Note

[1] GIS software and hardware used in the case studies: data acquisition was usually based on traditional, descriptive methods, with the exception of the archaeological assemblage discussed in Chapter 3 as well as archaeo-zoological and archaeo-botanical data in general. In these cases, both the recording and classification of data were carried out using computers. Dbase-type databases were subjected to data processing using software developed by the team. Integraph 4.0 software was utilized in the graphical handling of data. Presentation is possible in both vectorial and raster form. The three-dimensional surface modelling of the micro-region project in the Great Hungarian Plain is based on a 10 × 10 m grid. In order to emphasize otherwise minimal differences in altitude, a factor 20 distortion was introduced. These operations were carried out using an IBM 486 PC.

8

The spatial analysis of Bell Beaker sites in the Madrid region of Spain

J. Baena, C. Blasco and V. Recuero

8.1 Introduction

The dearth of studies showing the spatial distribution and the relative density of Beaker sites in the Madrid region has encouraged us to produce an analysis based on the use of a geographic information system (GIS). We suggest explanations for the patterns of site distribution, the economic potential of the territories where the settlements are found, the areas from where particular 'exotic' raw materials were obtained, the routes used for reaching these relatively nearby areas, and the relationships that existed between the various contemporary sites.

8.1.1 Cartographic data

One of the most important points of procedure when applying a GIS to spatial analysis is to use the most homogeneous and precise cartography that is available. In this respect the Community of Madrid has the advantage of many cartographic editions, although, in most cases they are syntheses of map series produced at different scales and based on older editions. The use of different scale maps does not present problems in computerized cartography, however, since this system allows the transfer of data to different scales, albeit at the risk of losing precision.

We have used cartographic data from various sources to produce the spatial database. Firstly, the thematic series edited by the Community of Madrid Board of Agriculture, in particular those relating to land-use capability, soil types (a synthesis of extant 1:100 000 scale maps) and the topographical maps of the province, at 1:200 000 scale. In addition, we have used two sheets of the 1:200 000 scale IGME geological maps, 1:200 000 scale maps showing the location of metal ores, also produced by the IGME, the lithological map of the Community (a synthesis of 1:50 000 scale maps), and the recent edition of the Community of Madrid map of cattle tracks, also to a scale of 1:200 000.

The rest of the information, relating to the river network and contour lines, derives from 1:25 000 digital data made available by various official bodies and whose subsequent processing has been carried out by the Autonomous University of Madrid's Cartography Service. The geographic elements shown in the maps have been digitized and ultimately combined with the archaeological data to produce the complete spatial database. Using the GIS capabilities an operating model for the management of territorial information that permits the integration of the graphic database with an alphanumeric database has been built. The data can be combined and analysed in various ways producing both two- and three-dimensional maps, of varying complexity. The analyses below show that the ability to ask spatially referenced questions is of great importance.

The hardware used to carry out this work were:

- a Calcomp 9.100 digitizing tablet for collecting the data;
- an AT 286 PC for storing information generated in the digitizing process;
- an IBM RT 6150 workstation for processing the information and for producing the digital terrain and the maps, and
- an HP 7596A plotter for producing the graphic outputs.

8.1.2 Archaeological data

As in most spatial archaeology based on site distribution maps, we are faced with the initial problem of having partial and uneven data. Some areas, such as river terraces, have been heavily exploited, bringing to light virtually all the remains buried in the subsoil, but in other areas the only discoveries known have been those resulting from casual finds in the course of farming, building or public works, and from the occasional appearance of artefacts. In order to overcome this problem as far as possible, we shall concentrate the greater part of our study on an area of the region south of Madrid where we have carried out intensive surveys and where quarrying has uncovered virtually all the archaeological remains in the Manzanares basin downstream from Madrid. These factors have provided intensive information for the area which is translated into a very high density of sites dating from various periods of prehistory including those belonging to the Bell Beaker horizon.

8.2 Settlements and environment

The excavation of some of these sites and the study of the artefacts and, in particular, the analysis of the raw materials used in their manufacture, has enabled us to gain an idea of some of the possible source areas of raw materials. It can be deduced from this analysis that some of these raw materials were obtained in the area immediately surrounding the sites while others, because of their exotic nature, came from areas outside the study area.

8.2.1 Distribution of settlements

Until new works appear to change the existing panorama, it seems that during the Neolithic the main habitat in the Madrid region was in caves in the foothills of the sierra (Antona, 1987: 53; Rubio, 1983: 11; Martinez Navarrete, 1987), while from the Chalcolithic onwards, particularly from the pre-Breaker phase, there was a relatively intense colonization of the middle and lower basins of the rivers. These Chalcolithic settlements are located both on the banks of the lower river terrace, dominating the entire flood plain, and on small rises, further away from the main river courses. More sporadically, cave dwellings were maintained, as in the case of Estremera (Sánchez Meseguer *et al.*, 1983).

This change could be related to the consolidation of a production economy based on farming which would more profitably exploit the middle and lower reaches of the rivers than in the area around the head waters. This does not mean that the area at the foot of the mountains was permanently abandoned, and there are indications of some human occupation there throughout later prehistory. This move is justified not only by the availability of fresh pasturage in summer, but also by the possibility of obtaining certain raw materials of vital interest for the industry practised by these people and their economy in general. These materials were granite for making mills and grinders, metamorphic rocks for making simple and trunnion axes and other polished tools and, in particular, the mineral ores needed for obtaining metals such as copper and tin.

This model of land occupation was definitively consolidated during the Beaker period and it appears to have persisted, with virtually no variations, until the end of the Bronze Age. It is only from the Early Iron Age that there appears to have been some movement away from the watercourses of the main rivers. If it is confirmed that the same pattern of settlement was maintained for an entire millennium, it must be assumed that such a model satisfactorily met the needs of the groups that populated the region and that their strategies cannot have varied substantially with the course of this time.

The first factor that stands out, in view of the large number of pre-Beaker and Beaker sites, in comparison with the limited presence of Neolithic groups, is that this is very possibly evidence of a progressive population increase beginning early in the Chalcolithic. This increase also meant progressive colonization and occupation of new lands suggesting, at least along the middle and lower Manzanares (and possibly also the Jarama and the Henares), almost complete occupation leaving very little space between settlements. These lands, as we shall see later, were particularly suitable for the development of agriculture and good pasturage, a possibility that did not exist in the upper basins of these rivers where the Neolithic sites are found since the gradients and the characteristics of the soils made farming difficult.

However, this apparently high level of population may be deceptive. The data furnished by most of the settlements excavated indicate that they consisted of small and superficial dwellings, often only a single hut, on very light soil, which can be interpreted as seasonal occupation for short periods by very small groups that may have consisted of no more than a single family. This mans that, in many cases, the sites represent successive occupations, even when the domestic artefacts are relatively similar, in which case the population density would not appear to have been excessively high.

What is certain is that we do not know what the internal layout of the settlement was like, or what relationship existed between the villages and the burial places. These are aspects of fundamental importance for interpreting sites less than 500 m apart, which is a common occurrence in the area of greatest population density along the banks of the Manzanares.

8.2.2 Settlements and soils

Although all of the middle and lower river basins should in theory be very attractive places, if we examine the distribution map of Beaker settlements closely we can see that there is a clear preference for occupying particular points of the first river terrace. Especially preferred are those overlooking the course of the river and its flood plain and, even more so, locations close to the confluence of streams and intermittent steams with the main basins, which coincide with areas where good quality soils predominate.

However, there are productive soils throughout the middle and lower basins of the Manzanares and Henares–Jarama, both arable and pasture (Figure 8.1). The best soils, referred to as type A (Monturici and Alcalá del Olmo, 1990: 8), are found in the Henares–Jarama basin, in an area now mainly dominated by industrial and urban use. It is possible that, due to their depth, type A soils were too heavy for the technology then available, although the fact that relatively few sites are known here could be due to the difficulties of conducting fieldwork in this area of extensive urban industrial development.

Other high and medium quality soils, types B and C (Monturici and Alcalá del Olmo, 1990: 9) are to be found along the Manzanares and around the courses of some secondary streams, and the vast majority of the settlements are sited on them. Type B soils are shallower than type A, a factor which may even have been an advantage in the period we are concerned with, and they are considered soils with a high land-use capability. Type C are defined as having a medium land-use capability. This is based on the useable depth, the risk of erosion (which was perhaps less serious in that period due to greater plant cover), poor permeability and its uneven texture due to a very sandy or clayey composition. None of these characteristics would have presented important problems for the type of agriculture developed by the Beaker groups.

Interestingly, there are virtually no sites located directly on soil types D and E, which are classified as of no use for arable farming, although there are some areas of type D soil very close to the greatest concentration of sites. Although unsuitable for growing crops, these soils could be used for grazing and are suitable for the growth of woodland, so they offer considerable advantages for an economy in which stock raising, hunting and possibly also gathering, played a very important part. It is relevant to note here that, particularly in the Manzanares basin, which is the area of current interest (Jimenez, 1991):

> the soils, despite the moderate gradient, were difficult to use because of the presence of varying amounts of chalk. Only in the areas closest to the river, on the terraces, were the soils worked to a small extent, even where they are deep, although the closeness of the water table improved its suitability. Where the containment of the river has allowed the development of terracing, as in the case of the area with the greatest concentration of sites, soils with clayey layers also occur, and these are more productive. In this case, permanent or semi-permanent pastures may have existed, encouraged by the proximity of water.

Analysis so far justifies the preference for the lands of the middle and lower basins of the Madrid rivers and, more precisely, the areas where terraces had developed with better irrigation. In their upper reaches, the main rivers of the region run through soils that would produce fresh pastures through most of the year together with large areas of forest that would have held good reserves of game, although little of the soil was suitable for growing crops and there were few flat areas with natural irrigation.

This strategy, adopted both by the Beaker groups with Ciempozuelos pottery and those with point-decoration pottery, appears to reflect a predominantly arable/pastoral economy carried on by a scattered population living in small groups, often

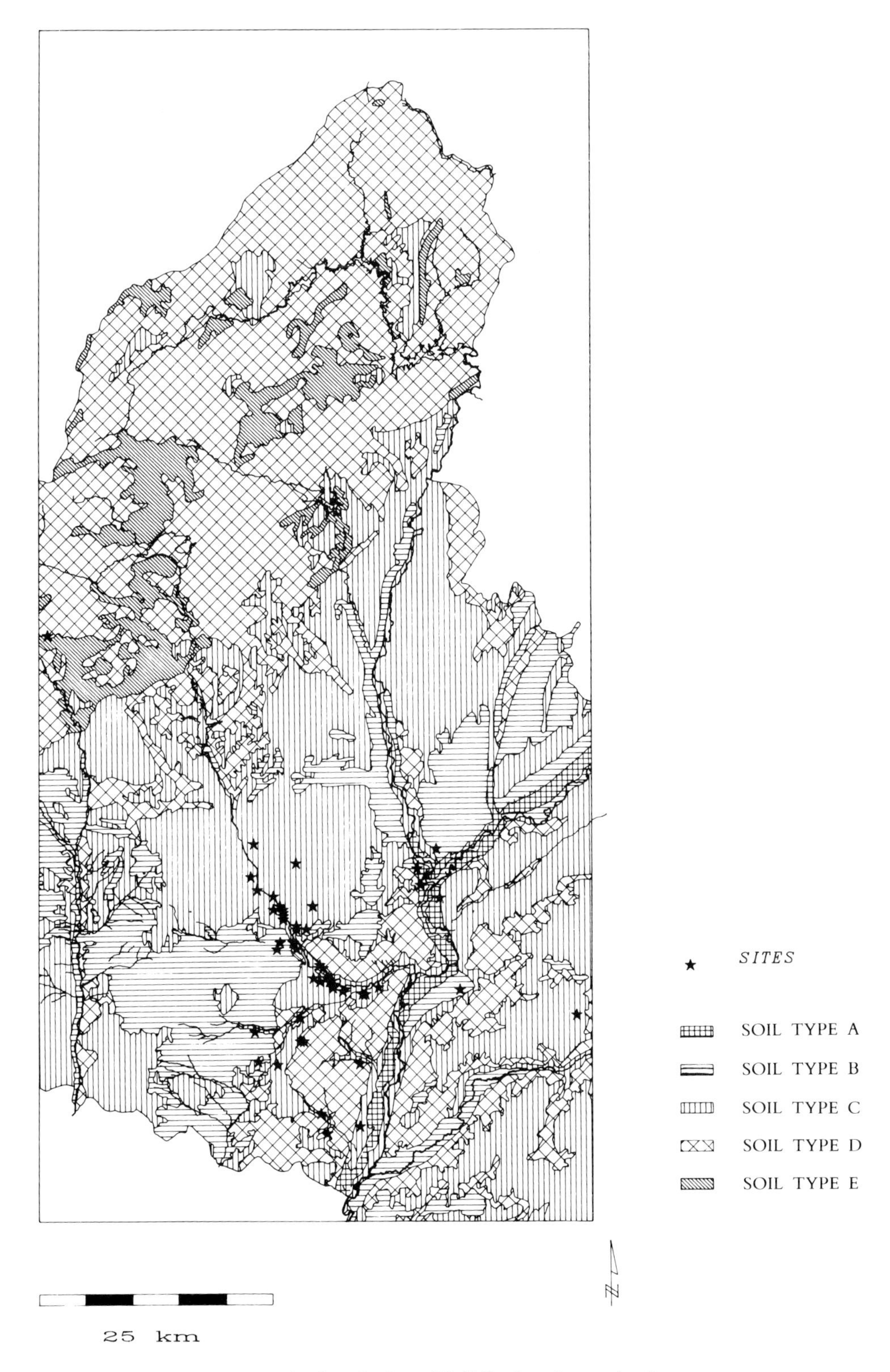

Figure 8.1 The distribution of Bell Beaker sites and soil types.

no larger than the family unit. It seems likely that a nomadic way of life was practised following an unknown cyclical pattern, probably consisting of relatively short cycles.

8.2.3 Settlements and mineral resources

These itinerant patterns may have been determined not just by the need to alternate pasture and arable land because of land exhaustion, but also to obtain particular raw materials used in industrial activities. In fact, if we look at Plate 1, showing only the area with the greatest concentration of sites, we see that the majority of Beaker sites are found very close (less than 3 km) to outcrops of good quality flint, in addition to the abundant nodules occurring below ground along all the terraces. Both of these sources provided material for lithic tools. Furthermore, many sites

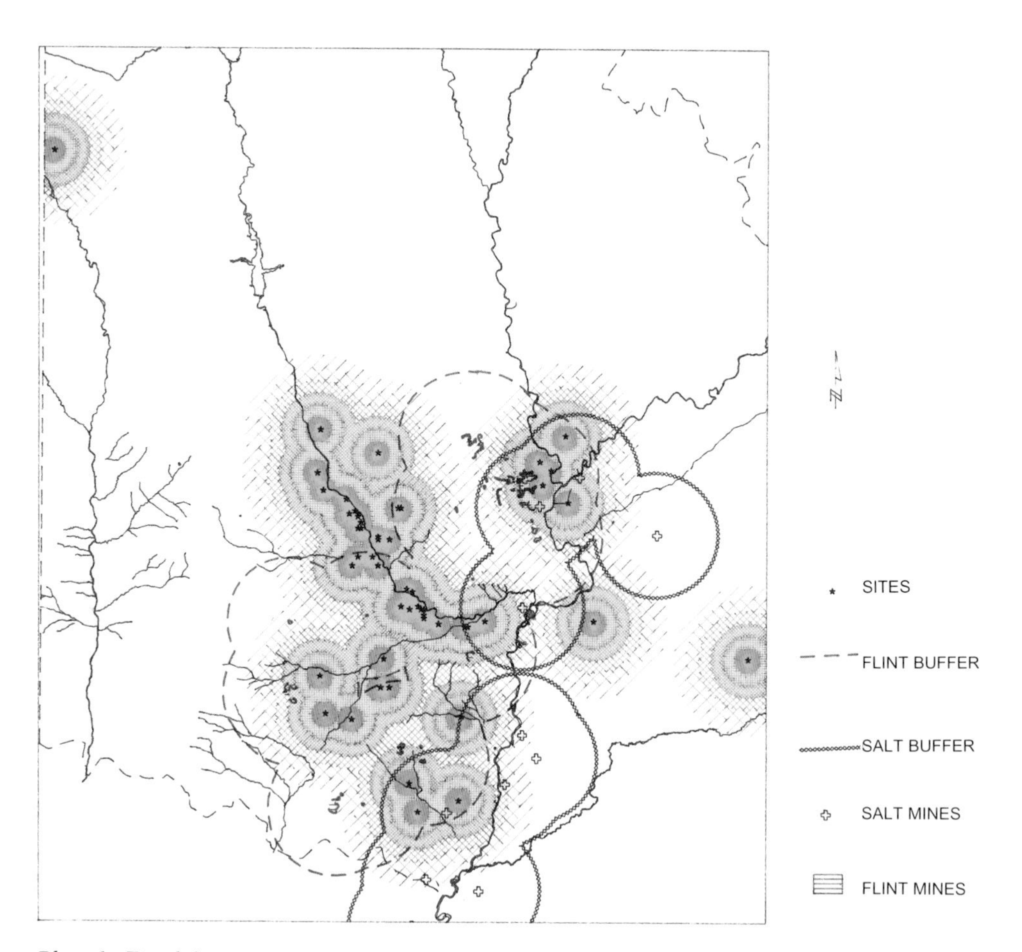

Plate 1. Five-kilometre salt and flint buffers together with variable buffers around Bell Beaker sites in the Madrid region of Spain.

within this same radius of 3 km have numerous salt deposits that would have been exploited, both for livestock and the preservation of food. The deposits in the area around Ciempozuelos–Valdemoro and San Martín de la Vega are of particular importance. In addition, there is a great abundance of clay that is found not only in fairly deep strata but also in numerous outcrops.

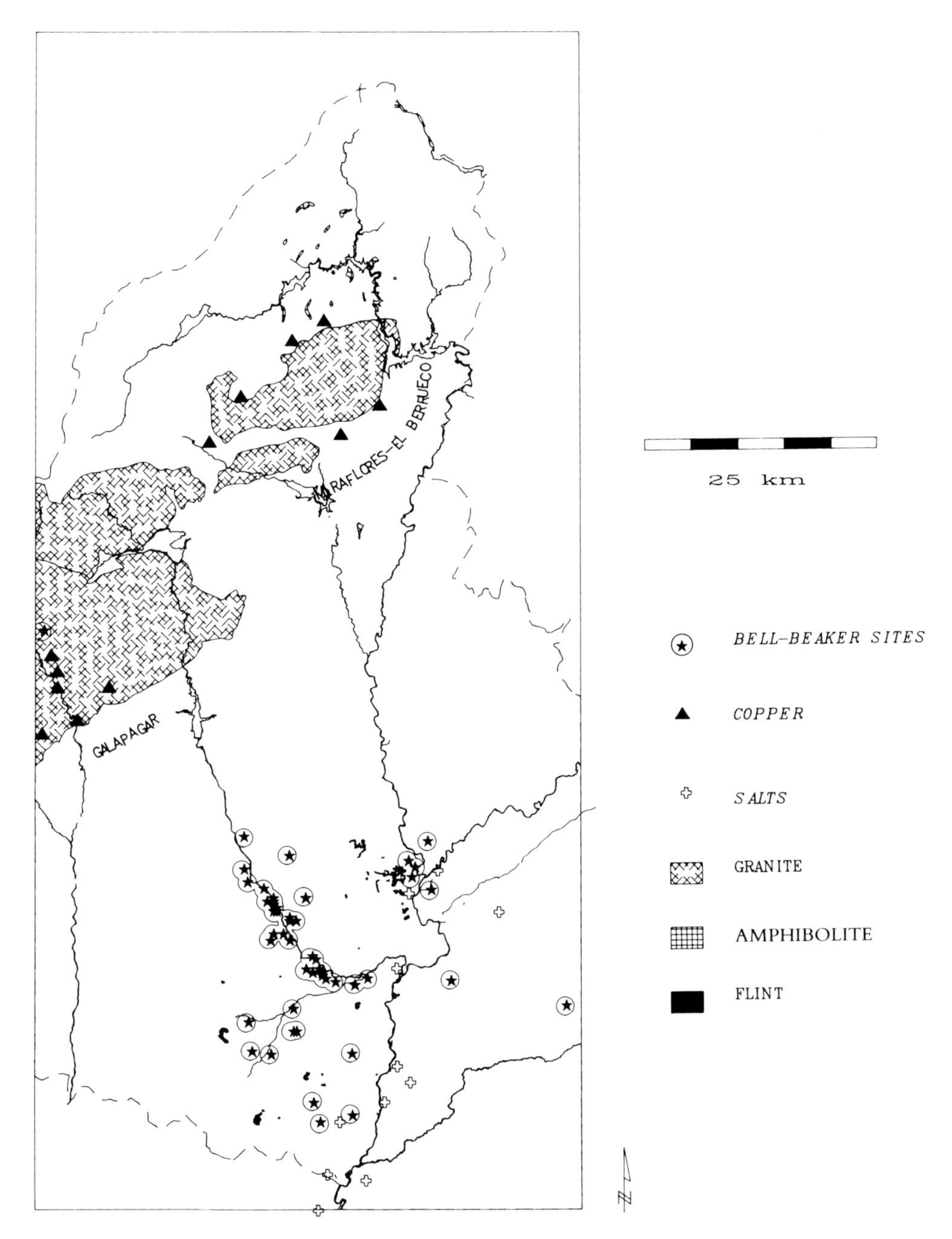

Figure 8.2. Bell Beaker sites in relation to exploited outcrops, showing two different copper zones.

Plate 1 also shows that the majority of sites have these three mineral resources (salt, flint and clay) within a maximum radius of 5 km together with being located in a region where the terrain is not difficult. Moreover, as shown below, some of the outcrops of these minerals were almost always visible. This is of considerable importance when considering that the majority of artefacts produced (ceramic and stone objects) were made from these materials and salt, too, was probably consumed in significant quantities.

Despite the fact that clay, flint and salt were mined on a considerable scale, they were not the only minerals exploited. Among the Beaker industries, we must not overlook the development of an incipient copper-working industry, which we know existed in the small communities in the lower basin of the Manzanares, shown by evidence from the sites at El Ventorro and at 8.8 km on the San Martín de la Vega road. In addition to metal, they also needed granite and metamorphic rocks for making polished tools. Figure 8.2 shows the salt, flint and copper mines, together with the Beaker sites. It can be seen that, with the exception of the Entretérminos dolmen, the sites are located outside the areas of copper ore outcrops and the solid granite with metamorphic rocks which occur with copper ore, since all these raw materials are abundant in the same places. In the case of the Madrid region, these are found in the Madrid mountains, around the headwaters of the rivers Guadarrama, Manzanares and Jarama at some 40–50 km from the area where most of the sites are concentrated.

The analyses carried out on some of the polished tools have shown that they are hard stones of a metamorphic type such as amphibolites, schists and others resulting from contact metamorphism. These are found at various points of the Central Range and the Toledo mountains, with the exception of the amphibolites, which are often intercalated in the gneisses of the Buitrago formation, close to the upper basin of the Jarama. These locations coincide with those provided by the minerals used in metal working, since the trace elements indicate their origin in the outcrops of the Miraflores–El Berrueco area, in the upper basin of the Jarama. To date, however, it has not been possible to prove the exploitation of the second of the Madrid areas with outcrops of copper ore: Galapagar and around the upper basins of the Guadarrama and Manzanares (Figure 8.4). Despite this, we see no reason why granite, other metamorphic rocks and copper should not have been equally well extracted both from the area of the upper basin of the Jarama and the headwaters of the Manzanares. Here, as in the case of the Jarama, its entire basin must have been a natural area for these Beaker groups, since there are no important impediments in the terrain.

8.3 *Mobility and axes of communication*

The mobility confirmed by the analysis of raw materials used in industrial activities could have been combined and co-ordinated with the requirements of livestock and the type of farming practised. This would explain the very high density of Beaker sites in the Madrid region, particularly in the stretch of the river Manzanares best known for its constant exploitation resulting in an average density 'between sites of 1–1.5 km' (Priego and Quero, 1992: 367). These figures, which relate to the lower basin of the Manzanares, are possibly very similar to the other Madrid river basins of the Henares and Jarama, but unfortunately this cannot be verified because of the dense modern population in the latter two areas.

It can be suggested that this mobility must have had two fields of action, local and longer distances. For the immediate area, both the main rivers and secondary water courses would have been followed and in particular it seems evident that, in order to obtain salt and particular types of flint, the secondary courses that cut across the main basin would have been utilized. In the case of longer journeys, both overland tracks and the river courses themselves may have been used to reach the mountain region close to the headwaters of the rivers.

In order to take a closer look at the routes followed to reach the mineral outcrop quarries we have produced maps of some of the well-attested cattle tracks in the region (Figure 8.3). In the case of the Manzanares it can be seen that they run parallel to the river, or even use the river itself as with the Jarama and although neither the rivers nor the cattle tracks are steep, they do go through gently rising terrain. The most direct routes present little difficulty because of their gentle profiles and in order to confirm this hypothesis a profile analysis was run on the course of the Manzanares and the cattle track that runs parallel to it (Figure 8.3). Neither route would have presented problems to the Beaker folk.

However, it would seem logical that it was the rivers themselves that were used as the axes of communication. It was possible to make the necessary stops along them with the guarantee of access to fresh pasture and vital water supplies. Also, the currents of the rivers could be used for moving heavy materials, such as large blocks of granite. As a hypothesis, we assume that both overland and river routes must have been used, and for the latter it would have been necessary to use rafts made from tree trunks. In contrast, Figure 8.3 also shows us how the courses of the most important Roman roads bore no relation to the interests of these earlier groups.

8.4 Settlement size and placement

This economic system, dictated in part by constant travelling, is consistent with the small settlements known, although there are, as yet, insufficient data which would allow us to generalize. Nevertheless, together with sites that have revealed a single hut, such as those at Fabrica de Ladrillos of Preresa and that of 8.8 km on the San Martín de la Vega road, there are others, El Ventorro for example, consisting of a group of various units which may or may not have been contemporaneous.

It remains to be seen if this clearly nomadic population was dependent on larger, more permanent settlements perhaps situated further away from the river basins or at a particular height. What is certain is that, co-existing with these lowland sites on the river terraces, in the same area, are other upland sites. In this study area, examples include La Loma de Chiclana and El Cerro de San Antonio in Madrid, Cerro Basura and the Motocross track site in Pinto and El Espartal in Valdemoro. Of these, the only Beaker site where it has been possible to prove relatively extensive occupation is Cerro Basura (Pinto). This is situated on the upper terrace of the river Jarama, some 20 m higher than the Arroyo Culebro plain, and is an authentic embankment settlement of considerable size, since the remains of artefacts are found scattered over an area of almost 9 ha. Its function could have been to unite and visually dominate not just some of the small settlements on the plain,

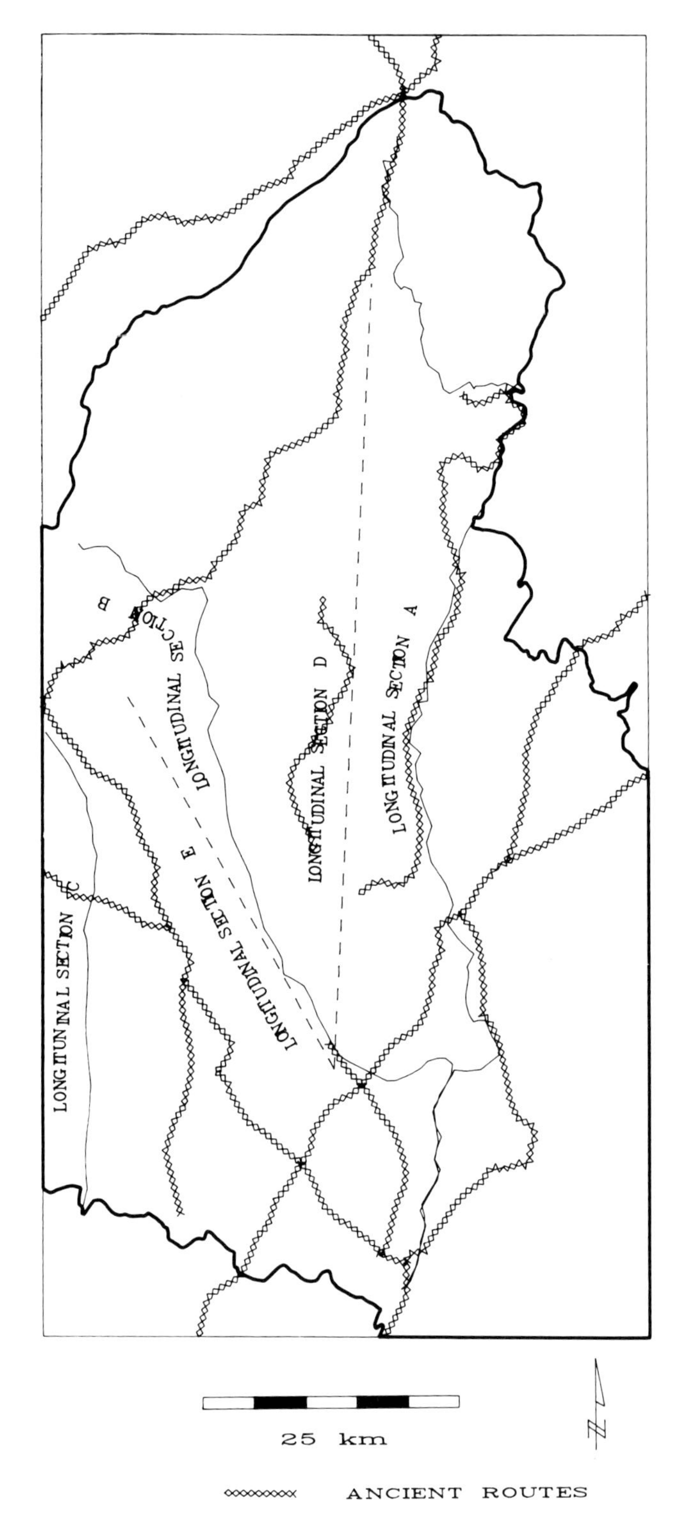

Figure 8.3. Profiles of ancient cattle tracks, direct routes and the rivers Manzanares, Guadarrama and Jarama.

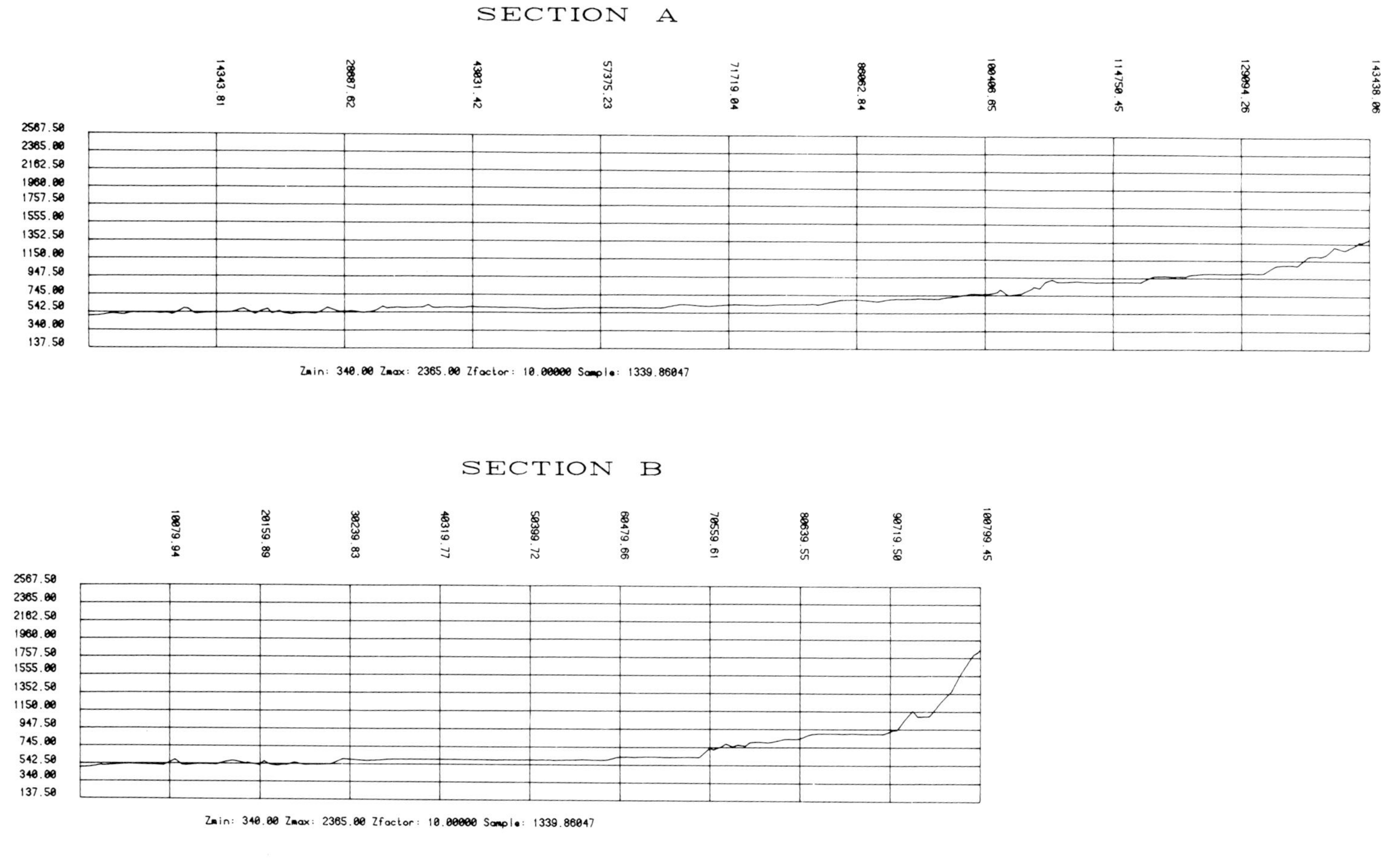

Figure 8.3 (continued). Profiles A and B (see map).

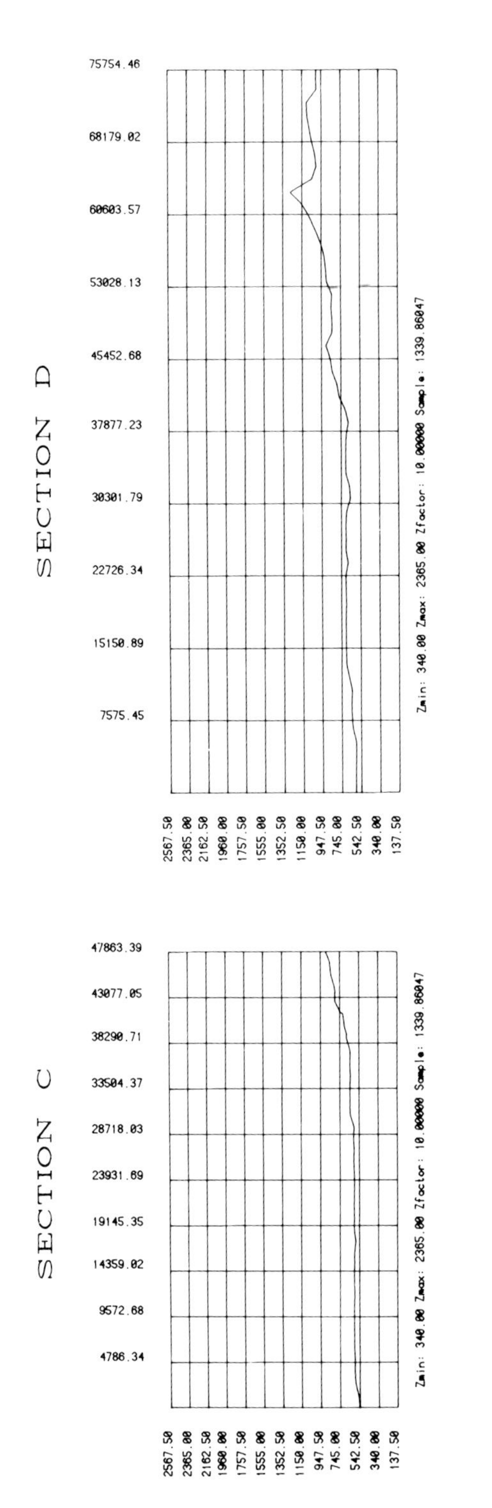

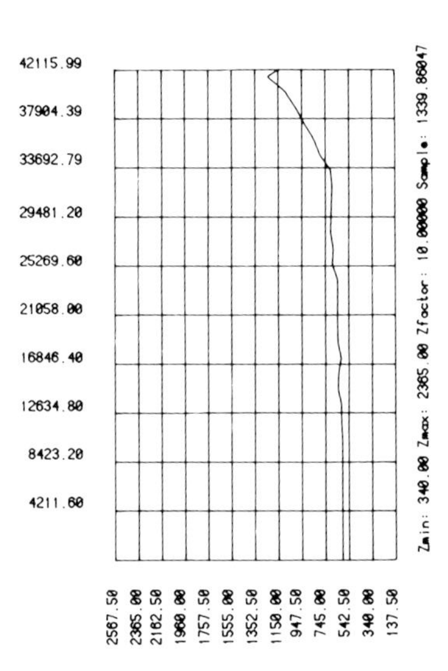

Figure 8.3 (continued). Profiles C, D and E (see map).

but also an important area of flint outcrops which includes Cerro de Los Angeles, certain points with salt deposits and numerous outcrops of solid granite in the Miraflores–El Berrueco area and Galapagar where, as we have said, there are easily exploitable veins of copper ore (Figure 8.4).

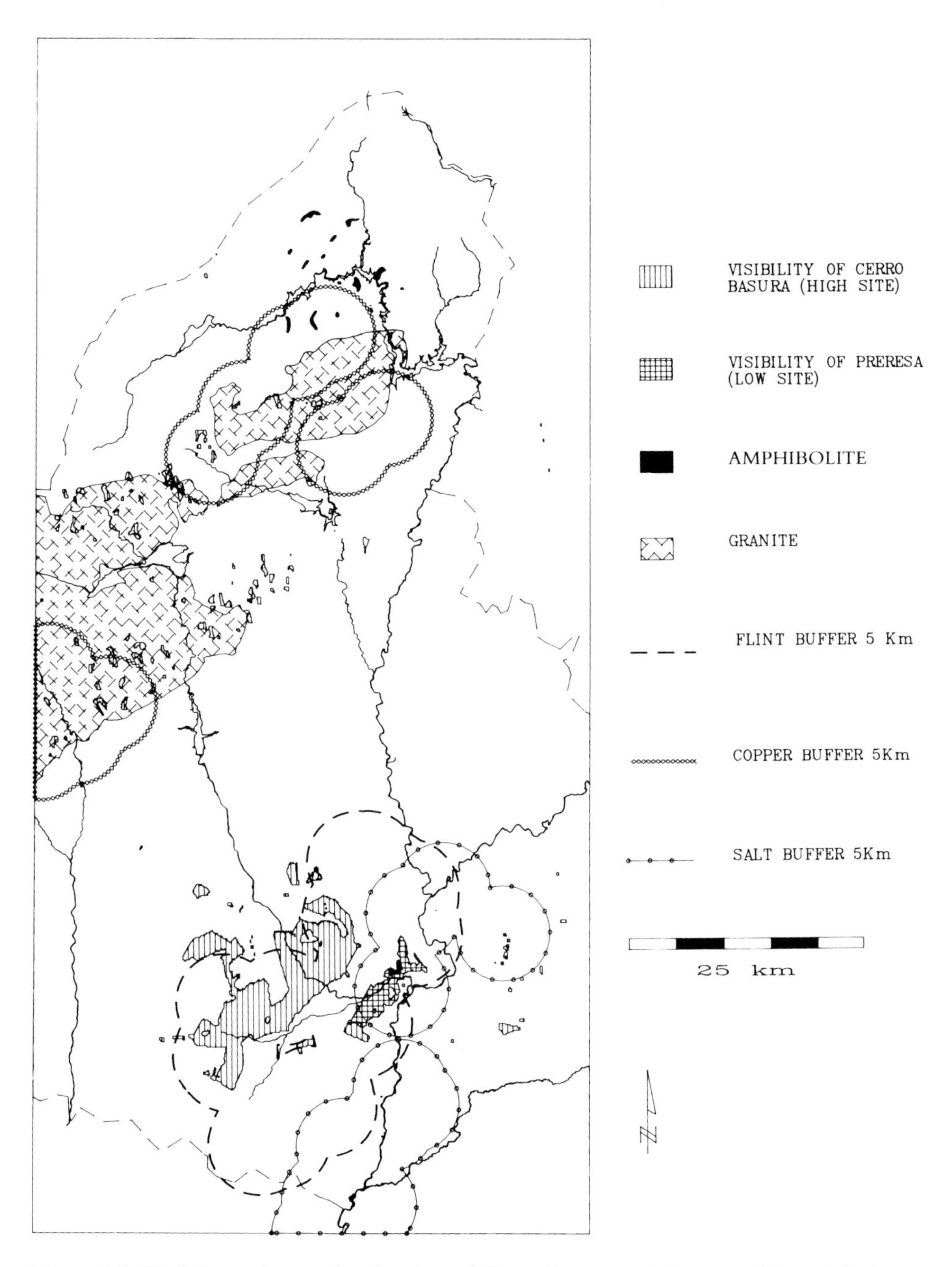

Figure 8.4. Visibility polygons for the sites of Cerro Basura and Preresa with exploited outcrops.

8.5 Visibility

The area visible from the Cerro Basura site is 16 247 ha which, as can be seen in Figure 8.4, contrasts with the area visible from the Preresa site, only 1911 ha, although the latter is of special strategic interest because it includes the low stretch of the Manzanares before it joins the Jarama, and also an area rich in salt.

A similar phenomenon can be seen in the area visible from the Motocross track, a total of 15 577 ha, clearly greater than that of the 8.8 km site located in a low position that looks out over only 4223 ha, although as in the case of Ventorro it includes points of mining interest in the mountains. This is particularly significant considering that evidence of metallurgical activity has been found in both settlements.

However, this type of upland site with great visibility constrasting with the lowland settlement from which little of the surrounding area was visible, does not seem to be a universal model, which the opposite of Ventorro. This is situated in the lowlands and has considerable visibility, including some points of the sierra with mining interest, totalling over 4070 ha. In contrast, Espartal in Valdemoro, situated on a small hill and clearly bigger than Ventorro since it has an area of 6 ha, only dominates a surrounding area of 2579 ha. Even so, this is an area of interest because of its rich salt deposits and outcrops of flint, which possibly justified its placement and would also explain the existence in the area of the only Beaker necropolis known in the entire region at Ciempozuelos.

Thus we cannot talk about a theoretical hierarchical ordering of territory, either on the grounds of location and visibility or on the basis of an altitude of site size correlation. The upland site at Cerro San Antonio, for example, is located on a rise on the right bank of the Manzanares and dominates a good part of the lower reaches of the river, some points of the sierra and, of course the terrace of the left bank on which there are a good number of small sites, giving a visible area of 9360 ha. Yet the site itself consists of little more than a few square metres in size and the lightness of the soil indicates little stability of occupation. Also, the finds, according to their discoverer (Fernando Lobo, personal communication) are concentrated in a dark layer barely a few metres in diameter and thus appear to correspond to the ground plan of a single hut constructed of organic material which has left little more than a few centimetres of archaeological deposit.

Another case which also fails to consolidate this very tempting hypothesis of a hierarchical order is the village of Loma de Chiclana (Andreu *et al*., 1992). This is situated at a point overlooking the courses of the Manzanares and the Olivar and Abroñigal streams, and excavations have revealed a series of temporary occupations belonging to various stages of the Chalcolithic and among them, a brief Beaker presence in a small area, thus suggesting a similar situation to that at Cerro San Antonio.

The reason for this dichotomy between terrace and upland settlements is not known, although the settlers' economic strategies would seem to have been comparable even if surveillance and defence requirements were different. The evidence provided by the two types is very similar and there is nothing that suggests they were devoted to different activities. Once again, however, Cerro Basura is an exception because it has provided a relatively large number of polished simple and trunnion axes which may be an indication of some farming activity, which would explain the site's greater size and stability. This example could be the first of a different type of site to that of the temporary sites located indifferently on the plain or in the uplands. To date, we have only Cerro Basura, although it is possible that other sites may be recorded, even in the mountains.

It seems sure that the special characteristics of both upland and terrace sites, combined with their small size and their very shallow soils, are evidence of their temporary occupation. It is even possible that at least some of them were not used for a complete farming cycle, suggesting they were camps used by shepherds who came from larger and more permanent settlements. If this were the case, it seems likely that regular contact would be maintained with the larger settlements, although it is impossible to determine if these were to be found in the same local area as the small settlements, or, alternatively, were in a more distant part of the river basin.

8.6 Conclusion

In short, with the help of the GIS, this work brings us closer to a better understanding of the relationship between these first metal-working groups of the Madrid region and their environment. It confirms that they had a good knowledge of the part of the natural region in which they lived, and from it they obtained the vast majority of their new materials, rationalizing and organizing their journeys to best effect, taking advantage of them to meet different needs on the same journey. Thus it is very probable that in the summer the mountains offered abundant fresh pastures for the livestock, and at the same time they could take advantage of the various minerals required for industry that were not available in the lower basins. At the same time, the possibility that specific journeys were made for prospecting, obtaining and transporting these raw materials cannot be ruled out.

Acknowledgements

This work has been undertaken within the framework of research project entitled 'Archaeological heritage of the lower basins of the Manzanares and Jarama (CO24/90)' subsidized by the Community of Madrid under the general plan of promotion of knowledge, and the cartographic application has been possible thanks to Project PS92–0023 of the DIGICYT of the Ministry of Education and Science. We would also like to thank the cartographic service of the Autonomous University of Madrid who produced all of the maps reproduced in this chapter, and especially its director, Dr Javier Espiago.

References

Audreu, M., Liesau, C. and Castaño, A., 1992, El poblado Calcolítico de la Loma de Chiclana (Vallecas, Madrid). Excavaciones de urgencia realizadas en 1987, *Arqueología, Paleontología y Etnografía* **3**, 31–116.

Antona, V., 1987, El Neolítico, *130 años de Arqueología madrileña*, Madrid, 44–57.

Jimenez, R., 1991, Informe sobre el uso de los suelos en el área del Bajo Manzanares. (texto mecanografiado).

Martinez Navarrete, I., 1987, Los primeros períodos metalúrgicos, *130 años de Arqueología madrileña*, Madrid, 56–81.

Monturici, F. and Alcalá del Olmo, L., 1990, *Memoria del mapa de capacidad potencial de uso agrícola de la Comunidad de Madrid*, Madrid.

Priego, C. and Quero, S., 1992, El Ventorro, un poblado prehistórico de los albores de la metalúrgica. *Estudios de Prehistoria y Arqueología madrileñas*, **8**. Madrid.

Rubio, I., 1983, Del Paleolítico al inicio de la Edad de los metales en Madrid, *Boletín de la asociación española de amigos de la Arqueología*, Dec. 1983, 4–14.

Sánchez Meseguer, J., Poyato, C. and Galan, C. 1983, El neolítico y la Edad del Bronce en la región de Madrid, *Arqueología y Paleoecología*, **3**, Madrid.

9

Prehistoric location preferences: an application of GIS to the Vinořský potok project, Bohemia, the Czech Republic

M. Kuna and D. Adelsbergerová

9.1 Vinořský potok project

The aim of the Vinořský potok project, the field part of which was accomplished by the District Museum of Brandýs n.L. in 1986–1991, was to study prehistoric population densities, settlement patterns and various types of interactions between past human occupation and the environment (Kuna, 1991). We present preliminary results concerning the correlation of settlement sites (from the Final Bronze Age to the Roman period) with local geomorphology. In particular, we want to test the hypothesis of non-random location preferences in individual archaeological periods in relation to several geomorphological characteristics of the landscape. If this hypothesis is correct, the locational preferences could be seen as displaying important aspects of past human behaviour with wider economic, social and symbolic implications. Also, we test the potential of GIS (PC ARC/INFO) for this kind of problem and provide the first application in Czech archaeology.

The landscape on the left side of the Labe river within the middle Labe lowland is quite monotonous and belongs to a large Cretaceous plateau of central Bohemia consisting mainly of various kinds of sandstone, mudstone and marlite. The plateau is slightly tilted towards the north-east, with the difference in altitude between the upper courses of small Labe tributaries and their confluences with the river being about 100 m and their usual length being about 10 km (Figure 9.1). Most of the area is covered by loess which gave rise to fertile soils of the chernozem type. The relief is very flat and does not impose any dramatic geomorphological features on the landscape causing particular constraints for human land use. This may represent an advantage for archaeological analysis since most of the observed settlement patterns can be explained by intentional human behaviour, rather than by external geographical determinants.

The main structural features of this landscape are the small streams that follow the tilt of the plateau to the north-east. Specific geological, tectonic and geomorphological processes created the characteristic shape of the valleys in this area: a relatively narrow valley floor, which gradually rises to the level of the surrounding plain on the western (left) sides of the valleys, while on the eastern banks the plain is usually separated from the valley floor by short steep slopes or terrace-like steps (Figure 9.2). The elevation of such terrace edges above the streams does not usually

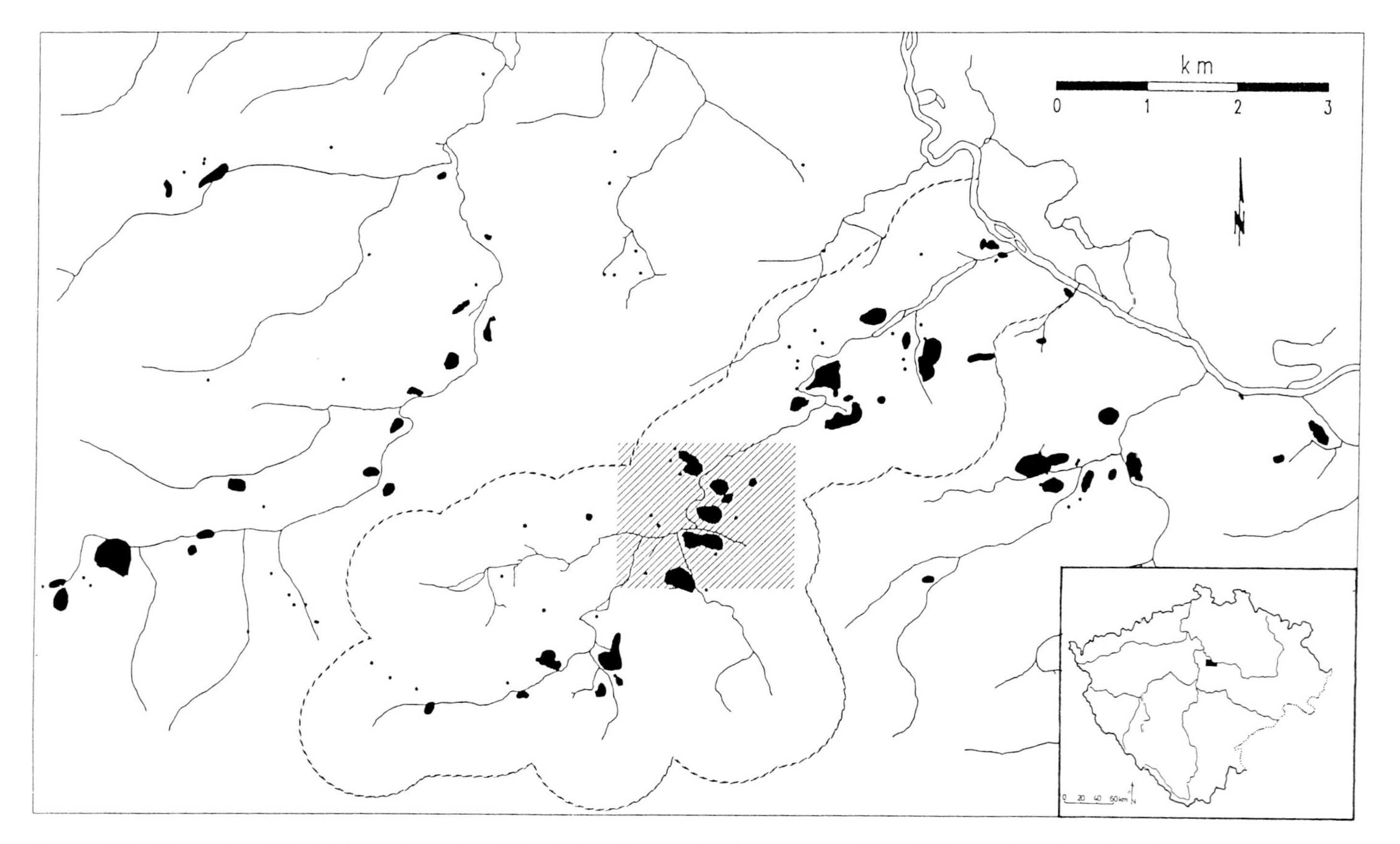

Figure 9.1. Map of the Vinořský potok project. The dashed line marks the core area (Vinořský potok valley) described in the text. The shaded box denotes the area shown in Figures 9.4 and 9.5. All archaeological sites from the final Bronze Age to the Roman period are mapped. The territory of Bohemia with the Vinořský potok project area is drawn in the insert.

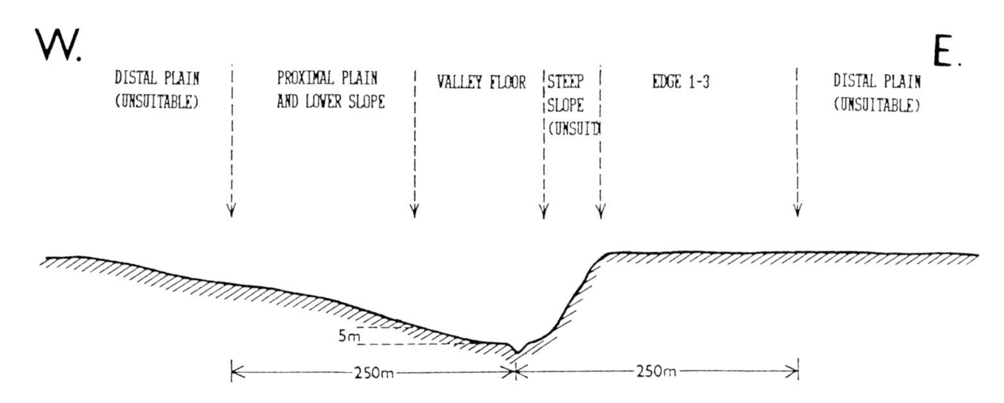

Figure 9.2. Vinořský potok project. Schematic valley section with marked relief categories.

exceed 20 m. Even so, in a relatively featureless landscape like this, the terrace edges form marked landscape features, which played an important role in the organization of the settlement pattern.

9.2 Description and classification of archaeological sites

The territory of the Vinořský potok project (190 km^2) was studied by a fieldwalking survey with all available earlier finds and sites being reclassified. This made it possible to build up a relatively reliable body of evidence for past settlement distributions reaching probably far beyond a 50 per cent sample of the existing settlement remains. Five archaeological periods were considered and all sites have been classified according to this time-scale (Final Bronze Age: 1000–750 BC; Hallstatt C: 750–600 BC; late Hallstatt–early La Tène period: 600–400 BC; La Tène B–D: 400–50 BC; Roman: 50 BC–400 AD.

We assume that most of the sites which were recognized as surface scatters represent settlement sites. There are several indicators suggesting that such an assumption is correct and any other interpretation (such as cemeteries or off-site activities) is quite improbable. Several burial sites known from earlier excavations were, however, also included. Settlement sites of individual periods differed in some aspects such as their total number or average size (Figure 9.3). This variability reflects, among other factors, a changing tendency towards either dispersed or nucleated settlement behaviour and it would certainly be an interesting question for further research.

Our analysis of archaeological sites was characterized by an approach that, as far as we know, has not been applied before and which cannot be realized without GIS techniques. In analysing the relationships between past settlements and their geographic backgrounds, we did not use site counts as means of quantifying the observed archaeological record but, rather, the surface extent of the past activity remains. During our field work, we established that the number of sites belonging to some periods was surprisingly high but, at the same time, their variability in size was extreme. Many sites were just small findspots represented by point-like distribu-

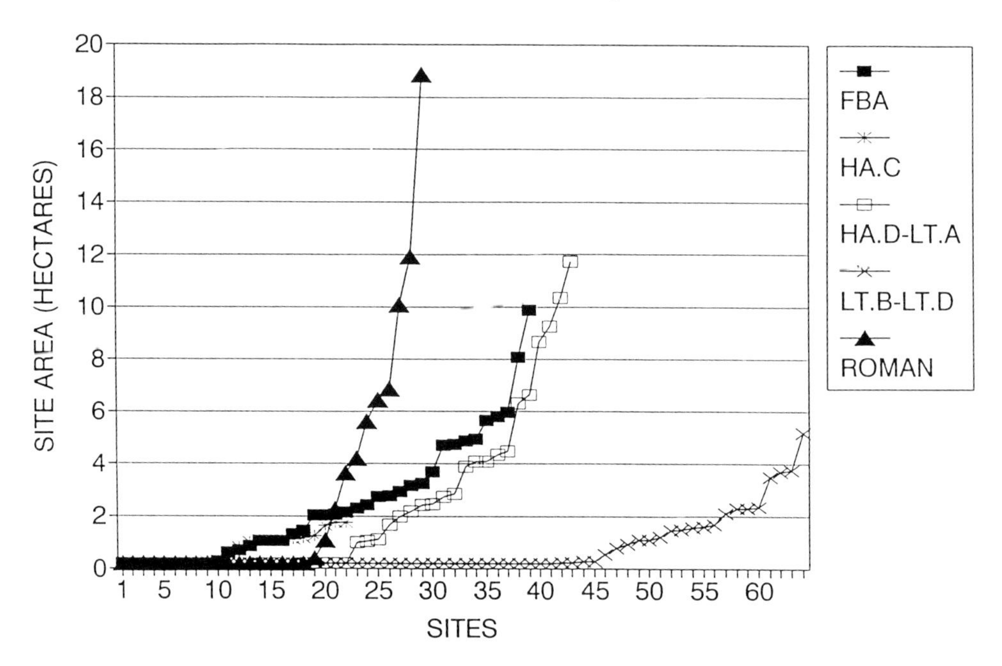

Figure 9.3. Vinořský potok project. Site counts and sizes according to archaeological phases.

tions of a few pottery fragments. In contrast to them, large sites were discovered representing either concentrated settlement activities during a shorter time-span or repeated (permanent) use of a habitation locus during an archaeological phase. Some such sites exceeded several hectares in size and the surface scatters consisted of thousands of pottery fragments per hectare. The largest site, belonging to the Roman period, covered some 18 ha (Figure 9.3). It seemed obvious that the larger a site appeared, the more value (weight) it should be given in further analysis.

Alloting different values to individual sites (habitation areas) according to their spatial extent (i.e. measuring their area instead of counting them) can also be reasoned in the theory of settlement processes. An isolated findspot could represent the remains of a single habitation event (e.g. a household on a given locus for a short period of time). Large sites, however, need not represent large communities behaving as co-ordinated social units: they could be explained as palimpsests of a larger number of household clusters (elementary sites or activity loci) accumulated over a longer time-span. This is why we can understand a large site not only as a result of a single location decision but as a product of (repeated) decisions of a relatively independent character made over over the time-span of a given archaeological phase. During our first analyses, we characterized each site not only by its spatial extent but also by the observed density of surface finds (or other kinds of archaeological data) discovered within its area. By multiplying the size of a site by a find density coefficient, we obtained a site 'score', which displayed a kind of relative value corresponding to the supposed sum of activities having occurred within the site during an archaeological phase. The use of this score was later abandoned for two reasons. First, it included the danger of a subjective bias in scoring sites because of the varying quality of archaeological observations. Second, although individual sites had different values when considering either their size only or their 'activity scores' (size multiplied by the find density index), general results of the two approaches regarding relative proportions of activities within individual landscape categories (the relief preference patterns) were nearly identical. As a result, we decided not to 'score' the sites but to work exclusively with their spatial extent, a more controllable variable.

The importance of considering the specific value (weight) of individual archaeological sites as a significant variable can be documented by the following example. It has been observed that Roman period activities were hardly ever located upon elevated edges above streams but that valley floors were intensively used. However, if we calculate percentages based upon site counts (including small findspots) there is very little difference between these two categories (32 per cent of sites are situated on valley floors, 29 per cent on elevated edges). This calculation is clearly missing the fact that most of the sites situated on edges are small, isolated findspots while most large sites are located on valley floors. When the site counts are substituted by the spatial extent of the sites, the results are very different (42 per cent of the total area of sites are located on valley floors, 3 per cent on edges). The latter result is obviously much closer to the real distribution of past human activities and may reflect real preferences of the past location behaviour. This example also shows the key problem of most traditional judgements using archaeological maps where sites are usually treated as homogeneous entities of the same interpretive value. Another problem of analysing sites as uniform units (points) is perhaps less important but also troublesome. Working with landscape features of a similar scale as archaeological sites, the probability that a site intersects more than one category of the background coverage is very high; it is often impossible to characterize a site by a single category of environment.

As already mentioned, we obviated these problems by using sites not as uniform point-like entities but as areas of variable size and, consequently, of variable analytical value. A GIS coverage was created for each of the archaeological phases. All the sites (197 in total) were digitized as polygons of variable size and shape: ranging from isolated findspots, marked as circles of 25 m radius, to larger sites, the size and shape of which reflected the available field evidence. For each of the archaeological periods the total surface extent of sites has been calculated and shows surprisingly similar figures for three of the periods (FBA, late Hallstatt and Roman period) while the values for the La Tène and Hallstatt C periods are somewhat smaller (Table 9.1). Whereas in the former case differences are presumably caused by the relative incompleteness of the available record, the latter case may reflect both a research bias and a real quantity of past settlement activities.

Table 9.1 Vinořský potok project. Site counts and total site extent

	Whole territory		Core area	
	counts	extent (ha)	counts	extent (ha)
FBA	39	96.4	19	49.2
Ha C	22	15.9	13	11.4
Ha D-LtA	43	97.4	23	59.3
Lt B-D	64	46.5	29	25.6
Roman	29	74.5	17	29.2

9.3 Classification of landscape elements

We assumed that, for the scope of our study, it was not necessary to consider and classify the relief of the whole territory, since large parts of landscape were not used for permanent settlement in the prehistoric past. Earlier investigations have shown that prehistoric settlement does not usually occur further than 300–500 m from a water source and certain other types of relief (such as slopes steeper than 4–5°) also remain unoccupied (Rulf, 1983; Kuna and Slabina, 1987). This has also been indicated by our field research during which we surveyed not only the areas close to streams but also a sufficiently large sample of the area outside those zones. In the case of the present analysis, we have established the distance of 250 m from a stream as the significant boundary. Our classification of geomorphological features concentrated on 500 m wide strips (250 m on each side) along water courses, henceforth called buffer zones. (The streams themselves must have been reconstructed using military maps commissioned by the emperor Joseph II in the 1780s.)

To support these ideas, we can point to the fact that only 21 sites (that is 10.7 per cent of the 197 sites in total), mostly smaller ones, were found outside the 250 m buffer zones. The remaining 176 sites appeared either totally or partly within the zones. If the total surface area of the sites is used, only 42.4 ha were situated outside the buffer zones, while the rest (287.7 ha) were included in them. These figures would be probably even more striking if we were able to reconstruct the prehistoric water streams with more accuracy as some smaller streams are obviously still missing from our maps since they were either already absent in the eighteenth century or they were neglected by the early military geographers. Not all the locations which are more distant from today's streams can, however, be explained by these factors. We might suppose that settlement sites of certain archaeological periods occur at a greater distance than 250 m from the nearest water source: a higher occurrence of sites outside the buffer zones could therefore also be a significant aspect of particular settlement patterns.

Buffer zones along reconstructed water streams were divided into polygons (more than 100 polygons in total) belonging to five main categories. The remaining area outside the buffer zones is the sixth category. The suggested categories of landscape relief are as follows (Figures 9.2–4):

1. **valley floor**—areas along the streams within the buffer zones, elevated less than 5 m above the water level;
2. **proximal plain and lower slopes**—plain terrain and moderate slopes (less than 5°) of the valleys, situated within the 250 m buffer zones and elevated more than 5 m above the stream;
3. **edge 1**—plain areas within the buffer zones, separated from the valley by a steep slope (steeper than 11.3°, which means an elevation of at least 10 m within 50 m horizontal distance);
4. **edge 2**—areas of the same type of relief as in the edge 1 category but with elevated edges on two sides;
5. **edge 3**—the same type of elevated plain as in the edge 1 category but enclosed on more than two sides creating a projected (promontory) position; and
6. **distal plain and other 'unsuitable' areas**—all areas either too distant from streams (more than 250 m) or 'unsuitable' for occupation for some other reason (e.g. slopes steeper than 5°: in the territory under study no prehistoric site displayed such a location).

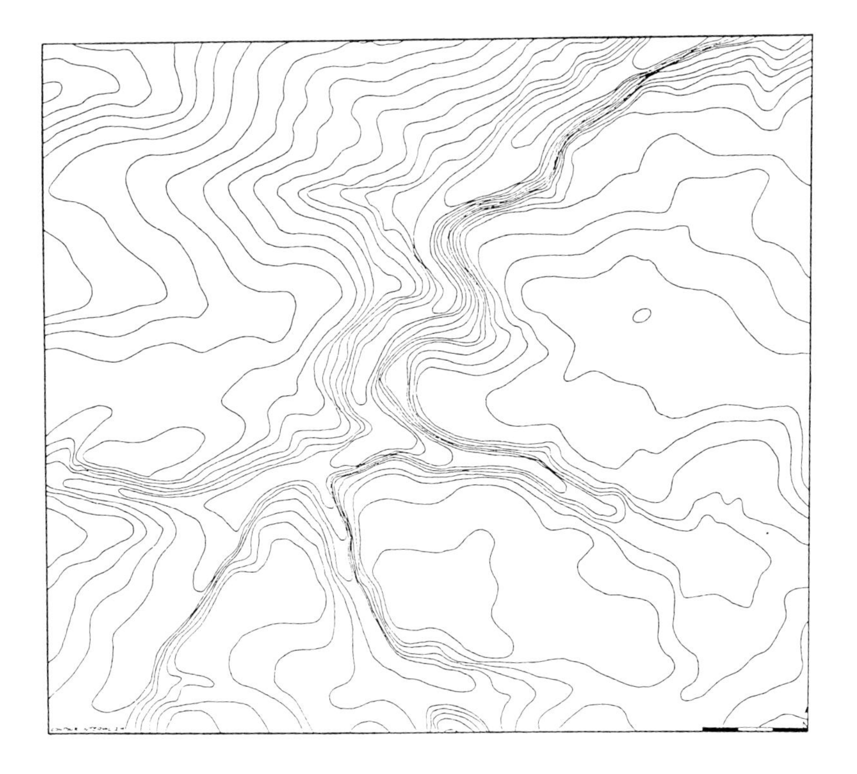

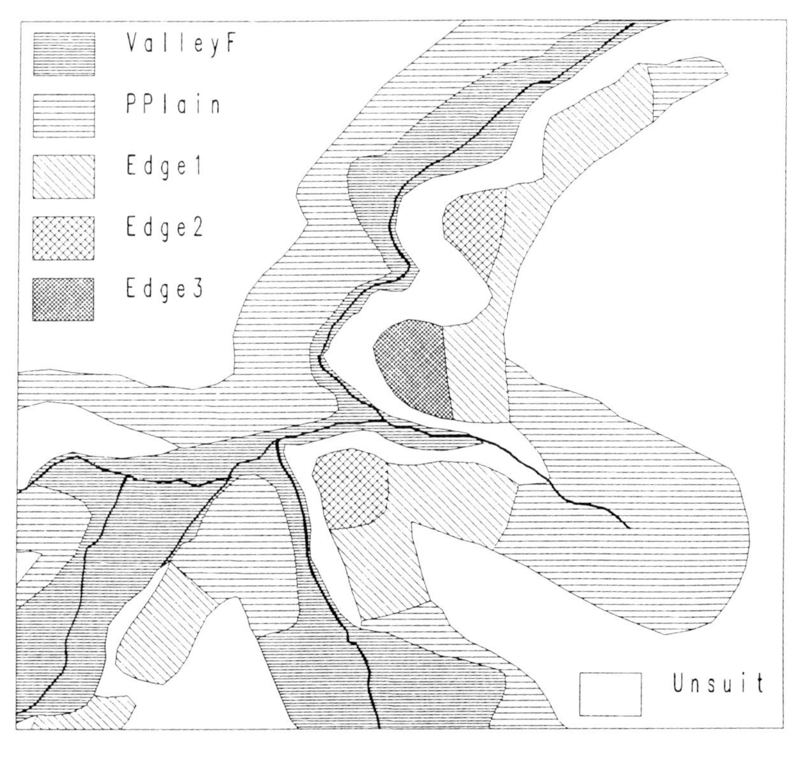

Figure 9.4. Contour map (2 m intervals) of a part of Vinořský potok valley (upper) and the corresponding relief categories (lower). Size of the box is 2.1 × 2 km (cf. Fig. 9.1).

For the present study we performed the classification of relief types by using a topographic map at 1:10000 and calculating elevations, slopes and distances by hand. The reason for this methodology was the absence of any digitized contour map at a suitable scale and the fact that the available PC ARC/INFO could not analyse contour lines in the way we required. Automatic procedures of analysing contour maps are, however, to be applied in the future.

9.4 Analysis

The map of relief types has been digitized as a separate coverage which was subsequently overlayed by and intersected with the five coverages representing settlement sites of individual prehistoric periods (Figure 9.5). Intersections of individual site polygons with individual relief polygons were summed according to relief categories and both their absolute and relative values (percentages) have been calculated. This procedure has been carried out twice: first, for the whole research territory and, second, for its core area, the central valley of Vinořský potok itself (Table 9.2). The reason for considering these two levels separately was the fact that the less frequent relief categories (edge 2 and edge 3) are represented relatively more often within the core area (the preference patterns were, therefore, less biased by the uneven distribution of the relief types themselves). The data from the core area were also given to E. Neustupný who we asked to test the results by means of formalized statistical methods. His methods and results are described in Chapter 10.

In the case of the core area, we also compared the observed preference pattern with the theoretical (expected) one. The expected pattern was defined as being identical to the relative proportions (percentages) of the individual relief types available within the area (Table 9.2). Since all the archaeological sites situated on this stream should, theoretically, be located within the 250 m buffer zone, the value of the 'distal plain' and other 'unsuitable' areas was defined as zero. We can expect that, if there were no intentional location rules or preferences, the preference pattern would correspond to the actual relative frequency of individual relief categories.

The calculated preference patterns of individual periods for the whole project territory show major changes in location behaviour during later prehistory (Table 9.3; Figure 9.6). The results from the core area displayed a similar process (Table 9.4; Figure 9.6). Comparison of the observed and expected preference patterns also

Table 9.2 Percentages of the relief type categories

	Whole territory		Core area (250 m buffer)	
	ha	%	ha	%
ValleyF	3148.7	16.4	421.3	30.3
PPlain	2812.2	14.7	736.5	52.9
Edge 1	324.2	1.7	187.8	13.5
Edge 2	16.5	0.1	14.7	1.1
Edge 3	34.5	0.2	32.2	2.3
Unsuitable	12813.1	66.9	—	0.0
Total	19148.8	100.0	1392.4	100.1

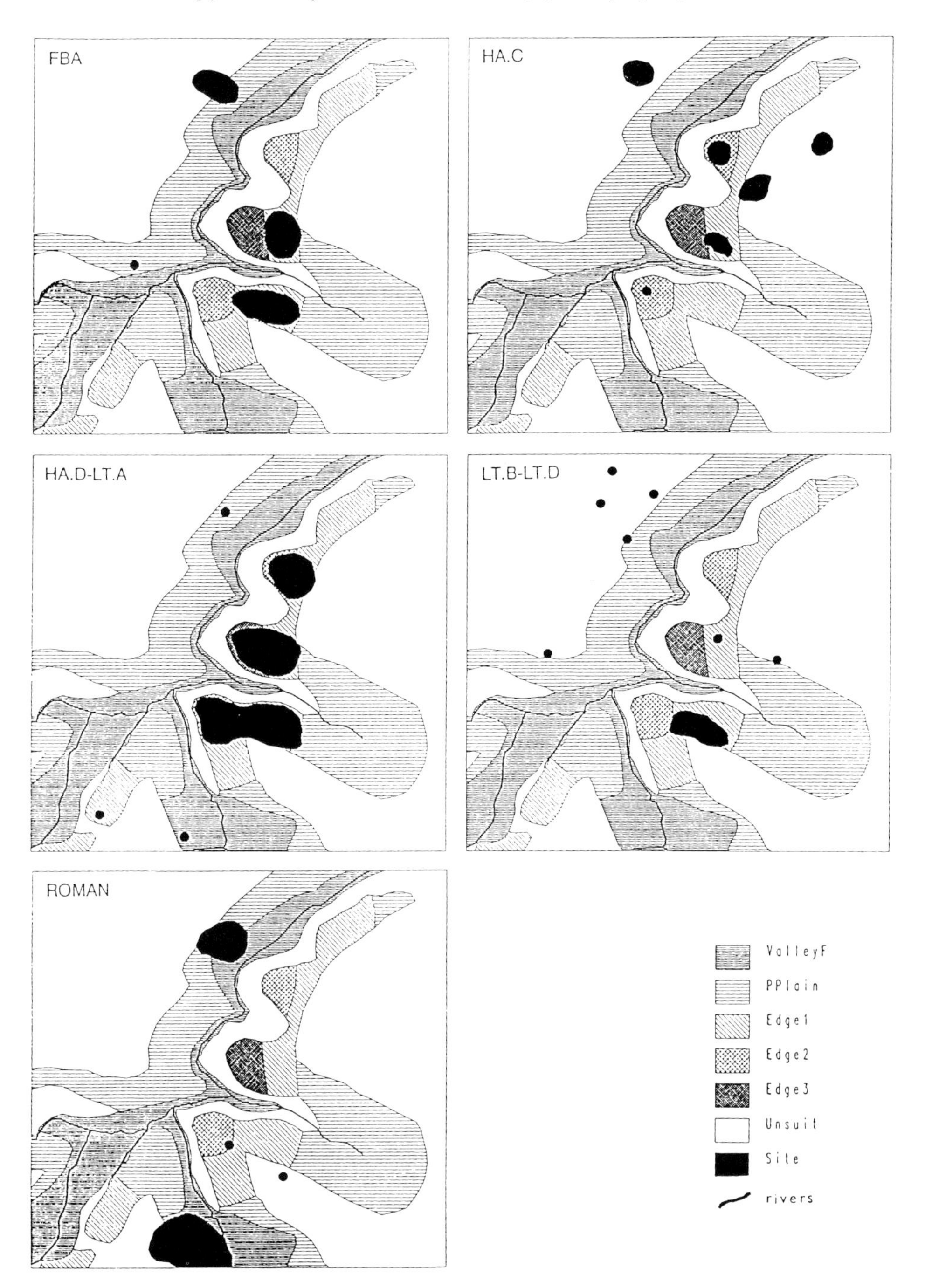

Figure 9.5. Vinořský potok project. Relief categories and site polygons. Size of the box is 2.1 × 2 km (cf. Fig. 9.1).

enabled us to approach the problem of the intentionality of the observed site distributions. Comments on the results and their initial interpretation are presented in the next section.

Table 9.3 Vinořský potok project. Intersections of archaeological sites with relief categories (the whole territory)

	Intersections (ha)				
	FBA	HaC	HaD/LtA	LtB-D	Roman
ValleyF	13.2	0.4	14.5	9.6	44.9
PPlain	27.6	2.6	17.4	6.5	18.8
Edge 1	44.5	2.6	32.7	15.5	0.4
Edge 2	1.9	1.5	8.7	1.7	0.1
Edge 3	1.9	1.1	15.4	4.5	0.4
Unsuitable	7.3	7.7	8.7	8.7	9.9
Total	96.4	15.9	97.4	46.5	74.5
	Intersections (%)				
	FBA	HaC	HaD/LtA	LtB-D	Roman
ValleyF	13.7	2.3	14.9	20.6	60.2
PPlain	28.7	16.5	17.9	14.0	25.2
Edge 1	46.2	16.6	33.6	33.3	0.6
Edge 2	1.9	9.2	8.9	3.7	0.2
Edge 3	1.9	7.0	15.8	9.7	0.5
Unsuitable	7.6	48.5	9.0	18.7	13.3
Total	100.0	100.0	100.0	100.0	100.0

Table 9.4 Vinořský potok project. Intersections of archaeological sites with relief categories (the core area)

	Intersections (ha)				
	FBA	HaC	HaD/LtA	LtB-D	Roman
ValleyF	0.8	0.0	1.5	1.6	12.2
PPlain	6.3	0.1	4.5	2.5	11.1
Edge 1	32.5	2.6	25.3	11.2	0.4
Edge 2	1.9	1.5	8.6	1.7	0.1
Edge 3	1.9	1.1	15.4	4.5	0.4
Unsuitable	5.8	6.1	4.0	4.1	5.0
Total	49.2	11.4	59.3	25.6	29.2
	Intersections (%)				
	FBA	HaC	HaD/LtA	LtB-D	Roman
ValleyF	1.5	0.0	2.6	6.2	41.6
PPlain	12.9	0.6	7.5	9.9	38.0
Edge 1	66.2	22.9	42.6	43.5	1.4
Edge 2	3.8	12.8	14.6	6.6	0.5
Edge 3	3.8	9.8	25.9	17.7	1.3
Unsuitable	11.8	53.9	6.8	16.1	17.2
Total	100.0	100.0	100.0	100.0	100.0

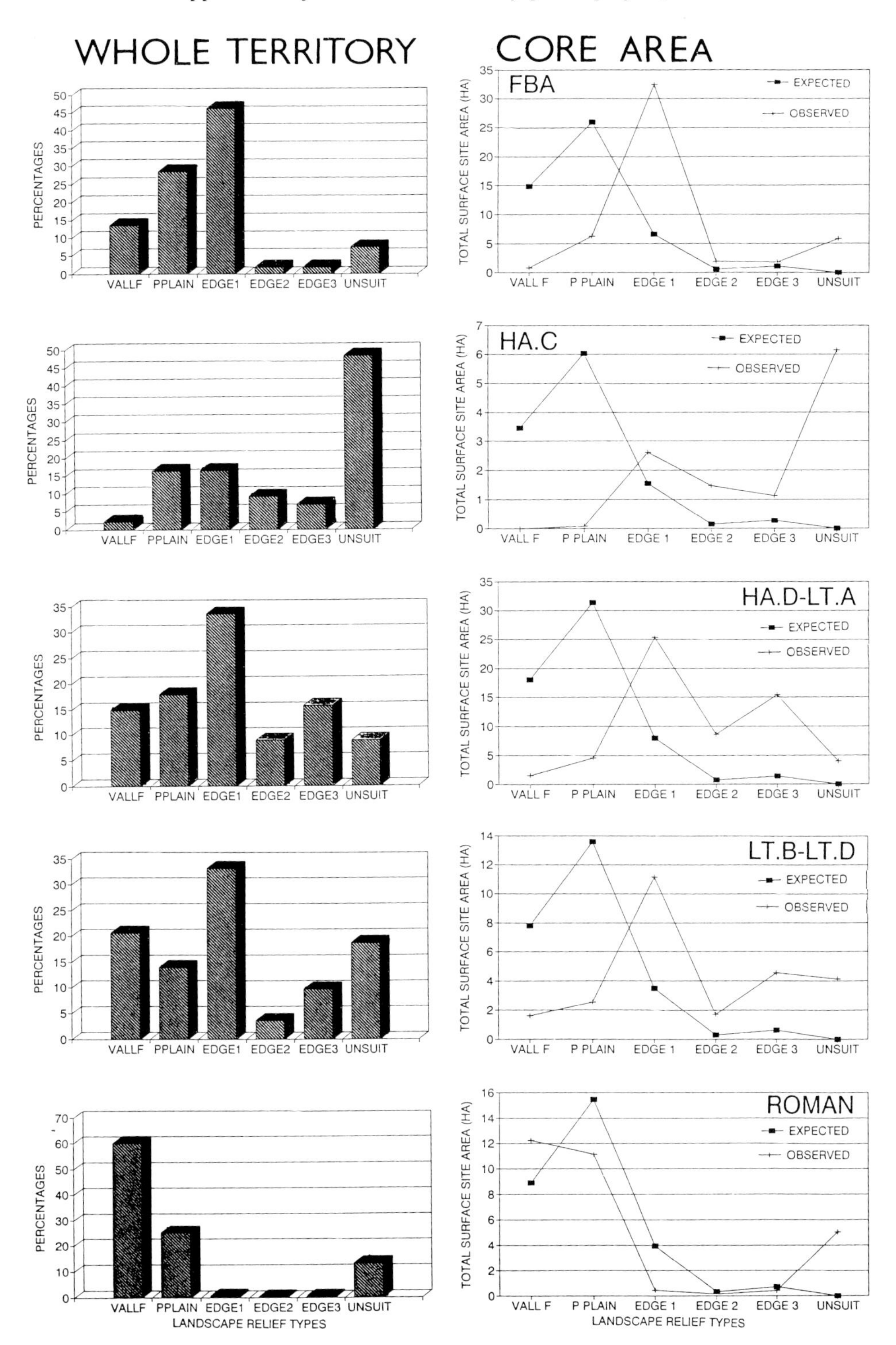

Figure 9.6. Vinořský potok project. Relief preference patterns within the whole territory (left) and within the core area including a comparison of the expected and observed preference patterns (right).

9.5 Results and interpretations

9.5.1 The Final Bronze Age (FBA)

The FBA clearly preferred the edge 1 type of location (46.2 per cent of the total area of sites within the whole territory are situated within this relief category). Considering the core area only, this preference is even more pronounced (66.2 per cent). The next relief category according to preference is the proximal plain, moderate slopes beyond the valley floor. The comparison of the relative values of proximal plain locations within the whole territory (28.7 per cent) and the core area (13 per cent) reveals that locating a site upon the proximal plain might have been only a secondary alternative, used in those community areas where no elevated edges were available (a higher preference for the proximal plain corresponds to the smaller extent of edge 1 within the whole territory). No less interesting is the fact that the FBA sites, although preferring elevated edges, avoided those parts which are usually described as 'strategic' positions: promontories and places surrounded by steep slopes on at least two sides (edges 2 and 3). This phenomenon has been noted by empirical observations in other regions as well (Slabina, personal communication). Valley floors (areas less than 5 m above the streams) were generally avoided but if no other relief was available they were settled as well (e.g. the area of Čakovice, Ďáblice, within the SW part of the project territory). Absence of the preferred relief type was obviously not a reason for abandoning the whole area. This may mean that site location in general was influenced more by cultural preference (which could have been ignored if necessary) than by strict functional constraints.

The pattern of the FBA sites suggests concentrated habitation activities with the lowest proportion of small findspots of all the periods (Figure 9.3). The percentage of 'distal plain' occupation (11.8 per cent within the core area) is not caused by separate locations but mostly by the fact that the edges of some of the larger sites extend beyond the 250 m buffer (Figure 9.5).

9.5.2 Hallstatt C

The subsequent Hallstatt C sites often appear beyond the 250 m buffer (48.5 per cent within the whole, 53.9 per cent within the core area) which is in itself quite interesting. This could be one of the reasons why settlement sites of this period have so far been rarely found. Sites located further from the stream than 250 m are rather exceptional in most prehistoric periods. This has many consequences: it means, for example, that sites located at such a distance from a stream are not a part of large multicultural clusters and, being small, isolated sites, have a smaller average visibility and, therefore, less chance of being identified.

The preference for the 'distal plain' is, however, partly a function of our own, arbitrary method of classification of relief types, namely of using the 250 m distance as the decisive boundary. A closer look at the map discloses that the settlement pattern is not so different from that of the preceding phase as it may seem from the calculations. Hallstatt C sites, similar to those of the FBA, are mostly situated upon the elevated banks of streams. The main difference between the two periods may be caused by the more dispersed pattern of the Hallstatt C settlements, allowing individual habitation loci to cross the 250 m boundary more often, thus shifting the distance tolerance to slightly higher values (perhaps 350 m or so). Even so, we do believe that our analysis has revealed an interesting feature of the settlement pattern of this period.

9.5.3 Late Hallstatt/early La Tène

Late Hallstatt/early La Tène sites show a clear inclination to be located upon edges (58.3 per cent for all three edge positions; 83.1 per cent within the core area). The location of settlement sites in 'strategic' positions (edge 2 and especially edge 3) reaches its peak (8.6 per cent and 15.4 per cent within the whole territory, 14.6 per cent and 25.9 per cent within the core area). The preference for these relief categories in HaD/LtA is clearly visible, especially if the absolute extent of the edge 2–3 relief covered by sites is compared to similar intersections of other periods (Tables 9.3; 9.4). Relative values (percentages) are, however, less convincing and this deserves special attention.

As mentioned above, there are no doubts about the correlation of HaD/LtA sites with promontories (edges 2 and 3) based on empirical knowledge of the area. It was, however, quite difficult to show this it by means of simple (non-formalized) methods of analysis as used so far. The reason for this being that the edge 2 and edge 3 categories are not present everywhere within the territory and, therefore, they were not available to all the communities of this period. These types of relief were unambiguously preferred wherever they were available; the fact that they are relatively less common and not evenly spread (when compared to the other types of relief) was probably the reason why the communities had to use other relief types as well (mostly edge 1). Of course, this makes the results less clear, especially if non-formalized methods of analysis are used. In contrast, the factor analysis used by Neustupný (factor 3–/3 proved our empirical knowledge of this correlation because it was able to compare the observed and the expected preference patterns including the quantitative relations among relief types themselves in the same analytical procedure (Neustupný, Chapter 10).

9.5.4 The Late Iron Age

The Late Iron Age (LtB–LtD) does not display unambiguous preferences for a certain relief type but the 'edge locations' (edges 1, 2 and 3) still largely prevail (46.7 per cent within the whole, 67.8 per cent within the core area). Considering these figures, the relatively large proportion of 'unsuitable' locations (18.7 per cent and 16.1 per cent), and the pattern of site sizes (Figure 9.3), we may presume that the distribution of sites was influenced by factors similar to those during the Hallstatt C period, namely, the dispersed character of habitation activities.

9.5.5 The Roman period

The situation changes during the Roman period when the occupation tends to towards larger sites again. It is very interesting that the location of large settlement sites seems to be very specific in being located by preference on the valley floors (60.2 per cent in total, 41.6 per cent within the core area). This specific feature of the Roman period settlement sites has already been noticed previously (Kuna *et al.*,

1989). It seems that there was no precedent for such a preference during the prehistory of Bohemia and it could perhaps be understood as a profound change in the spatial behaviour of the population.

The comparison of the expected and observed preference patterns within the core area (Figure 9.6) enabled us to consider the question of intentionality of past locational behaviour. It is supposed (and supported by the statistical tests of Neustupný) that, for example, the edge locations for FBA, HaD/LtA and LtB-D represent results of intentional past behaviour. The high preference for 'unsuitable' areas (distal plain) during the Hallstatt C period is certainly also significant but it should be tested on other regions before a generalization is made. The bias, caused by the arbitrary definition of relief categories and by indirect effects of different settlement processes (higher mobility and dispersion of activities), should also be considered in this case. Although the Roman period displays a clear tendency towards site concentration within the valley floors, the intentionality of this appears to be questionable. The observed preference pattern can be successfully explained even without presuming any intentionality because it more or less copies the expected pattern, which reflects the relative proportions of available relief categories within the core area. Consequently, the observed situation could be interpreted in terms of random shifts over the existing relief categories without applying any intentional location rules at all.

If this is correct the obvious change in site distribution during the Roman period need not be explained as the exchange of one intentional principle for another but it could be understood as the abandonment of existing intentional location rules. The results of a linear correlation of the expected and observed patterns reveal a striking difference between the Roman period and all other archaeological phases, although the coefficient is not significant at the 95 per cent level (Neustupný, Chapter 10).

Summarizing the available data we can infer that a profound change of location principles in the Roman period has been identified. This change could mean the adoption of new cultural principles of site location but it may also be interpreted as a departure from any intentional preferences. In the latter case the relief categories, as classified in this paper, would not have been significant for the adopted location strategy. The apparent preference for valley floors during the Roman period could thus have been just a function of the abundance of this relief type within the given territory. This problem cannot be solved without considering other regions with different relief compositions. Valley floors have obvious advantages from the economic viewpoint, based on their immediate proximity to water, and their frequent use during the Roman period could be understood as a shift towards a more economics-based location strategy. In contrast to the Roman period, most preceding cultural phases clearly preferred elevated valley edges, a preference that may consequently appear as to be caused more by social and symbolic factors than by economic considerations. It should also be mentioned that the valley floor location was also the most preferred one during the subsequent early medieval period and later, resulting in the development of stable medieval villages. Such an assumption would, however, require more careful consideration.

The present analysis of location principles in a micro-region of central Bohemia during later prehistory has revealed, we feel, new ways of interpreting distribution maps not so far used by archaeologists. While empirical data of this type have often been used in archaeology, the exploitation of GIS techniques seems to open a new chapter in this particular area of analysis. Such procedures can become a very effective tool for predicting and analysing site locations both in the field of theoretical research and archaeological resource management.

References

Kuna, M., 1991, The structuring of prehistoric landscape, *Antiquity*, **65**, 332–47.

Kuna, M. and Slabina, M., 1987, Zur Problematik der Siedlungsareale (in der Bronzezeit), in, Černá E. (Ed.), *Archäologische Rettungstätigkeit in den Braunkohlengebieten und die Problematik der Siedlungsgeschichtlichen Forschung*, Prag: Institute of Archaeology.

Kuna, M., Waldhauser, J. and Zavřel, J., 1989, *Říčany 1986. Záchranný archeologický výzkum sídliště doby laténské a železářského areálu starší doby římské* [Archäologische Rettungsgrabung einer latenezeitlichen Siedlung und eines Eisenverhüttungsareals der älteren römischen Kaiserzeit], Brandýs n.L.–Stará Boleslav: Okresní muzeum.

Rulf, J., 1983, Přírodní prostředí a kultury českého neolitu a eneolitu [Naturmilieu und Kulturen des böhmischen Neolithikums und Äneolithikums], *Památky archeologické*, **74**, 35–95.

10

Beyond GIS

Evžen Neustupný

10.1 Introduction

GIS is a major invention for archaeology perhaps comparable to radio-carbon dating in its importance. Following initial over-optimistic claims for GIS, the time for sober evaluation has come. To understand fully the role of GIS in archaeology, it is necessary to delimit its field not only positively, but also negatively, in other words it is necessary to realize what lies beyond the capabilities of GIS.

This chapter uses the case of the Vinoř brook in central Bohemia (Kuna and Adelsbergerova, Chapter 9) to analyse the role of GIS in archaeological reasoning. This is because the Vinoř case is simple and transparent, based on the descriptive system devised by Kuna using powerful structures allowing both non-formalized and mathematically oriented methods to be used to achieve the same results. In comparing the two methodologies, I shall follow the three principal steps outlined elsewhere: analysis, synthesis and interpretation (Neustupný 1993).

10.2 The non-formalized approach to the Vinoř case

10.2.1 Analysis

Generally, the term 'analysis' should be returned its original meaning by using it for the decomposition of archaeological contexts into meaningful constituents. The goal of analysis conceived as decomposition is the establishment of a descriptive system which includes the definition of sets of objects, sets of descriptors (variables) and sets of mappings between objects and descriptors.

The decomposition can be performed either on the basis of simple observation unaided by any tool, or it can use complicated aids, tools and research procedures as, for example, in the case of radio-carbon dating. Specialized software, such as GIS, able to transform the original information into a form which is easily exploited by archaeological synthesis clearly belongs to the same class of aids.

However, I do not believe that even a highly sophisticated software package can replace the theoretical judgement of an archaeologist. In the particular case of Vinoř the decisive factor was, without doubt, Kuna's ability to select the right set of descriptors. His understanding of the prehistoric landscape in Bohemia is in terms of edges—strips of sloping terrain which structure the landscape vertically. In devising this concept, he has identified something that seems to have had a real symbolic meaning for the prehistoric people of that area.

10.2.2 Synthesis

Archaeological synthesis means the generation of archaeological structures, i.e. looking for order, regularity or pattern hidden in the record. Synthesis, always being based on the previous analysis, is performed in either a non-formalized way or by means of mathematical algorithms.

In the case of Kuna and Adelsbergerová, the synthesis of archaeological structures has been accomplished either by considering the raw data furnished by GIS techniques or by their transformation into percentages or indices. All of these structures have been obtained either by observation or by simple calculation, and the role of various types of graph in discovering and assessing individual oppositions certainly should not be underestimated.

It is clear that the generation of archaeological structures, whether done in the traditional way or by means of mathematical methods as shown later, have nothing to do with GIS. The only possible exception to this is the intersection of coverages containing geomorphological types of relief and maps of prehistoric sites which could possibly be understood as elementary structure formation.

10.2.3 Interpretation

Generally, interpretation proceeds by way of modelling, that is, by comparing archaeological structures to models derived either directly or indirectly from non-archaeological cultural contexts. Some models are abstract and yet quite productive; for example, among those formulated by Kuna and Adelsbergerová one was used to interpret the differential presence of individual culture groups in the classified types of relief.

Such a situation, however, could not be observed in instances other than the Roman period when the occupation of valley floors could be explained as a consequence of non-selectivity in site location. Kuna and Adelsbergerová could argue that the abandonment of locational preferences in the Roman period meant a departure from the earlier rules of site location thus introducing the importance of social and symbolic factors rather than economic considerations. This argument is supported below.

10.3 A formalized approach to the synthesis of structures

10.3.1 Principles

Exploiting the descriptive system described by Kuna and Adelsbergerová and the data generated by them using GIS, I have attempted to synthesize archaeological structures by means of formalized, mathematical algorithms. Specifically, I have used a matrix of data containing the area of intersection between sites of the five culture groups and the six types of relief. The rows of the matrix are types of relief and the columns are culture groups. The individual cells of the matrix contain numbers expressing the area in hectares.

Another view of the same problem is to use entities (defined as areas with the same type of relief), described by standard archaeological variables such as culture groups. This view reduces the Vinoř case to the most usual type of archaeological descriptive system (items in rows, attributes in columns). In consequence, there is no reason why the rows should not be considered as objects, and the columns as attributes within a descriptive system. As soon as this is realized, it becomes possible to calculate linear correlation coefficients between each pair of culture groups and to treat the matrix of correlations by means of some orthogonalizing procedure from the factor analysis family. I have chosen the Varimax rotated principal component solution, mainly because of its theoretical properties, which gives the results shown in Table 10.1.

The number of factors to be extracted from the correlation matrix is based on the consideration of the pattern of latent roots. If there is a sharp relative decrease in the values of two neighbouring latent roots (i.e. a pronounced 'spring' between them), it is usually taken as an indication of the number of factors. In Table 10.1 the biggest relative spring occurs after the third root, and it is almost absolute, i.e. the values of the remaining two roots are negligible, suggesting three factors.

10.3.2 The case of three factors

The so-called communalities almost equal unity, which is an indication that the information contained in the correlation matrix has been maximized by the solution. The Final Bronze Age, La Tène period and, to a lesser degree, the Late Hallstatt/La Tène transition have high positive coefficients (so-called factor loadings) in respect of factor 1 (Table 10.2). Hallstatt C (the Bylany culture) has a high loading on factor 2, while the Roman period is diagnostic for factor 3 (its positive pole). It should be noted that the Late Hallstatt/La Tène transition has a high negative loading in respect of factor 3 (this can also be described as factor 3^-). The findings so far indicate that the Final Bronze Age, La Tène and the Late Hallstatt/La Tène periods form one group, while Hallstatt C forms another group

Table 10.1 The Vinoř case: latent roots of the correlation matrix

1	3.250
2	0.993
3	0.707
4	0.047
5	0.003

Table 10.2 The Vinoř case: rotated factor loadings

	factor 1	factor 2	factor 3
1. Final Bronze Age	0.976	0.113	−0.107
2. Hallstatt C (Bylany group)	0.111	0.984	−0.134
3. Hallstatt D/La Tène A	0.777	−0.093	−0.615
4. La Tène B to D	0.938	0.180	−0.277
5. Roman period	−0.220	−0.203	0.951

and Roman period a third one. The Hallstatt/La Tène transition is the one which maximally opposes the Roman period.

The factor solution in terms of factor loadings (Table 10.2) can be recalculated into a 'dual' solution in terms of so-called factor scores (Table 10.3). While factor loadings, similar to correlation coefficients from which they are derived, take values from −1.000 to +1.000, factor scores may theoretically acquire any real value. It can be seen in Table 10.3 that edge 1 has a high score in respect of factor 1, the distal plain scores high on factor 2 while the valley floor and proximal plain have a slight but probably not statistically significant tendency to oppose this factor. Valley floor and proximal plain, however, have significant scores on factor 3, while edge 2 and edge 3 oppose it.

A summary of the results so far is given in Table 10.4 which shows four structures:

1. Factor 1: Final Bronze Age, La Tène, Hallstatt D/La Tène A and edge 1
2. Factor 2: Hallstatt C, distal plain
3. Factor 3^{+} (i.e. the positive end of factor 3): Roman period, valley floor, proximal plain
4. Factor 3^{-} (i.e. the negative end of factor 3): Hallstatt D/La Tène A, edge 2, edge 3.

A high factor loading on a variable (a descriptor) does not indicate how frequently the variable occurs; rather it shows how typical it is of the factor. Factor analytical

Table 10.3 The Vinoř case: factor scores relating to factors 1, 2, and 3

1. Valley floor	−0.418	−0.704	1.254
2. Proximal plain	−0.021	0.731	−1.149
3. Edge 1	2.151	0.052	−1.149
4. Edge 2	−0.942	−0.222	−1.127
5. Edge 3	−0.384	−0.542	−1.264
6. Distal plain	−0.387	2.147	0.379

Table 10.4. Summary of the factor analytical solution of the Vinoř problem

	Factor 1	Factor 2	Factor 3
Evidence of factor loadings:			
Final Bronze Age	typical		
Hallstatt C (Bylany c.)		typical	
Hallstatt D/La Tène A	typical		opposing?
La Tène B to D	typical		
Roman Period			typical
Evidence of factor scores:			
Valley floor			typical
Proximal plain			typical
Edge 1	typical		
Edge 2			opposing
Edge 3			opposing
Distal plain		typical	

methods have the advantage that no descriptor and no object from the initial descriptive system is exclusively typical of one resulting structure. This, of course, is in full agreement with what we intuitively feel.

It is immediately clear that our factor solution of the Vinoř brook problem yields almost exactly the same structures as those obtained by Kuna using largely non-formal means, except for factor 3^-, which does not seem to have any counterpart presumably because it was combined with the seemingly similar factor 1.

10.3.3 Statistical tests

What should now follow are statistical tests to demonstrate that the supposed structures are not a play of random numbers; this applies equally to both the non-formal and the factorial solutions.

Testing the significance of the data used by Kuna and Adelsbergerová is not easy, as most statistical tests dealing with nominal variables rely on counting frequencies. Once they are replaced by measurements, the choice of statistical methods becomes rather restricted. One of the testing tools that can be used is the linear correlation, which tests the independence of two sets of measurements.

Table 10.5 shows the linear correlation between the area occupied by late prehistoric culture groups within the defined types of relief and the maximum area of the same type of relief available in the region. There is high positive correlation coefficient (0.808) for the Roman period; however, because of the very few degrees of freedom, it is not statistically significant at the 5 per cent level. If we forget statistical significance, there is another fairly high (negative) coefficient relating to the Hallstatt C Bylany group (−0.607). It could be concluded, therefore, that the distribution of Roman period sites tends to match the distribution of individual types of relief while that of Hallstatt C sites tends to oppose it.

10.3.4 Advantages of the formal solution

This chapter has attempted to show that when the structures contained in the record are strong, when the archaeologist involved is specialized in the field and, mainly, when the extent of the descriptive system is small, reliable results can be obtained by means of non-formal methods for the generation of archaeological

Table 10.5 Vinoř brook. Linear correlation between the area occupied by late prehistoric culture groups within types of relief and the maximum area of the same relief type available in the region

	linear correlation coefficient
1. Final Bronze Age	−0.013
2. Hallstatt C (Bylany group)	−0.607
3. Hallstatt D/La Tène A	−0.339
4. La Tène B to D	−0.216
5. Roman period	0.808

structures. Kuna's treatment of the Vinoř case is a good example. However, when the structures contained in the record are weak, when the archaeologist lacks intuitive knowledge, and the case passes a certain level of complexity, the formal solution is much better and much more productive. Even in the case of Vinoř, the factorial solution has clearly revealed factor 3^- which remained in the background of the non-formal treatment.

10.4 What is beyond GIS?

10.4.1 GIS as an analytical tool

The spell of GIS, which has now captivated so many archaeologists, lies in the fact that it is a tool which drastically widens the possibilities of archaeological analysis by the decomposition of archaeological contexts. GIS is a tool for greatly enhancing the number of possible descriptors available to archaeologists. It furnishes a means of achieving descriptions of mainly unlimited territories using such descriptors.

Despite the fact that GIS has so far been mostly used as a graphical software, its full potential is not to produce maps or other kinds of graphical output but to analyse spatial properties of the archaeological record. However, applications which would reveal archaeological structures as defined above seem to be lacking. In this sense GIS does not belong among the tools of archaeological synthesis, i.e. tools for the generation of order, rules or patterns contained in the archaeological record. Similarly, it does not belong among tools appropriate for the interpretation of structures. These two fields of archaeological method clearly lie beyond GIS.

10.4.2 Is GIS also a synthetical tool?

It may be asked whether GIS could not be extended with some basic tools for archaeological synthesis as described above, dendrograms for example. Whether such hybrids would be useful is not obvious as, technically, joining anything to GIS does not mean that it becomes logically integrated. Yet, this is a field worthy of development as shown by the simple comparison between vector synthesis as defined above and what could be done with descriptors and objects produced by GIS, as discussed below.

The descriptors created by GIS are coverages—if a coverage is well designed, its role corresponds to the role of a vector in vector synthesis:

1. A coverage may consist of real numbers, for example a coverage of contour lines, which in some cases may be very dense similar to a field in physics. In vector synthesis such coverages correspond to vectors filled with values of cardinal variables.
2. The coverage may consist of polygons, for example those representing soil types. Such a coverage is analogous to generalized vectors whose elements are individual states of a nominal variable.
3. Sets of coverages, taken as descriptors in the above mentioned sense, describe two-dimensional objects such as points or areas on the surface of the Earth. The areas can be defined in a variety of ways one of which is exemplified by the Vinoř

case. Another, perhaps more natural way, would be to use an arbitrary but regularly spaced grid forming a raster.

Typically, a descriptive system based on GIS will consist of

1. a set of areas,
2. a set of coverages, and
3. a set of mappings between the areas and the coverages.

This is a scheme in which archaeological items are mapped into archaeological attributes, generating matrices in which structures can be sought by methods from the factor 'analysis' family. The obvious question is whether archaeological and, indeed, any other sort of structures can be formally generated on this basis.

The Vinoř case suggests that this is indeed possible, as coverages can be correlated and the correlation matrices analysed for hidden structures. However, in addition to the usual difficulties in handling nominal variables, missing values, etc., there is another difficulty to be overcome, caused by the individual areas being prearranged in two dimensions. It is theoretically possible to ignore this arrangement during the synthesis of structures (as in the example above) but it clearly means a loss of information.

In my view, all of the procedures that generate archaeological structures, to say nothing of their interpretation, lie beyond the scope of GIS, which should concentrate on analysis of the record. Anyway, it should not be forgotten by users of GIS that this 'beyond' exists, and that it is what gives sense to the archaeological endeavour.

Acknowledgements

I am grateful to Martin Kuna for putting his unpublished primary data at my disposal and for discussing various aspects of this chapter with me.

References

Neustupný, E., 1993, *Archaeological Method*, Cambridge: Cambridge University Press.

11

Perceiving time and space in an isostatically rising region

M. Nunez, A. Vikkula and T. Kirkinen

11.1 Introduction

Finland has undergone a series of major environmental changes during the last 12 000 years. From a glaciated country to a fairly temperate region where certain central European plants thrived (e.g. waterchestnut) and then climatic deterioration to the somewhat mild subarctic climate of today. At the same time, the surrounding Baltic basin evolved from a vast ice-lake, to an arctic sea, to a 'great lake', and finally to a sea that has become increasingly brackish in the past 3000 years. Moreover, as all these changes took place the country was experiencing powerful isostatic rebound.

It is therefore not surprising that for over 80 years Finnish researchers have paid attention to environment and space (e.g. Ailio, 1909; Ramsay, 1920; Auer, 1924). This preocupation led to the creation of geographically anchored archaeological databases in the 1980s and to the adoption of GIS techniques in the 1990s. GIS are thus a very recent addition to Finnish archaeological research and we are barely beginning our attempts to model the ever-changing environment and, particularly, the isostatic uplift and related phenomena that have influenced Finland's settlement and culture during the past 10 000 years.

This chapter deals with some of the basic problems and the preliminary solutions that are being implemented in two regions affected by different types of processes related to isostatic upheaval: the Saimaa water system in the eastern lake plateau and the Åland Archipelago in the southwest.

11.2 Prehistoric Finland and the isostatic rebound

During the last Ice Age Fennoscandia was depressed by the weight of a 3 km thick ice-cap. By 13 000 BC the ice-sheet had begun its slow retreat from the North European Plain and by 9000 BC its border had reached southern Finland, where it stood still for *c.*1000 years forming the Salpausselkä end moraines. After this last stadial episode, the ice melted rapidly and by 7500 BC it was virtually gone.

When Mesolithic people reached Finland *c.*8200 BC they found a dynamically changing world. The remnants of the ice-sheet were being rapidly wasted by relatively mild preboreal climates, which in turn stimulated the spread of flora and fauna into deglaciated areas. Ice-free Finland, which formed then an archipelago within a 'great lake' (Lake Ancylus), was reacting with powerful isostatic rebound. Furthermore, the level of the isolated Baltic basin was high due to massive glacier ablation and, despite the powerful uplift, much of ice-free Finland lay under water.

Although the uplift has been slowing down since deglaciation, the country is still rising at rates of 9 mm per year in the northwest and 3 mm per year in the southeast (Figure 11.1). Corresponding rates were *c.* 120 and 30 mm per year early in the eighth millennium BC. Obviously this continuous land rise has been

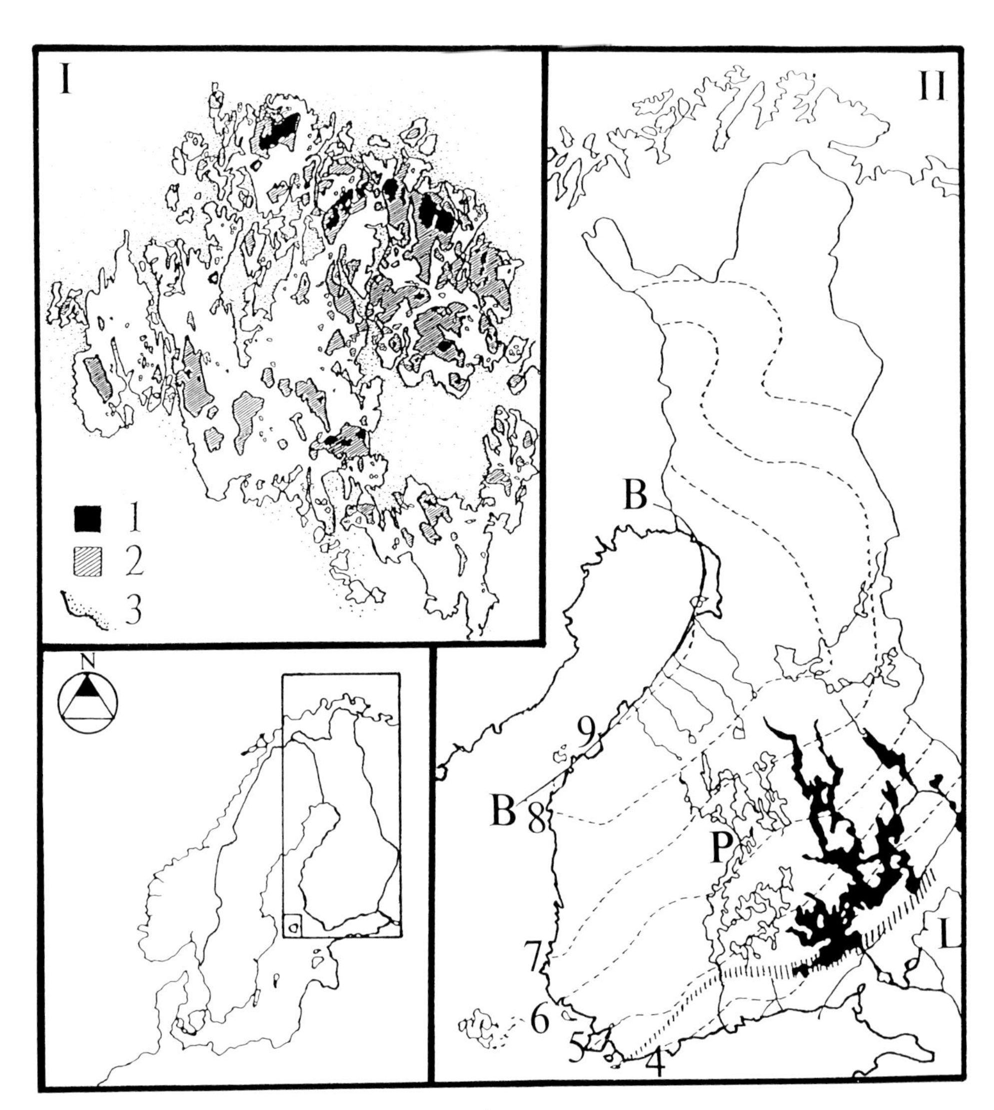

Figure 11.1 The two study areas: I, Main Åland during the last 7000 years—(1) dry land when humans reached the archipelago c.5000 BC, (2) dry land c.1500 BC, and (3) present coastline; and II, Finland with the present uplift isobases of 4–9 mm per year: B represents the baseline from which the horizontal distances for the gradients in Figure 11.2 were measured. Observe the present water systems of Päijänne (P) and Saimaa (in black) with its Vuoksi outlet flowing southeast through the natural dam created by the Salpausselkä end-morains (striped) into Lake Ladoga (L) (Siiriäinen, 1974; Suutarinen, 1983; Nunez, 1986).

responsible for the gradual emergence of the country from the waters of the Baltic. Postglacial shore displacement has been generally regressive, but there were episodes when the water rose faster than the local uplift and led to transgressive events. No such transgressions have taken place after the stabilization of world ocean levels some 8000 years ago.

The immediate implication of isostatic uplift is the increase of land area, but other uplift-related phenomena have had an even more direct effect on settlement and culture: first of all, the ever-changing shoreline and the closely connected never-ending environmental sequences of sea–bay–lagoon–lake–bog and seabottom–island–pennisula–mainland; in general the continuous transformation of coastal environments to inland ones. It should perhaps be stressed that environmental zones were seldom destroyed, they merely shifted. In other words, as islands and bays were turning to peninsulas and lakes, new islands and bays were being formed. The effects of isostatic upheaval are not restricted to the coastal zone, however. The fact that the land rose faster in the northwest than the southeast causes the continuous tilting of lake basins and considerable environmental changes inland (Figure 11.2).

Despite the problems connected to changing environments, the gradual and regular nature of such uplift-related phenomena as land tilting and shore displacement has allowed the establishment of a chronological framework for Finnish prehistory, particularly the Stone Age. The relative chronology of the 1930s has more recently received absolute values thanks to radio-carbon dates from both archae-

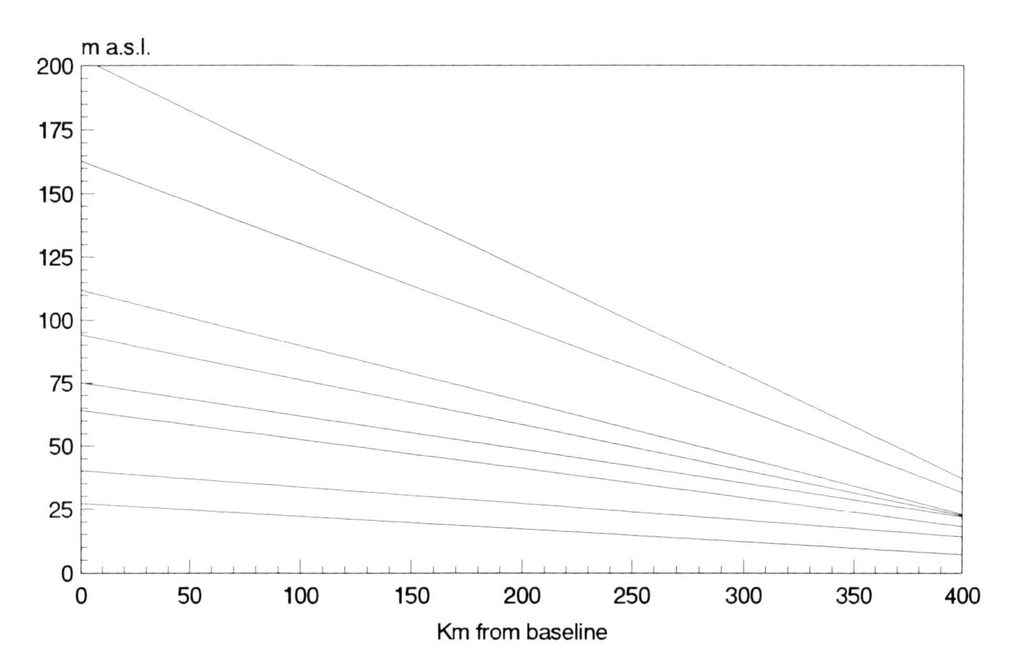

Figure 11.2 Tilted planes of radio-carbon-dated geological and archaeological events measured on the basis of altitude above present sea level and distance from baseline B in Figure 11.1. From top: end of Preboreal period, c.8000 BC; end of Boreal period, c. 7000 BC; highest Litorina shore, c.6200 BC; end of preceramic period, c.5200 BC; end of early Comb ware phase (CC1) c.4200 BC; end of CC2, c.3600 BC; end of Kiukais culture, c.1500 BC; end of Sub-Boreal period, c.800 BC. (Siiriäinen, 1978).

ological sites and isolation levels of basins (Europaeus-Äyräpää, 1930; Meinander, 1971; Siiriäinen, 1974; 1978; Nunez, 1978, Nunez and Storå, 1994). One important feature of isostatically rising areas is that the *z*-coordinate of any point, i.e. its position above present sea level, is a function of time.

11.3 The Lake Saimaa project

This research project of the Archaeology Department of the University of Helsinki uses SOAR, a geographically referenced relational database, and IDRISI to study prehistoric ecology and culture of the Saimaa region concentrating on four main subjects:

1. shoreline chronology based on sites with datable pottery found at known elevations above present sea level;
2. development of a typological and chronological sequence for the Comb and Asbestos wares in the eastern Lake District;
3. determination of the ecological needs of prehistoric human populations through analysis of dwelling-site environments and their comparison with ecofact and soil data (Figure 11.3); and
4. comparison of these data with the spread patterns of innovations.

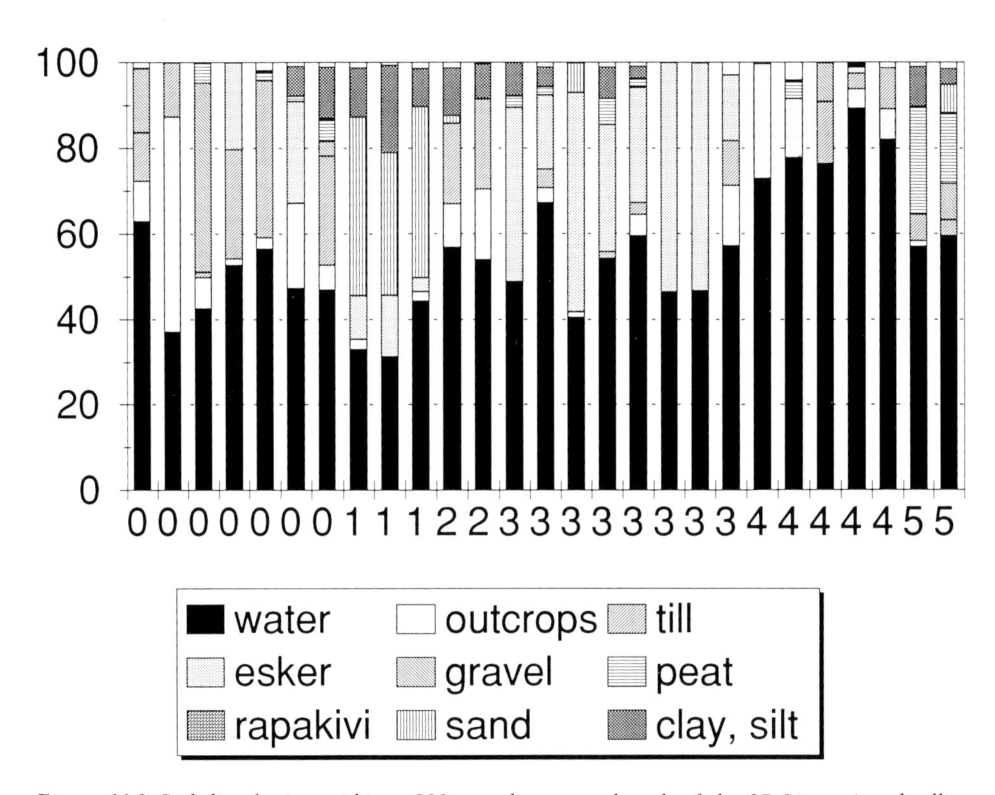

Figure 11.3 Soil distribution within a 500 m radius around each of the 27 Stone Age dwelling sites within a selected study area of 30 × 40 km by the shores of the ancient Lake Saimaa. Categories 0–5 were obtained by a cluster analysis.

The idea is to create a computer simulation model to explore the mechanisms affecting interaction networks, trade routes and year cycles of the prehistoric inhabitants (Vikkula 1994a, 1995).

Finland's 56 000 lakes are grouped into some 70 water systems (Rikkinen, 1990). They tend to be small, but four are major lakes: the Näsijärvi and Oulujärvi flow to the Gulf of Bothnia, the Päijänne to the Gulf of Finland, and the Saimaa runs via the Vuoksi river into Lake Ladoga in Russian Karelia (Figure 11.1). The largest of these water systems is that of Lake Saimaa at 52 390 km^2. The prehistory of the lake region is imperfectly known. There have been some major excavations and geological and settlement histories (Saarnisto, 1970; 1971; Siiriäinen, 1974; Matiskainen, 1979), but a comprehensive archaeological study of a water system with its environmental history and the ecology and lifestyle of its prehistoric population has never been made.

The Saimaa region contains material from all periods, being particularly rich in Stone Age sites. The southern part of Ancient Lake Saimaa was isolated from the Yoldia Sea Baltic stage around 8500 BC. The process was complete around 5850 BC, when the Saimaa and Päijänne systems flowed northwest into the Gulf of Bothnia. Tilting due to differential uplift rates (Figure 11.2) eventually led to the blocking of this early outlet and to a transgression in the southeast end of the basin. Contact with the Päijänne system to the west was reached via the Matkuslampi pond around 4900 BC and for a couple of centuries these two lakes formed a vast water system with a single northwest flowing outlet. This stage ended with the formation of the Kymi river as a southern outlet of the Päijänne around 4700 BC. The Saimaa discharged into the Gulf of Finland via Lake Päijänne from *c*.4700 to 3800 BC, when it broke through the Salpausselkä ridges forming the Vuoksi channel and began to discharge southeast into Lake Ladoga (Saarnisto, 1970; Donner, 1976; Rikkinen, 1990).

Although no major changes have taken place after the formation of the Vuoksi river, the uplift-induced tilting of the basin has continued to cause the water level to recede at the northern end of the lake system. In the southern part of the study area, the Saimaa's highest raised shore lies at 80 m, which corresponds to the youngest pre-Vuoksi level and dates to immediately before the formation of the new outlet. But due to the tilt, this shore becomes higher towards the northwest: 85 m at Savonlinna, 87 m at Joroinen, 102 m at Kuopio and 111 m at Iisalmi (Saarnisto, 1970).

The pronounced tilting phenomenon creates certain problems when using GIS in the Saimaa region. The 5800-year-old Ancient Saimaa shoreline is over 30 m higher at the northwest end than in the southeast, and even within the 30 × 40 km area of a regular topographic map sheet the difference is around 5 m. Since the available topographic maps are based on 5 m contour lines, a contour line can be representative of a shoreline for a specific time point only within a very limited area (Vikkula, 1994b). To approximate the ancient shorelines we have interpolated the elevation value of each cell in a raster-based digital elevation model (DEM) of the study area. We know the land uplift isobases, which can be used to interpolate a coefficient for corresponding cells in another file. Ancient elevation values ($h_{x,y}$) can be calculated for each cell as a function of time:

$$h_{x,y} = H_{x,y} - L_{x,y} * t$$

where $H_{x,y}$ represents the present elevation of cell x, y in metres, $L_{x,y}$ the interpolated value for the land uplift in the same cell in metres per century, and t is the time in years. There is an error source due to the deceleration of the uplift rate through time, but it can be ignored as long as reconstructed elevation models are based on a relative time-scale. To use an absolute time-scale, local deceleration algorithms could be controlled using the radio-carbon-dated gradient data in Figure 11.2.

Tilting the landscape back to the position it had during the Stone Age, or any other prehistoric period, requires an elaborate operation. It involves the reconstruction of water levels in an area with thousands of small lakes that were connected or isolated from the water system during past millennia. We must determine the lowest threshold of each basin and fix its level at that elevation—a time-consuming task that has to be done manually. Not until all these levels are correctly defined, can the other types of GIS data be applied and the typical conditions and differences in soil and terrain formations be analysed.

Since the sites of the hunting–fishing population were located close to or on lake and sea shores, we can use the above methods to obtain a fairly reliable reconstruction of ancient shoreline configuration for the time the site was in use. Unfortunately, dates carry error margins of up to several centuries, which restricts the interpretation of spatial conditions as a function of time and forces us to work with broad intervals corresponding to cultural phases.

A total of 2409 sites and find locations have been registered in the study area. Not all areas have been thoroughly surveyed, but surveys are still in progress, and new sites are likely to be discovered in the near future. The sites belonging to the period of hunting–gathering economy fall within the Stone Age (SA), 10 000–1500 BC. Other local sites belong to the early Metal Age (EMA, 1500 BC–AD 500) and the late Iron Age (LIA, AD 500–1250). The various types of sites in the study are listed in Table 11.1.

Most of the Metal Age sites of the Saimaa region belong to the Viking and Crusade periods (*c.* AD 800–1250). Slash-and-burn agriculture and animal husbandry were practised in the study area during the Iron Age (Simola *et al.*, 1985; Ukkonen, 1993), though hunting and fishing were important as well. Interestingly, agriculture indicators have been reported also from areas with no known Iron Age settlement.

The spatial relationship of Iron Age sites to important environmental factors such as their contemporaneous water systems, soil types, eskers, outcrops, sub- and

Table 11.1 Stone Age (SA), early Metal Age (EMA) and Late Iron Age (LIA) sites from the Saimaa region

Features	SA	EMA	LIA
Dwelling sites	697	27	14
Cairn cemeteries	—	22	19
Cremation cemeteries	—	—	7
Inhumation cemeteries	—	—	6
Hoards	1	—	3
Hillforts	—	—	9
Rock paintings	12	2	—
Cup stones	—	—	88
Stray finds	1206	6	78
Total	1916	57	224

supra-aquatic areas was analysed with IDRISI. The results showed that Iron Age sites are not as closely related to the shore as Stone Age ones (Figure 11.3). About 45 per cent of the sites are situated within 100 m of the shore and 80 per cent of them are within 1 km. Closest to shore are cremation fields and dwelling sites.

Iron Age sites tend to be situated in the vicinity of terrain rich in fine soils. This does not apply only to the area as a whole, the same preferential distribution can be observed also within a 1 km belt along the ancient shore. It was also observed that Iron Age agriculturalists were willing to give up the advantages of shore settlement for areas with nutrient-rich fine soils situated over 500 m away.

11.4 The Åland Archipelago project

The Åland Archipelago is situated at the entrance to the Gulf of Bothnia, halfway between the Swedish and Finnish mainlands. It occupies a surface of *c.*7000 km^2, with *c.*1500 km^2 of total land area. About two-thirds of this is concentrated in what is known as Main Åland (Fasta Åland) and the remaining land is shared by thousands of smaller islands and skerries. However, the total land area was less than 200 km^2 when the Stone Age population first reached the islands around 4900 BC. This difference is due to the isostatic uplift, which has been responsible for continuously changing the coastline and a six-fold increase in land surface area during the past 7000 years. Åland is still rising at the rate of 5–6 mm per year.

The Åland archaeological GIS project began in December 1992, when prehistoric settlement data from a previously compiled geographically referenced database were incorporated into the GIS of the Provincial Government of the Åland Islands. The system, Miljöflex (from milieu), is custom-made for the Provincial Planning Bureau and is basically a planning and resource management tool. It is an object-oriented GIS with a detailed vector image of the archipelago. The map image may be combined at will with the various object files: lakes, roads, buildings, pipes, sewers, wells, dumps, purification plants, cultivation fields and many other land-use features. All these data, including the recently added information on archaeological sites from Main Åland, are related to the basic map image, and each other, through a database. In the brief time it has been in use, Miljöflex has shown great potential as a CRM tool.

Although the capabilities of Miljöflex are somewhat marginal for archaeological research, the relational database with which the system interacts, Paradox 3.5, allows complex query and statistical operations as well as export for even more sophisticated treatment. It is through this Paradox end that we have been able to carry out our preliminary spatial analytical work. At this early stage, the research has been directed to two important aspects of Åland's prehistory:

1. the spatial organization of late Iron Age settlement and its evolution into the late medieval period and;
2. the effects of the isostatic uplift on the environment and settlement of the islands through time.

Tilting is not as noticeable in the small E–W oriented archipelago as in the larger NW–SE oriented Lake Saimaa, but the uplift-induced increase in land surface area is much more dramatic on the islands than on the mainland (Figure 11.1). Despite an ever-changing nature that complicates the modelling and interpretation of prehistoric settlement, the archipelago offers excellent possibilities for the controlled study of the responses of settlement and culture to environmental changes through time.

A hypsographic curve of Main Åland (Figure 11.4) reveals an uneven distribution of land surface at different elevations above sea level, particularly in those points above 80 m and below 10 m, which correspond respectively to approximately 3 and 50 per cent of the total land surface. An interesting feature possibly relatable to the mode of shoreline regression is the distribution of soil deposits with respect to their position above sea level. Bedrock predominates at points higher than 65 m while clay and silt are dominant in areas below 10 m. The terrain between 10 m and 65 m consists mainly of till, sand and outcrops (Jaatinen *et al.*, 1989; Nunez, 1993).

When the hypsographic curve from Main Åland is placed within the time frame provided by an exponential fit to 19 radio-carbon-dated isolation levels from local basins (5260 BC–1260 AD), we see that although the isostatic land rise has been a fairly smooth process (Nunez and Storå, 1994) the rate of areal increase has differed considerably in different periods (Figure 11.4): *c*.7 km^2 per century during the Stone (*c*.4900–1500 BC), *c*.4 km^2 per century during the early Metal Age (*c*.1500 BC–500 AD), *c*.18 km^2 per century during the late Iron Age *c*.500–1000 AD), and *c*.40 km^2 per century during the historic period.

The continuous regression of the shoreline affected the location of dwelling sites, which were successively shifted to follow the retreating shore. This was apparently the case during the hunting–fishing phase of the Stone Age occupation (*c*.4700–3200 BC) as well as after the adoption of primitive farming practices during the late Stone Age (*c*.3300–1500 BC) and the early Metal Age (*c*.1500 BC–AD 500).

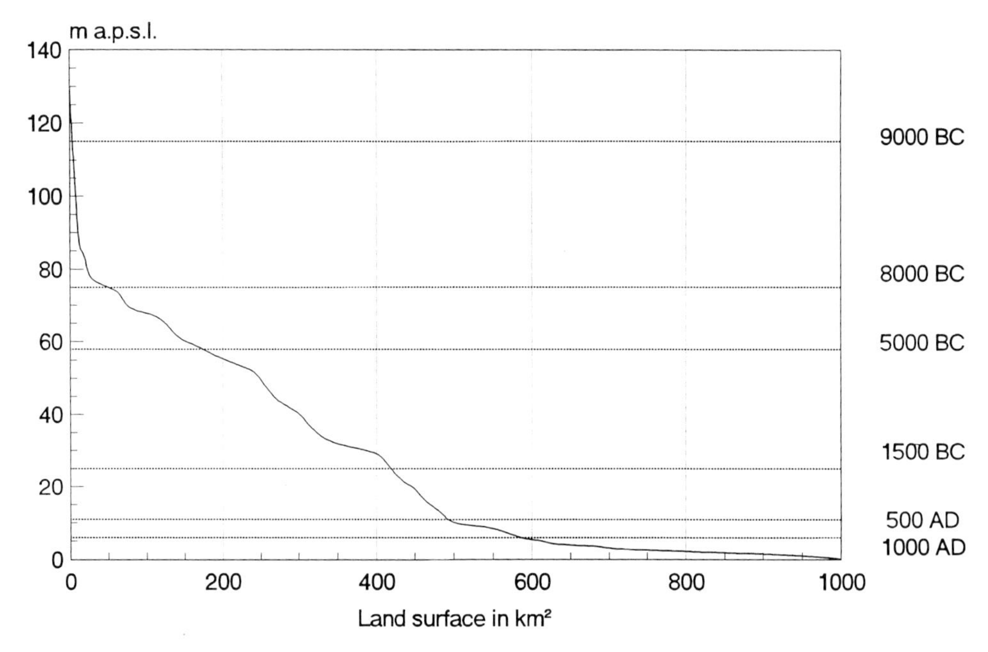

Figure 11.4 Areal increase of Main Åland (currently 1035 km^2) during the last 11 000 years. The diagram combines the local hypsographic curve with dates obtained from an exponential fit based on 19 radio-carbon-dated positions of the sea level. (Glückert, 1978; Jaatinen et al., 1989; Nunez, 1993; 1994).

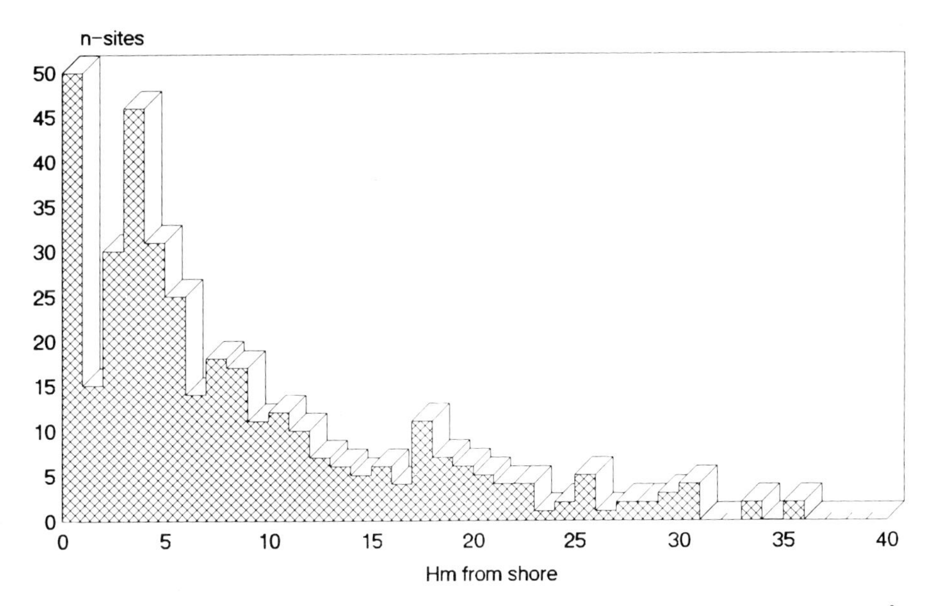

Figure 11.5 Frequency of distances from the major late Iron Age farmstead sites of Main Åland to their contemporaneous shore. In contrast, all Stone Age sites appear to have been located within 50 m from shore

The apparent stability of settlement during the latter period may well be related to the low rate of areal increase between 1500 BC and AD 500 (Figure 11.4) and not necessarily to farming practices.

During the late Iron Age (*c.* AD 500–1000) proximity to the sea was advantageous due to seafaring and the excellent pasture provided by coastal meadows (Nunez, 1993); but a greater dependence on an agrarian economy bound the farmsteads to the ard-tillable light sandy soils, which may have been close though not necessarily in the immediate vicinity of the shore (Figure 11.5).

Table 11.2 Spatial relationship between 404 late Iron Age sites, 6 hillforts and 8 medieval churches from Main Åland

Size number of graves	N sites	NNM km	NFM km	NCM km
All	404	0.4	4.7	3.5
>10	282	0.5	4.5	3.4
>30	130	0.9	4.4	3.4
>50	60	1.5	4.7	3.3
>70	28	2.4	4.2	3.1
>90	14	3.9	4.0	3.5
>110	9	5.0	3.9	2.7
>130	5	9.2	3.2	2.0
>150	2	29.4	2.3	0.2

Sites are grouped by size (all, >10 graves, >30, etc.). Nearest neighbour values (NNM) tend to increase with cemetery size (due to the decreasing number of sites?), while the mean distances from each site to its nearest hillfort (NFM) and church (NCM) decrease.

Since the population density on Main Åland seems to have been rather high during the late Iron Age (5–9 people per km^2). the spatial relationship between the farmstead sites of this period was studied through near-neighbour analysis. The nearest neighbour mean (NNM) for the 404 known farmstead sites was calculated on the basis of geographical coordinates, and the same procedure was applied to the distances from each site to their nearest hillfort (NFM) and medieval church (NCM). The results were 0.44, 4.7 and 3.5 km, respectively. When the sites were grouped according to their size (measured on the basis of the number of graves) the NNM increased with the site size whereas the NFM and NCM decreased (Table 11.2). The latter suggests that larger and more important communities, which had more to lose from raids, needed to be closer to hillforts, and that early missionary activity was directed to the larger centres, which could best support the spiritual and economic needs of the Church (Nunez, 1993; 1995).

11.5 Concluding remarks

Although the application of GIS techniques to archaeological research is at a very early stage in Finland, it is obvious that GIS are very suitable for the modelling of our complex environmental history, particularly the uplift-related phenomena. We foresee a surge of archaeological applications of GIS in Finland during the 1990s. However, it must be borne in mind that the adoption of any new method requires a certain degree of caution. We should be critical about the nature of our raw data, most of which were never meant for GIS purposes, and about the adequacy of using them in some GIS routines. Furthermore, there is a substantial danger of falling into the trap of creating exciting graphic images of our data. We should go beyond, combine, test, simulate, predict. Regardless of whether or not these goals can be achieved, GIS will be of invaluable help in the elucidation of the complex phenomena related to isostatic uplift in Finland.

References

Ailio, J., 1909, *Die steinzeitlichen Wohnplatzfunde in Finland*, Helsinki.

Auer, V., 1924, Die postglaziale Geschichte der Vanajavesisees, *Bulletin de la Commission Géologique de Finlande*, **69**.

Donner, J., 1976, Suomen kvartäärigeologia, *Helsingin yliopisto, Geologian ja paleontologian laitos*, moniste, **1**.

Europaeus-Äyräpää, A., 1930, Die relative Chronologie der stenzeitliche Keramik in Finland, *Acta Archaeologica*, **1**.

Glükert, G., 1978, Österjöns postglaciala strandförskjutningen och skogens historia på Åland, *Publications of the Department of Quaternary Geology of the University of Turku*, **34**.

Jaatinen S., Peltonen A. and Westerholm. J., 1989, Ålandskulturlandskap 1700-talet, *Bidrag till kännedom av Finlands natur och folk*, **137**.

Matiskainen, H., 1979, Päijänteen arkeologinen rannansiirtymiskronologia, *Lahden museo-ja taidelautakunnan tutkimuksia*, **16**.

Meinander, C. F., 1971, Radiocarbon dateringar till Finlands stenålder, *Societas Scientiarum Fennica*, **58**B, 5.

Nunez, M., 1978, A model to data Stone Age sites within an area of abnormal uplift in southern Finland, *Iskos* **2**, 25–52.
Nunez M., 1986, Om bosättningen på Ålandsöarna under stenåldern, *Åländsk Odling*, **46**, 13–28.
Nunez, M., 1993, Searching for a structure in the late Iron Age settlement of the Åland Islands, Finland, *Karhunhammas*, **15**, 61–75.
Nunez, M., 1995, Agrarian colonization of the Åland Islands in the first millennium AD, *Fennoscandia Archaeologica*, **11** (in print).
Nunez, M. and Storå, J., 1995, Settlement, environment and shore displacement in the Åland Archipelago 7000–4000 bp. Acts of the Symposium Time and Environment, Helsinki, Oct 1990, *PACT* **36** (in print).
Ramsay, W., 1920, Litorinagränsen i sydliga Finland, *Geologiska Föreningens i Stockholms Förhandlingar*, **9**, 217–39.
Rikkinen, K., 1990, *Suomen aluemaantiede*, Lahden tutkimus-ja koulutuskeskus. Vammala.
Saarnisto, M., 1970, The Late Weichselian and Flandrian History of the Saimaa Lake Complex, *Commentationes Physico-Mathematicae*, **37**.
Saarnisto, M., 1971, The history of the Finnish lakes and Lake Ladoga, *Commentationes Physico-Mathematicae*, **41**, 71–8.
Siiriäinen, A., 1974, Studies relating to shore displacement and Stone Age chronology in Finland, *University of Helsinki Department of Archaeology Stencil*, **10**.
Siiriäinen, A., 1978, Archaeological shore displacement chronology in Northern Ostrobothnia, Finland, *Iskos*, **2**, 3–23.
Simola, H., Grönlund, E., Huttunen, P. and Uimonen-Simola, P., 1985, Pollen analytical evidence for Iron Age origin of cup-stones in the Kerimäki area, *Iskos* **5**, 527–531.
Ukkonen, P., 1993, Unpublished report of faunal analyses, Archaeology Department, University of Helsinki.
Suutarinen, O., 1983, Recomputation of land uplift values in Finland, *Reports of the Finnish Geodetic Institute*, **83**.
Vikkula, A., 1994a, Ecological approaches to the Stone Age of the Ancient Lake Saimaa, *Muscovicaston Arkeologian Osaston Julkaisu*, **5**, 167–179.
Vikkula, A., 1994b, Stone Age environment and landscape changes at the Eastern Finnish Lake District, *Proceedings of the UISPP Inter-Congress Meeting*, Sydney-Mt. Victoria, Series 2.
Vikkula, A., 1995, Settlement and environment in the Southern Saimaa Area during the Stone Age, *Fennoscandia Archaeologica*, **11** (in print).

12

GIS on different spatial levels and the Neolithization process in the south-eastern Netherlands

M. Wansleeben and L. B. M. Verhart

12.1 Introduction

The Neolithization process, the cultural change to an economy based primarily on agriculture, is not simply a transformation in prehistory. After a very long period where humans could sustain a living by hunting, gathering and fishing, at some moment a transition to the much more labour-intensive production of food occurred. The course of this process and the underlying causes remain of interest to many archaeologists, at both regional and European levels, the topic attracts a lot of attention (Barker, 1985; Bogucki, 1988; Madsen, 1987; Louwe Kooijmans, 1993; Zvelibil, 1986). Various 'prime movers' ranging from ecological to social causes have been put forward to explain this change. In the south-east of The Netherlands (province of Limburg) research is being carried out into the Neolithization process as well (Figure 12.1). Characteristic of this area is that over a distance of 100 km insight can be gained into the cultural processes of both the central European loess area and the north European plain. The research area is rich in archaeological information, thanks to the presence of many amateur archaeologists and a long tradition of scientific research. This makes it eminently suitable for an investigation into the rise of agriculture and animal husbandry, hence the Meuse Valley Project and the use of GIS. Apart from more methodical research, GIS is specifically used as a research tool in the archaeological investigation of the Neolithization process.

12.2 The Meuse Valley project

First, there is a well-documented picture for the south of the research area of the first agricultural communities in The Netherlands. Settlers from the Linear Bandceramic culture appeared in the loess area at around 5300 BC. From a large number of excavations (Modderman, 1958/59; 1970; Waterbolk, 1958/59) it is known that these people lived in villages of clustered farms, kept cattle and cultivated cereals and vegetables (Bakels, 1978; 1982; Clason, 1972; Stampfli, 1969), they were farmers in the traditional sense of the word.

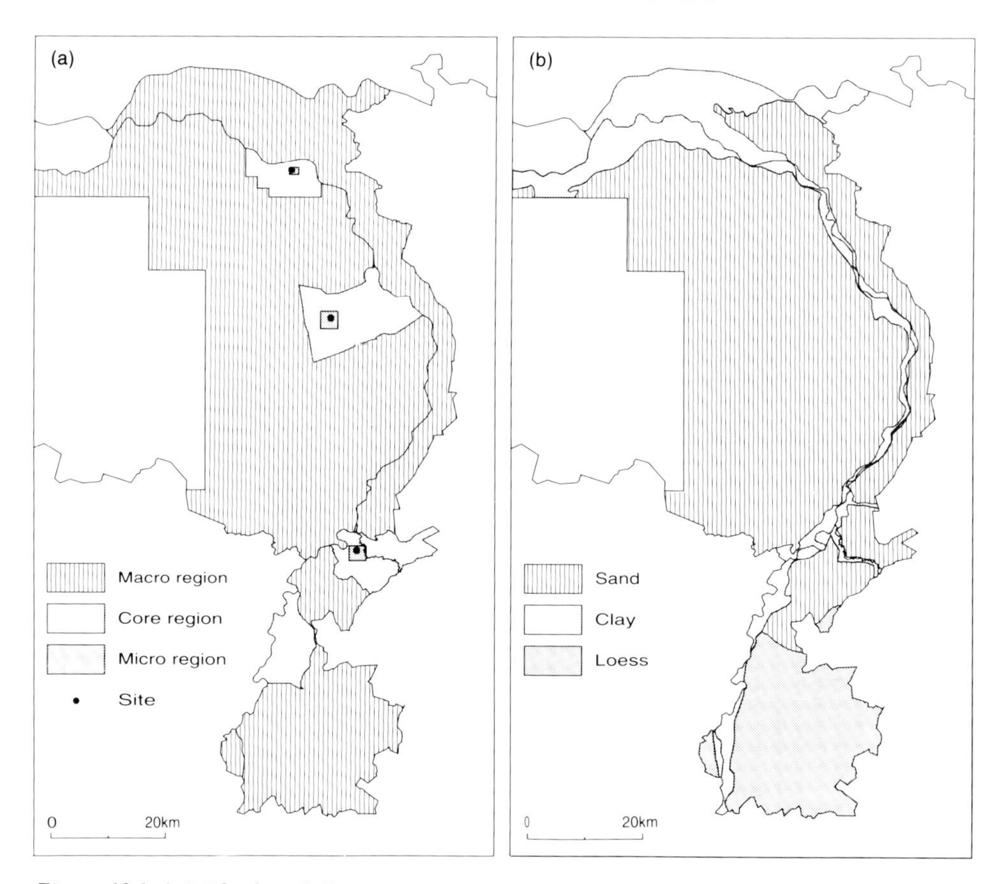

Figure 12.1 (a) The four different spatial levels of investigation in the Meuse Valley Project; and (b) and the simplified geology of the research area.

Second, there is a lot of information from the northern part of the Meuse Valley, especially in the river area, where a number of sites are known, dating from the period around 4100 BC. These were small, short-lived encampments on sand ridges and crests, belonging to the Michelsberg culture (Louwe Kooijmans, 1974; 1980; Louwe Kooijmans and Verhart, 1990; Verhart and Louwe Kooijmans, 1989). These camps probably consisted of just one or a few huts and apart from a lot of game, the diet was composed to a limited degree of products from livestock and agriculture (Bakels, 1981; Louwe Kooijmans, 1993). The archaeological information from the two parts of the research area is clearly widely divergent, although two things should be taken into account:

1. the areas are almost 100 km apart and in completely different geographical settings—loess as opposed to river deposits; and
2. there is a gap of over 1000 years between the sources of information.

So both chronologically and spatially there is a gap in our knowledge; the Meuse Valley Project has targeted that lacuna. To us, the Neolithization process is primarily an economic change and therefore the project is aimed in the first instance at reconstructing the (food) economy. The radical changes in the food supply are of course not isolated facts; many other facets of society have changed as well or themselves caused changes in the food economy.

The Meuse Valley Project is a regional archaeological investigation and is not primarily aimed at excavations and extrapolation of those data to larger areas, but instead attempts to obtain a survey of as many sites as possible over an entire region. The basic data are provided mainly by surface finds that mostly yield archaeological information of a lower quality than excavations. Many Stone Age sites in the Meuse Valley consist exclusively of a few handfuls of flint, sometimes with some pottery as well. The loss of quality however is compensated by a gain in quantity; there are many sites.

However, it is a laborious process to try to deduce economic information from such disparate archaeological material. For the Meuse Valley Project, three archaeological correlates have been defined to allow a connection to be made from surface sites to food economy:

1. the nature of the site, as it can be deduced from the size and artefact composition;
2. the spatial distribution of sites, as visualized on distribution maps; and
3. the geographical location of the sites in relation to environmental background information.

These three correlates are not mutually independent, but strongly interconnected. In fact, the Meuse Valley Project attempts to reconstruct the settlement system as a whole for various phases, concerning for example, the distinction between clustered villages of large wooden farms on the fertile loess and small, short-lived camps of a few huts along brooks.

The original settlement system corresponds only to a very limited degree to the present-day archaeological distribution pattern. A large number of human and natural factors strongly distort the modern image. These depositional and post-depositional processes act like a kind of sieve, allowing only part of the archaeological information to pass, either in the original state or disturbed. So, in a regional archaeological investigation, large problems in data reliability occur, as perceived for a long time by many archaeologists (Fokkens, 1991; Hamond, 1980; Shiffer, 1976). To approach this problem, it was decided to use a stepped geographical strategy, with research on four different spatial levels (Figure 12.1(a)). Each level has its own research method, detail and types of conclusions (Table 12.1). These four levels,

- macro region,
- core region,
- micro region and
- site

provide the opportunity to assess the distorting factors. A lower level, where more detailed investigations have taken place, verifies the reliability of the find patterns at a higher level.

The importance attached to the geographical location of sites in reconstructing the economy has been mentioned above. The usefulness of applying a GIS to site location analysis, at present almost taken for granted, has been clear to us from an early stage (Wansleeben, 1988). However, the use of GIS has not been limited to this as the principles, methods and programs have turned out useful at each level of research. GIS allows, in a relatively simple way, the processing of all kinds of spatial

Table 12.1 The different spatial levels of research within the Meuse Valley Project

Spatial level	Research area	Data collection	Archaeological results
Macro region	Limburg (4500 km^2)	Literature and national database	Large-scale cultural patterns
Core region	4 regions (100 km^2 each)	Museums and private collection	Settlement patterns
Micro region	One per core region (5 km^2 each)	Field survey, geological mapping and trial excavations	Detailed exploitation pattern
Site	One per micro region	Excavation	Local activities

data and their presentation in such a way that meaningful archaeological conclusions can be drawn. The procedures necessary for the archaeological problems posed can, quite often, not be done by using standard GIS procedures and adaptations of existing software modules or even entirely new ones are necessary. This GIS concept, specifically that of a grid-based GIS, does offer a framework for thinking and programming where archaeological reasoning or ideas can be tested and implemented.

The use of GIS at the four different spatial levels will be illustrated below and although the examples are related to the study of the Neolithization process in the south-east of The Netherlands, they were mainly chosen for methodological reasons. The archaeological conclusions are, due to the early stage of the investigations at some levels, still very preliminary, although the GIS procedures used are the methodological framework for the future processing of the extensive databases.

12.3 Four spatial levels of research

12.3.1 The macro region of the Meuse Valley

The macro region is such a large research area (4500 km^2) that it is impossible to obtain archaeological data through individual research, even in a relatively long-term project. The data available at this level have been gathered from the literature and from the files of the Dutch State Service for Archaeological Investigations at Amersfoort. This data file is at present being automated and standardized under the Archis project (Wiemer, Chapter 22). Based on both sources of information, data could be compiled on over 3800 Stone Age sites (Figure 12.2(a)). The large number of sites precludes verification of the data for each site within the available time. For the descriptions we are completely dependent on others, such as the observer (primary source), or other people who have recorded the archaeological information. In that case the Meuse Valley Project can only draw upon a secondary or tertiary source, which, not surprisingly, has caused differences in the quality of the archaeological information. A description of the sites can range from a very general designation like 'Stone Age site' to a highly detailed description of the number of

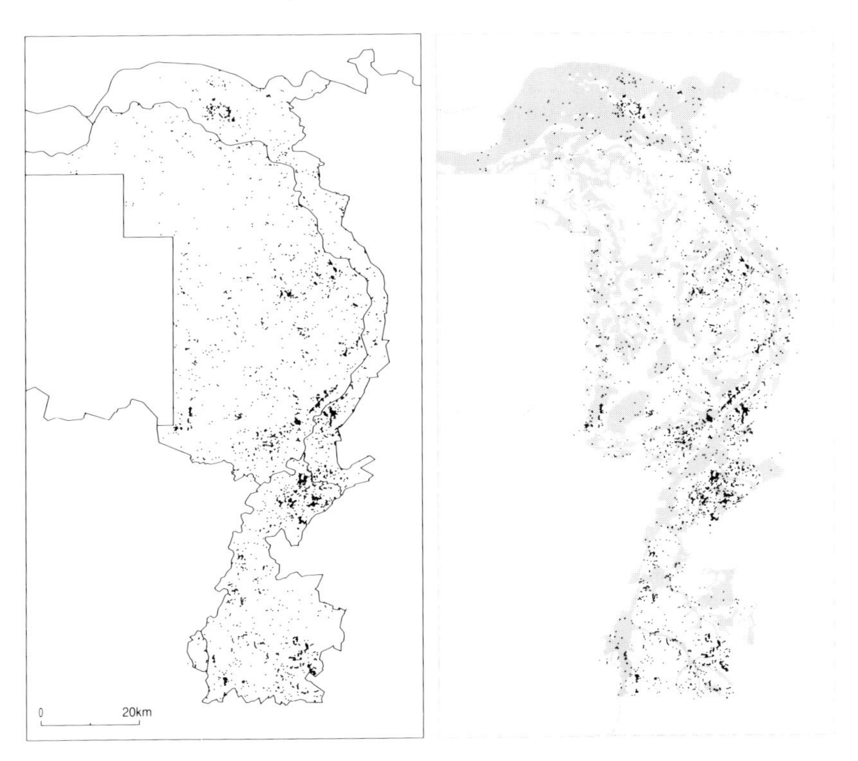

Figure 12.2 (a) The distribution of all Stone Age sites, and (b) superimposed with the areas where the presence or absence of dots does not reliably reflect the true archaeological distribution (woods, towns, river deposits, peat and sod-manuring areas).

artefacts, divided by type and raw material. For the investigation at the macro-regional level it was however very important to standardize the data files to some degree. We therefore decided to reduce all information to presence/absence data (Wansleeben and Verhart, 1990). This decision does have serious drawbacks. Not least, this database does not shed any light into the nature of the sites. So the macro region does not appear to be the most appropriate level for economic reconstructions, giving only a general overview of the settlement system at specific stages.

But the data at this macro-regional level are perfectly suited to gaining insights into the distribution of 'cultural units', i.e. where to find the material remnants of a specific archaeological culture. Although the aims of such a cultural–historical approach are now considered antiquated, in particular by too close a link between material culture and population groups and the attendant ideas on migration, distribution maps do provide a valuable first impression of the cultural succession in space and time, especially in a long diachronous investigation like the Meuse Valley Project, where it offers a general chronological framework for studying the Neolithization process.

The archaeological distribution map (Figure 12.2(a)) is by no means as simple as it seems at first sight. The quality of the information on such a dot map is, due to recovery and post-depositional processes, very irregular and fragmented, with the pattern being more or less influenced by factors that are completely unrelated to the occupation pattern in the past. The way the data are usually presented causes a kind of 'false accuracy' and a more realistic way is to indicate in which parts of the research area distorting factors should be taken into account, together with the distribution map (Figure 12.2(b)). In this way, it is easier to judge the merits of a map.

Of the observed distribution only the general distribution pattern should be considered a more or less exact reflection of the past habitation pattern. GIS proves to be a good way of handling an investigation into a general distribution. For example, in many GIS standard tools are included to describe distribution patterns by means of mathematical trend surfaces. In practice, however, this approach proves to be too crude for archaeological investigations. The distribution patterns are reduced too much to just an increase or decrease in one or two directions. Improvement is provided by moving average techniques, displaying both the average trend and local details. Depending on the distribution pattern and the archaeological problem posed the manner of visualization can be adapted by varying the moving average parameters (the size of the mask used for averaging and the use of a weighted or unweighted average). To gain insight into the cultural distribution of the late-Mesolithic and the Bandceramic culture in the south-east of The Netherlands we used such a moving average technique. In view of the size of the research area, we decided upon a relatively large mask of 20 × 20 km and a weighted (inverse distance) average.

The late-Mesolithic distribution pattern displays a relatively even spread, both in the sand and in the loess areas (Figure 12.3(a)). An increase in density of sites occurs, that can be attributed to a varying research intensity. This became apparent when the same areas showed clusters in distribution patterns of other archaeological phases as well. The distribution map of the early farmers shows a remarkably different pattern with a strong clustering of the well-known Bandceramic settlements on the loess plateau of the Graetheide (Figure 12.3(b)). At the same time, however, a slight fanning out of sites into the immediate adjoining sand area is visible. Based on the character of the surface material and the data from a single excavation, these are tentatively considered to be encampments used for transhumance (Bakels, 1982) or hunting.

Before formulating models about the cultural succession and interaction, based on a comparison of the two distribution maps, one should stop to consider that these maps are based exclusively on typological dating. The late-Mesolithic sites have been defined on the basis of the occurrence of trapezia flint arrowheads. These are commonly placed in the period between 7000 and 5000 BC. The map of the Bandceramic sites in turn, represents the sites where Bandceramic pottery has been found or a combination of adzes and asymmetrical Bandceramic arrowheads. Unfortunately, it is not possible to determine which sites are really contemporaneous on the basis of the typological attributions. At the moment, we assume that the north Limburg sand area was not depopulated during the Bandceramic phase, as supposed by Arts (1987). Beside the early farmers there were still hunter/gatherers in this area, so a number of the late-Mesolithic settlements in the sand area should really be displayed on the map together with the Bandceramic settlements. After the arrival of the Bandceramic people the Mesolithic population appears to exploit the south Limburg loess area rarely or not at all. The same could be true for the middle Limburg sand area, although this region would appear to be eminently suited for contacts between hunter/gatherers and farmers to occur. As yet there is no clear answer as to what form these contacts took—peaceful or hostile.

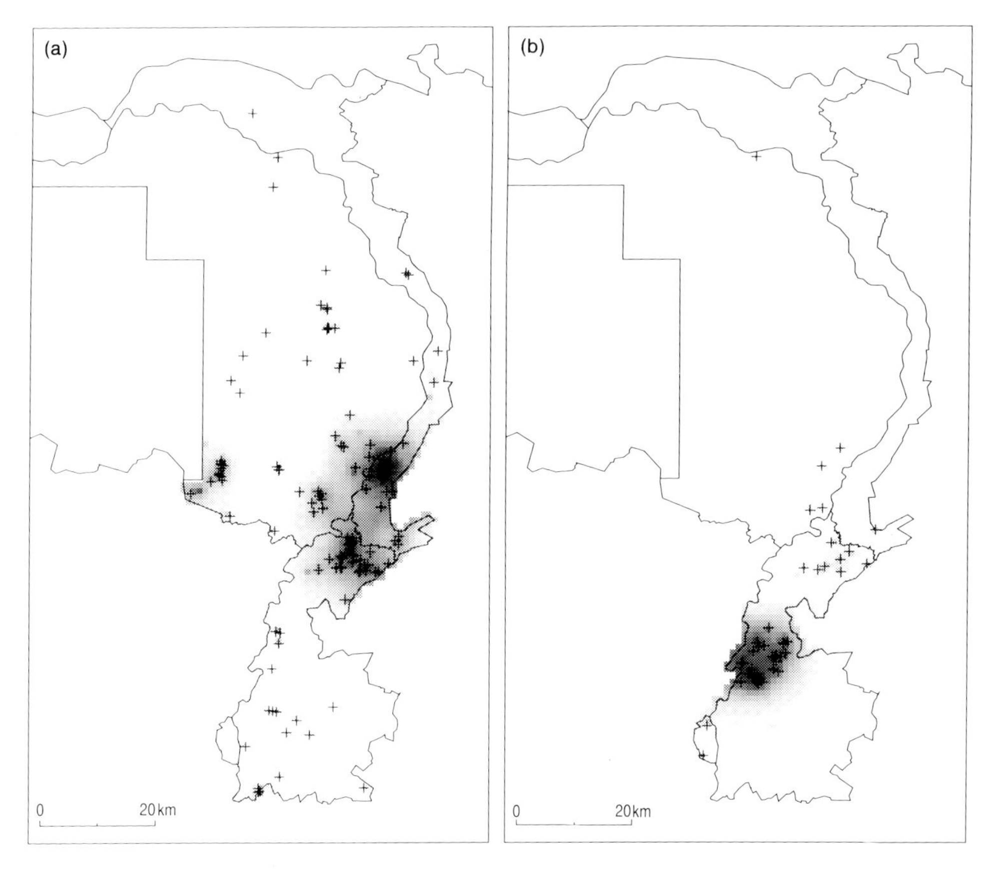

Figure 12.3 (a) The distribution of late-Mesolithic sites, and (b) Bandceramic sites in the macro region Meuse Valley, as well as the weighted average trend surfaces.

12.3.2 The core region of Venray

One spatial level lower is the core region, an area of 100–150 km^2, where it was possible to make an extensive inventory of the majority of the collections of amateur archaeologists and local museums. We ourselves could describe the Stone Age finds from the core region, both the flint and the pottery, in a standard way. The large differences in quality of the macro region are less pronounced at this research level. The number of artefacts at each surface site has been recorded by type and raw material; however, this does not mean that the sites are clearly known as there are still a large number of distorting factors. The relationship between the surface finds and all finds still present in the soil is for example still completely obscure. One or two surface finds could be the proverbial tip of an immensely rich site, but just as likely these could be sole scattered remnants of a short-lived activity and only an excavation can provide a decisive answer to this. It should be pointed out that the spatial distortions are here more limited than at the macro-regional level. Of course, some subareas have not been investigated, but interviewing the amateur archaeologists allows a much better estimate of them. Many amateur archaeologists have been active in a certain areas for dozens of years and have searched fields under all kinds of circumstances. The finds collected in this way are much more representative than systematic surveys by professional archaeologists. A field survey usually lasts only one or two seasons and is planned in advance, meaning that the work takes place no matter what the limitations of ground cover and weather.

The core region especially is the spatial level at which both insight into the nature of sites can be gained and the geographical location of the sites can be investigated. Site location analysis is the more or less traditional GIS application within archaeology, although many GIS contain limited procedures for statistically summarizing the relationship between sites (point observations) and geographical variables (areas, lines). Usually the chi^2 test is applied and it is characteristic of the present state of GIS that a separate module had to be written for the popular DOS-GIS IDRISI to allow chi^2-calculations. However, the chi^2 test has to meet a strict number of conditions, for example, even the small geographical units should have an expected value above the threshold of five and the total number of sites from a certain period or of a certain type should be more than 40. Combining the legend units into larger geographical types is often undesirable from an archaeological point of view and application of Yates' correction of continuity or the calculation of Fisher's exact probabilities are statistical solutions to this problem (Siegel, 1956; Thomas, 1986). Basically, the result of a chi^2 test is simply a statistical answer (yes or no) to the question: is the difference found between the observed and the random distributions of sites over the legend units large enough not to have been caused by chance? In itself such an answer is valuable, but to understand the location choices of a certain group of people it may be more important to indicate which geographical units were preferred or avoided. The choice of relevant legend units turned out to be difficult with the aid of the high chi^2 values, since in that case the choice of units is a subjective process, and while such a subjective choice does not detract from the value of the investigation, it is not clearly defined and cannot be repeated, which is precisely what is required by formal statistical techniques.

A formal method for selecting units has been described by Atwell and Fletcher (1985; 1987), where for each geographical unit a weight factor is calculated that can be considered an estimate of the relative importance of that unit. Atwell and Fletcher proposed using a test to determine whether the weight factors found deviate significantly from the expected values. However, this test should probably be modified (Wansleeben and Verhart, in press) to offer a formal way of indicating which units seem to have been avoided and which units preferred, producing an important addition to the chi^2 test.

Another alternative has been developed based on cultural resource management (CRM) projects where there is a need to manage the soil archives as efficiently as possible by attempting to model the site location pattern. On the one hand, this model has to consist of an area as small as possible, but on the other it has to comprise as many sites as possible. This principle has been used to develop a site location parameter (K_j) (Wansleeben and Verhart, in press) which attempts to make such a choice from the legend units that would create the optimum model (smallest area, most sites). The effect of K_j, the selection of relevant units, is similar to the results of the Atwell–Fletcher test. Furthermore the absolute value of the parameter, between 0 and 1, also yields an estimate of the predictive value of the geographical variable concerned by testing whether the observed K_j value deviates enough from the expected value. In this respect, the result is similar to the chi^2 test and the site location parameter K_j appears to be a useful combination of both previous tests.

In the Meuse Valley project, the visual inspection of the distribution maps is the initial, important way to gain insight into the site location pattern (Figure 12.4). Besides this, a new module for IDRISI is important, which includes the three statistical techniques, chi^2, modified Atwell–Fletcher and K_j, to application at the same time.

The distribution of the late-Mesolithic sites in the core region of Venray over the soil moisture classes, which is a factor of both geomorphology and natural vegetation, is presented in Table 12.2. Even after reducing the 13 classes of the original map (Stiboka soil map of 1:50 000) to seven moisture regimes, this in no way meets the statistical requirements for the chi^2 test, where the chi^2 statistic 23.235, is significant at the 0.05 level, but unreliable. The Fisher exact probability however, indicates that the value found is indeed significant and a high chi^2 value only occurs for moisture regime 4. According to the Atwell–Fletcher weight factors ground water regime 4 is most important, followed by 3 and 7. Only regime 4 is significantly preferred, while regimes 5 and 6 have been significantly avoided. The K_j value of 0.531 is significant, but not very high. This model with 91 per cent of the sites in 60 per cent of the area consists of the legend units 3, 4 and 7. Comparison of the results of all three statistical tests proves a useful way of getting to know the trends in the data. The late-Mesolithic sites are either located in the dry areas (7) or in the moist ones (3, 4) and the distribution map demonstrates that the sites commonly occur at the transition between these areas, just in the dry or in the moist area. This high

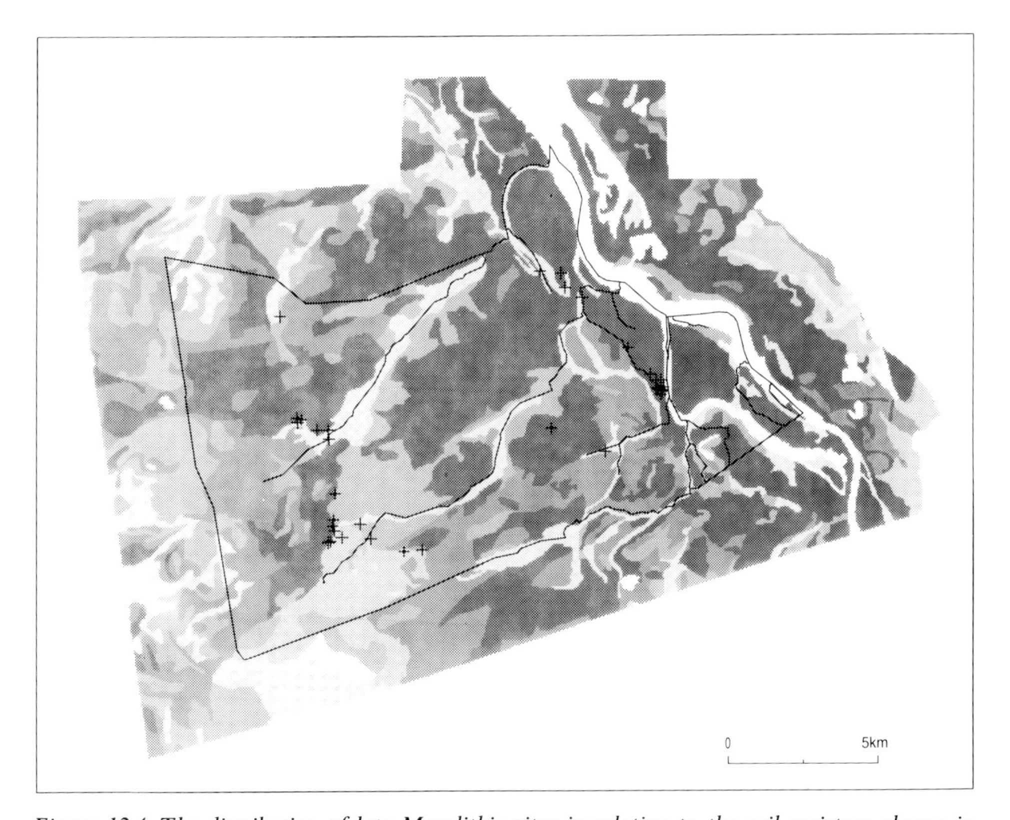

Figure 12.4 The distribution of late-Mesolithic sites in relation to the soil moisture classes in the core region Venray. Darker shades of grey indicate drier areas.

Table 12.2 Site location statistics for the distribution of late-Mesolithic sites over the various soil moisture classes (higher numbers indicate drier areas) in the core region of Venray (50 × 50 m grid resolution)

Site Location Statistics
Landscape image file: Soil moisture classes
Site image file: Late-Mesolitic sites

Lnd	Cells	%	Sites	%	Expct	Chi^2	At–Fl	(prob)	K_j
1	1307	0.018	0	0.000	0.6	0.594	0.000	(0.535)	0.516
2	792	0.011	0	0.000	0.4	0.360	0.000	(0.689)	0.522
3	7208	0.099	6	0.182	3.3	2.271	0.251	(0.136)	0.531*
4	3488	0.048	6	0.182	1.6	12.312	0.519	(0.001)	0.407*
5	11750	0.162	1	0.030	5.3	3.523	0.026	(0.011)	0.410
6	15314	0.211	2	0.061	7.0	3.529	0.039	(0.010)	0.394
7	32810	0.451	18	0.545	14.9	0.645	0.165	(0.372)	0.226*
	72669	1.000	33	1.000	33.0	23.235	1.000		

Simulated Fisher exact probability: 0.006
Atwell–Fletcher expected value: 0.143
Highest K_j value: 0.531 ($p < 0.000$), 90.9 per cent sites in 59.9 per cent of area
* = included land units

diversity of the terrain around the Mesolithic settlements is in good agreement with current ideas about the exploitation strategies of these last hunter/gatherer communities.

Using the grid GIS we have produced a derivative map showing the geographical diversity in ground water levels. The degree of diversity was calculated by determining for each grid cell how many different legend units occur in a radius of 500 m around it. From the late-Mesolithic onwards there is a gradual increase (Michelsberg through Single Grave cultures) in the homogeneity of the terrain around settlements (Figure 12.5). An increasingly homogeneous terrain may indicate a growing importance of agriculture in the food economy and the fact that there appears to be a gradual transition and not a sudden change indicates, we feel, that there was not a colonization by a completely agricultural community in this sandy area. A gradual change in the lifestyles of the local hunter/gatherers seems more plausible, possibly resulting from a long period of contacts with the agricultural communities of the Bandceramic and subsequent Roessen cultures, in an acculturation into farmers. Over 1000 years after the Bandceramic culture the Neolithization process in the south-east of The Netherlands has progressed to a point where, with the Michelsberg culture, we can speak of a Neolithic society, at least in a material sense. Economically, however, there appears to have been more of a continuation of the traditional exploitation strategy with botanical and zoological data from the river area, pointing to game, domesticated animals and agricultural produce, seemingly applying to the sandy area as well. Only near the end of the Stone Age does there seem to exist a completely agrarian economy (Louwe Kooijmans, 1993).

12.3.3 The micro region of Linden

At a lower level still is the micro region; a relatively small area (approximately 5 km^2) where we can do the basic archaeological survey ourselves. This limits the

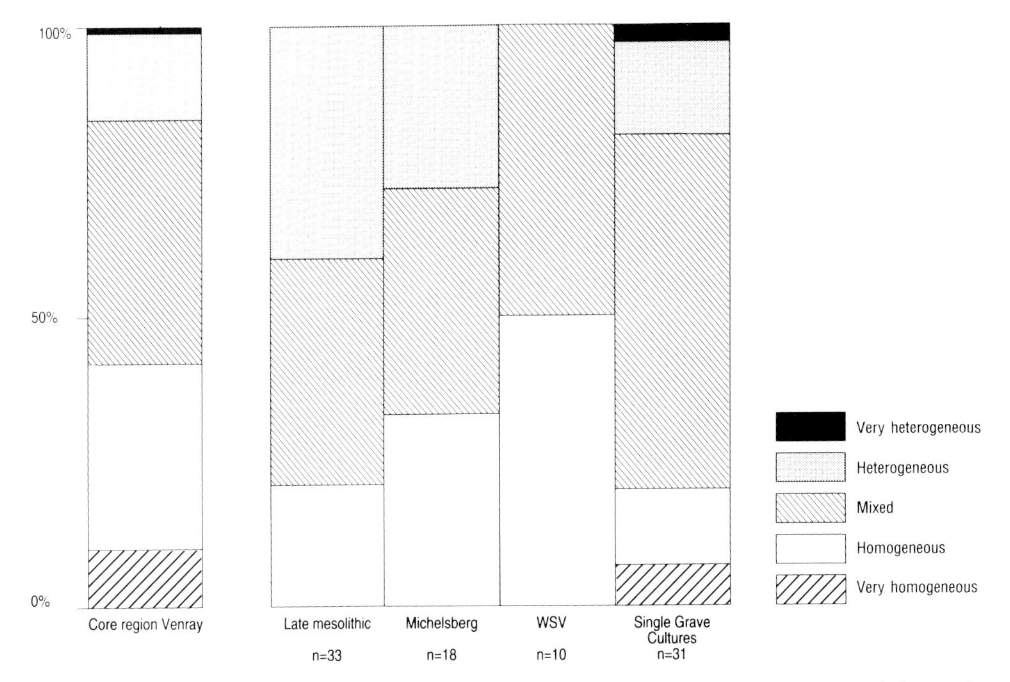

Figure 12.5 A histogram of the variation of soil moisture classes for the Late Mesolithic and three successive phases of the Neolithic, for the core region of Venray.

influence of the spatially distorting factors that are still active at the level of the core region. However, such investigation is so labour-intensive that it can only be done for a small area. In the micro region Linden, part of the core region of Cuijk, there was an additional problem. An archaeological survey includes walking the fields and gathering surface finds, but as Linden lies in the river area, almost the entire research area is covered with a layer of recent clay. A field survey in the traditional sense of the word is therefore barely possible, so an alternative was sought. A method was used whereby on a 25 × 25 m grid, a 10 cm diameter hole was bored. Below the clay a sample of approximately 1 litre of sand was collected from the top of the ancient surface, which was then sifted on a sieve with a mesh of 1 mm and the residue searched macroscopically for the occurrence of small pieces of flint, pottery, charcoal, stone and bone—'micro-debris' which is an indication of human activity. In Linden an area of about 70 ha was investigated, although this method is more labour-intensive than the traditional field survey, but it has the added advantage of allowing a survey of otherwise almost inaccessible terrain (clay deposits, sod-manuring areas and woods).

The plots, geology and logistics in practice resulted in a more or less irregular distribution of bore points. Automated interpolation methods to transpose irregularly distributed data into a regular grid have been available for some time and there is a lot of discussion about the advantages and disadvantages of the various interpolation techniques, specifically in the area of generating digital elevation models (DEMs) (Kvamme, 1990). A good understanding of the interpolation algorithm and the meaning of the parameters, as well as checking the results against the original data are important in this respect. The possibilities for interpolations are relatively limited in many GIS and explicit mention of algorithms in the manuals is rare. Therefore, in the Meuse Valley Project, the software package Surfer is considered to be a highly useful tool.

Within GIS the results of the interpolation can be integrated easily with other sources and here two are important:

1. the ground bores which are in an irregular distribution—for each bore hole the numbers and types of micro-debris and the depth below ground level where the sample was drawn were recorded; and
2. the elevation map of 1:10000 (Dutch Topographical Service)—irregularly distributed observations at about ground level have been recorded with the average observation density approximately 100 observations per km^2.

The data have been converted by means of interpolation techniques into a single, uniform, regular grid and stored in the GIS. This allowed a number of research possibilities that would otherwise have been almost out of our reach; for example, from the height of the ground level the depth of the sand in the bore sample could be subtracted, yielding a DEM of the prehistoric terrain. Based on the number of pieces of flint per sample it was possible to establish an approximate map of flint densities, to indicate Stone Age sites and by varying the parameters in the interpolation, it was possible to even out irregularities in the spread of the flint. By combining the two maps a clear picture emerges of the location of the Stone Age sites as related to the terrain at that time.

In the area under investigation some twenty small sites were discovered on a late-glacial parabolic wind-blown river dune, situated on a Meuse terrace. The rims of the dune seem to have been preferred site locations with the larger sites somewhat more towards the central part of the sand body.

A number of Stone Age sites almost certainly date from the Michelsberg phase, as shown by additional 20 cm diameter borings. Together with data from excavations, both in the past and from the Meuse Valley Project (Verhart, 1989; Wansleeben and Verhart, 1992) the bore data provide a relatively clear picture of the settlement system of the first Neolithic communities in this relatively wet river area. The Michelsberg people settled for a short time on the dry sand crests with the length of stay varying from several days to several months. These sites were not, however, permanent settlements, but rather base camps and hunting camps. This was a Neolithic society with a mobile way of life, at least for part of the year and/or part of the group and these data appear to confirm the gradual acculturation of the Mesolithic population in the sand and river areas. This process was still under way during the Michelsberg phase and was by no means complete.

12.3.4 The site of Merselo-Haag

Finally, the site is the lowest spatial level of research. In the core region of Venray a late-Mesolithic camp was excavated near Merselo-Haag, west of Venray (Verhart and Wansleeben, 1989; 1991). At the source of the Loobeek, a brook draining in a north-easterly direction into the Meuse, some twenty encampments are located. In this micro region, a traditional field reconnaissance has supplied additional information about the sites already known to amateur archaeologists and almost all the sites are located at the rim of the valley which displays very few variations in height.

This geographical location and the relatively small number of surface finds per site suggest these were hunting camps from the late-Mesolithic and one of the sites was excavated to test this hypothesis.

Traditionally, Mesolithic camps in The Netherlands are excavated by trowel, making three-dimensional measurements of all finds—a very time-consuming process, which is considered acceptable because of the few remaining indications of human activity. However, this approach was impossible for the investigation at the Loobeek as the excavation area was only available for a single field season, because of re-allotment. The excavation was performed by collecting blocks of soil of 25 × 25 × 15 cm (length × width × depth) and sifting these on a sieve with a mesh size of 3 mm. All finds were recorded in a regular grid during the excavation, producing data ready for the application of GIS technology for a number of different purposes.

First, it was possible to generalize the data from the original 25 × 25 cm grid to a grid of 1 × 1 m (Figure 12.6). This was important to allow comparisons with published excavations, where in spite of the fact that with traditional excavations the exact location of the finds is known, quite often distribution maps are published showing find densities per square metre providing a highly useful global impression of the spatial distribution.

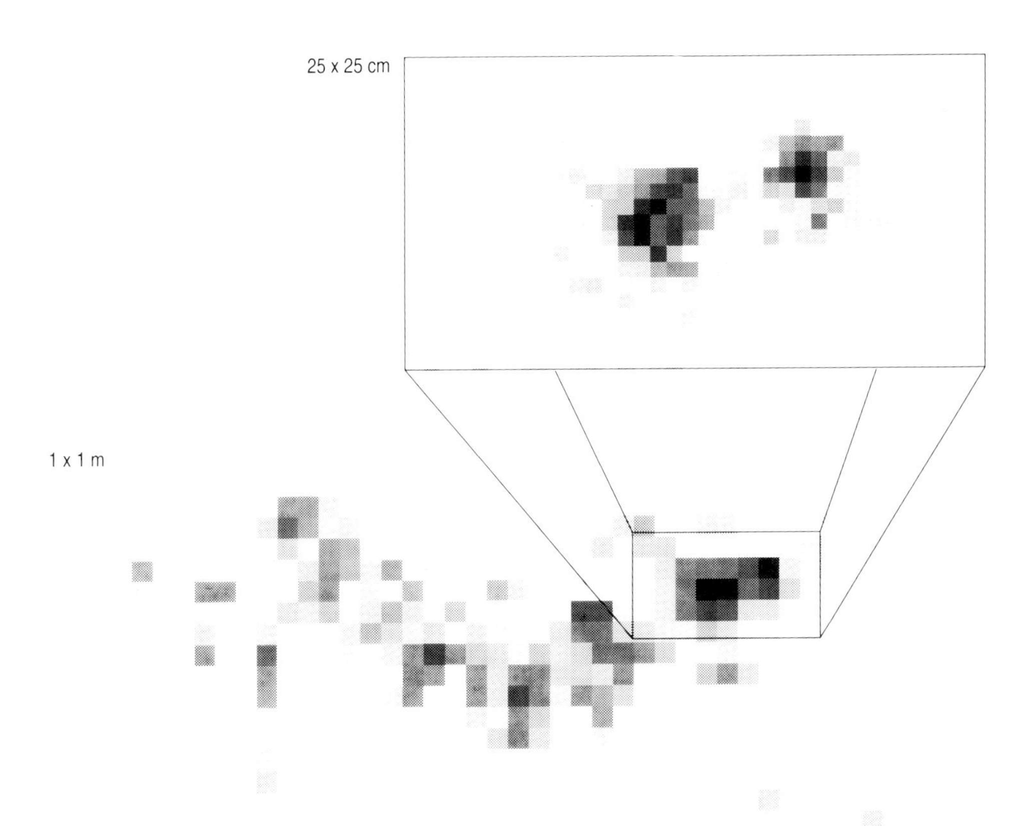

Figure 12.6 The excavation of Merselo-Haag: densities of flint artefacts per m^2, and an enlargement of the main concentration with squares of 25 × 25 cm. Darker shades of grey indicate higher artefact densities.

Second, GIS was used to reclassify the find densities many times, which is useful since different classifications yield widely differing distribution pictures. Several reclassifications are necessary to show certain facets of the spatial distribution to their full advantage and this is one of the basic procedures in GIS.

Third, sometimes very large individual differences occur in the find densities of adjoining squares, there are even instances where rich squares adjoin completely empty ones. This could be caused by particular post-depositional processes, such as tree falls, the accuracy of the person doing the sieving, mistakes in the find administration or by chance. These kinds of differences can be removed by using moving average techniques giving distributions which are easier to visualize since the image is no longer distorted by small individual differences.

Finally, GIS was used to determine the spatial association of various artefact types. There are a large number of different association parameters, of which the Jaccard, Bray and Driver coefficients were most useful. These parameters are very robust, since only the presence or absence of an artefact type in a cell matters and squares without finds are not included in the calculation. In many GIS no procedures are available for such analyses, although because many parameters are based on grid information, the creation of new GIS procedures is relatively simple.

The excavation in Merselo-Haag has yielded a late-Mesolithic camp where three or four spatially separated clusters of flint artefacts could be distinguished. Furthermore, all over the excavation area small pieces of flint were dispersed. Most concentrations consisting specifically of many small artefacts. These are mainly waste-flakes, caused by working the flint and few implements occur. The main concentration (Figure 12.6) is somewhat larger and contains many more large artefacts and implements; it is also remarkable that in this concentration both Wommersom quartizite, a specifically late-Mesolithic raw material, and arrowheads occur much more frequently than in the rest of the excavation. For this concentration we are at the moment tentatively considering the function of dwelling place, but nothing definite can be said about the presence of a hut or a tent, based on the find distribution. This site has yielded small amounts of finds, compared to published data on Mesolithic camps, together with a relatively high percentage of arrowheads which do not preclude an interpretation that this was a short-lived hunting camp. The source of the Loobeek with its large geographical variation and the presence of water would probably have harboured a wealth of game.

In connection with the Neolithization problem there is another remarkable find: the presence of an arrowhead stylistically closely resembling the characteristic asymmetric Bandceramic arrowhead, which may indicate contacts between these hunter/gatherers and the early farmers. At present, it is hard to determine whether this is a case of (in)direct exchange of goods or imitation.

12.4 Conclusions

To allow further analyses of the data of the Meuse Valley Project, a preliminary working model has been postulated for the course of the Neolithization process. When the South Limburg loess area was colonized by the Bandceramic farmers, the first contacts occurred with the hunter/gatherers in the sand and river areas. The agricultural system of the Bandceramic culture was eminently suited to the loess, but could not be applied on the sand without adaptations. This can be deduced from the nearly total concentration of Bandceramic villages and the loess. The Bandceramic farmers and the hunter/gatherers do not appear to have initiated the necessary changes quickly. After the Bandceramic culture, there is no apparent continuity in the use of the Dutch loess area; their successors of the Roessen culture used it a lot less intensively. At the same time, there are hardly any changes in the way of life of the hunter/gatherers, although an increase in 'Neolithic' objects is visible in the sand and river areas. These artefacts, adzes, Breitkeile and pottery point to an intensification of the contacts between hunter/gatherers and farmers generating a gradual process of mutual influence, dependence and/or exchange of goods and knowledge.

The gradation of the Neolithization process indicates that there was probably no direct or immediate need for change; the hunter/gatherer existence managed to supply the basic necessities of life very well. The agrarian products appear to have been introduced at first as a supplement to the economy of the hunter/gatherers producing a 'very broad spectrum' economy. This might have been accepted more or less voluntarily, as a way of lowering risks or to enhance social status; maybe the agricultural products were bartered at first and later cultivated on a small scale. Without clear 'prime movers' probably all kinds of interwoven social and economic changes occur in such a society, for example, a decrease in mobility, increase in population size, redistribution of duties, changes in social relationships and an increase in the need for food. Together these processes form a sort of spiral, in which cause and effect can no longer be clearly distinguished. Small changes have their effects on other processes and in this way are reinforced, the hunter/gatherer communities could eventually evolve into a completely agricultural community. This is a process that certainly happened by fits and starts, only to reach its economically temporary ending at the end of the Neolithic.

From the examples presented here in connection with the Meuse Valley Project, it is evident that many archaeological questions and hypotheses can be tested by the spatial properties of the archaeological data. Apart from the finds themselves, the spatial context is an important source of information for reconstructing the past and GIS is a general computer tool specifically designed for processing spatial information. For a large number of different archaeological problems GIS tools may be useful in visualizing, processing and analysing spatial patterns so the soaring use of GIS in archaeology is not a complete surprise. The number of applications, at each spatial level and within the various subdisciplines of archaeology, can only increase further.

Still it is important to remember the origins of GIS and that many procedures have been developed for and by geographers and geologists. By no means all GIS possibilities are useful for archaeological research, because of the particular characteristics and peculiarities of archaeological data. Suffice it to recall the often qualitative nature of the data, the frequent occurrence of incomplete data sets, large spatial distortions and non-systematic collection of data. Many existing GIS-procedures will have to be adapted to this and for many particular archaeological questions completely new tools will have to be written. A critical use of GIS as an automated tool for spatial questions should be emphasized in future. Allen, Green and Zubrow were in this respect quite right to conclude (Allen *et al.*, 1990: 383):

> We caution against the use of GIS as an end in itself. Good research and management is based on asking good [archaeological] questions—something GIS does not do for us.

Good archaeological questions and the adaptation of GIS to our needs offer a chance to exploit fully the potential, and archaeology as a spatial science can only gain by it.

References

Allen, K. M. S., Green, S. W. and Zubrow, E. B. W., 1990, Interpreting Space, in Allen, K. M. S., Green, S. W. and Zubrow, E. B. W. (Eds) *Interpreting space: GIS and Archaeology*, pp. 383–6, London: Taylor & Francis.

Arts, N., 1987, Mesolithische jagers, vissers en voedsel- verzamelaars in noordoost België en zuidoost Nederland, *Oude land van Loon*, **XLII**, 27–85.,

Atwell, M. R. and Fletcher, M., 1985, A new technique for investigating spatial relationships: Significance testing, in, Voorrips, A. and Loving, S. H. (Eds), *To pattern the past*, Pact Vol. **2**, 181–9, Strasbourg.

Atwell, M. R. and Fletcher, M., 1987, An analytical technique for investigating spatial relationships, *Journal of Archaeological Science*, **14**, 1–11.

Bakels, C. C., 1978, *Four Linearbandkeramik Settlements and their environment: a Paleoecological study of Sittard, Stein, Elsloo and Hienheim*, Analecta Praehistorica Leidensia 11, Leiden: Leiden University Press.

Bakels, C. C., 1981, Neolithic plant remains from the Hazendonk, Province of Zuid-Holland, The Netherlands, *Zeitschrift für Archäologie*, **15**, 141–8.

Bakels, C. C., 1982, Zum Wirtschaftlichen Nutzungsraum einer Bandkeramischen Siedlung, in *Siedlungen der Kultur mit Linearkeramik in Europa*, Sonderdruck Internationales Kolloquium Nitra, pp. 9–16, Nitra.

Barker, G., 1985, *Prehistoric Farming in Europe*, Cambridge: Cambridge University Press.

Bogucki, P. I., 1988, *Forest Farmers and Stockherders. Early Agriculture and its Consequences in North-Central Europe*, Cambridge: Cambridge University Press.

Clason, A. T., 1972, Some remarks on the use and presentation of Archaeozoological data, *Helinium*, **12**, 139–53.

Fokkens, H., 1991, 'Verdrinkend landschap. Archeologisch onderzoek van het westelijk Fries–Drents plateau 4400 BC tot 500 AD', unpublished PhD thesis, University of Groningen.

Hamond, F., 1980, The interpretation of archaeological distribution maps: biases inherent in archaeological fieldwork, *Archaeo-Physika*, **7**, 193–216.

Kvamme, K. L., 1990, GIS algorithms and their effect on regional archaeological analysis, in Allen, K. M. S., Green, S. W. and Zubrow, E. B. W. (Eds), *Interpreting space: GIS and Archaeology*, pp. 112–25, London: Taylor & Francis.

Louwe Kooijmans, L. P., 1974, *The Rhine/Meuse delta. Four studies on its Prehistoric Occupation and Holocene Geology*, Analecta Praehistorica Leidensia 7, Leiden: Leiden University Press.

Louwe Kooijmans, L. P., 1980, De midden-neolithische vondstgroep van het Vormer bij Wijchen en het cultuurpatroon rond de zuidelijke Noordzee circa 3000 v. Chr., *Oudheidkundige mededelingen uit het Rijksmuseum van Oudheden te Leiden*, **61**, 113–208.

Louwe Kooijmans, L. P., 1993, The Mesolithic/Neolithic Transition in the Lower Rhine Basin, in Bogucki, P. I. (Ed.), *Case Studies in European prehistory*, pp. 95–145, Boca Raton: CRC Press.

Louwe Kooijmans, L. P. and Verhart, L. B. M., 1990, Een midden neolithisch nederzettingsterrein en een kuil van de Stein-groep op de voormalige Kraaienberg bij Linden, gemeente Beers (N.-Br.), *Oudheidkundige mededelingen uit het Rijksmuseum van Oudheden te Leiden*, **70**, 49–108.

Madsen, T., 1987, Where did all the hunters go? An assessment of an epoch-making episode in Danish Prehisory, *Journal of Danish Archaeology*, **5**, 229–47.

Modderman, P. J. R., 1958/59, Die Bandkeramische Siedlung von Sittard, *Palaeohistoria*, **6–7**, 33–121.

Modderman, P. J. R., 1970, *Linearbandkeramik aus Elsloo und Stein*, Analecta Praehistorica Leidensia 3, Leiden: Leiden University Press.

Schiffer, M. B., 1976, *Behavioral Archaeology*, New York: Academic Press.

Sheenan, S. J., 1988, *Quantifying Archaeology*, Edinburgh: Edinburgh University Press.

Siegel, S., 1956, *Nonparametric statistics in behavioral sciences*, New York: McGraw-Hill.

Stampfli, H. R., 1969, Die Tierreste der Grabung Müddersheim, Kr. Düren, in: Schietzel, K., Müddersheim: eine Ansiedlung der jüngeren Bandkeramik in Rheinland, *Fundamenta*, Reihe **A**, Band **1**, 115–24.

Thomas, D. H., 1986, *Refiguring Anthropology*, Prospect Heights: Waveland Press.

Verhart, L. B. M., 1989, Nederzettingssporen uit het Midden-Neolithicum langs de Pater Berthierstraat te Grave, *Westerheem*, **38**, 190–7.

Verhart, L. B. M. and Louwe Kooijmans, L. P., 1989, Een midden-Neolithische nederzetting bij Gassel, gemeente Beers [N.-Br.], *Oudheidkundige mededelingen uit het Rijksmuseum van Oudheden te Leiden*, **69**, 75–117.

Verhart, L. B. M. and Wansleeben, M., 1989, Een laat-Mesolithische nederzetting te Merselo-Haag, Gemeente Venray, Nederland, *Notae Praehistoricae*, **9**, 29–30.

Verhart, L. B. M. and Wansleeben, M., 1991, Het Maasdalproject en de activiteiten van mesolithische jagers en verzamelaars in het dal van de Loobeek bij Merselo, gemeente Venray, *Archeologie in Limburg*, **46**, 48–52.

Wansleeben M., 1988, Applications of geographical information systems in archaeological research, in Rathz, S.P.Q. (Ed.), *Computer Applications and Quantitative Methods in Archaeology 1988*, BAR International Series 446 (ii), 435–51.

Wansleeben, M. and Verhart, L. B. M., 1990, Meuse Valley Project: the Transition from the Mesolithic to the Neolithic in the Dutch Meuse Valley, in Vermeersch, P. M. and Van Peer, P. (Eds), *Contributions to the Mesolithic in Europe*, pp. 389–402, Leuven: Leuven University Press.

Wansleeben M. and Verhart, L. B. M., 1992, Beers, Linden- de Geest, in Verwers, W. J. H. (Ed.), Archeologische Kroniek van Brabant 1991, *Brabants Heem*, **44**, 144–8.

Wansleeben, M. and Verhart, L. B. M., in press. Meuse Valley Project: GIS and Site Location Statistics, *Analecta Praehistorica Leidensia*, **25**.

Waterbolk, H. T., 1958/59, Die Bandkeramische Siedlung von Geleen, *Palaeohistoria*, **6–7**, 121–61.

Zvelibil, M. (Ed.) 1986, *Hunters in Transition*, Cambridge: Cambridge University Press.

13

Cumulative viewshed analysis: a GIS-based method for investigating intervisibility, and its archaeological application

D. Wheatley

13.1 Introduction

This chapter introduces a general method, which may be called cumulative viewshed analysis (CVA), which can be used to make inferences about the relationships of intervisibility between related sites within a landscape. The method is not statistically complex, and therefore can be considered to be quite robust. All stages of the method may be implemented using currently available GIS and statistical software, requiring only that a suitable elevation model be available, together with the locations of the sites. Although cumulative viewshed analysis may well find wider application, the method has been developed in response to a particular research problem, namely an attempt to understand the spatial relationship between archaeological monuments.

13.2 Cumulative viewshed analysis

13.2.1 Visibility from single sites

The calculation of a line of sight map or 'viewshed' for a point location, given a digital elevation model (DEM) is a relatively trivial computing problem and is available within the current functionality of many GIS. In a raster system the calculation requires that, for each cell in the raster, a straight line be interpolated between the source point and each other cell within the DEM. The heights of all the cells which occur on the straight line between the source and target cells can then be obtained in order to ascertain whether or not the cell exceeds the height of the three-dimensional line at that point. Figure 13.1 shows diagramatically how this operates: the example top right shows the case for the existence of a line of sight, while the example bottom right shows the case for no line of sight. It is usual, in most cases, to add an additional value to the source cell to account for the height of the human eye above the surface.

The result of each of these discrete calculations is either a positive (Figure 13.1, top) or negative (Figure 13.1, bottom) result, conventionally coded as a 1 for a visible cell or a 0 for a cell which is not visible. When performed for the entire raster,

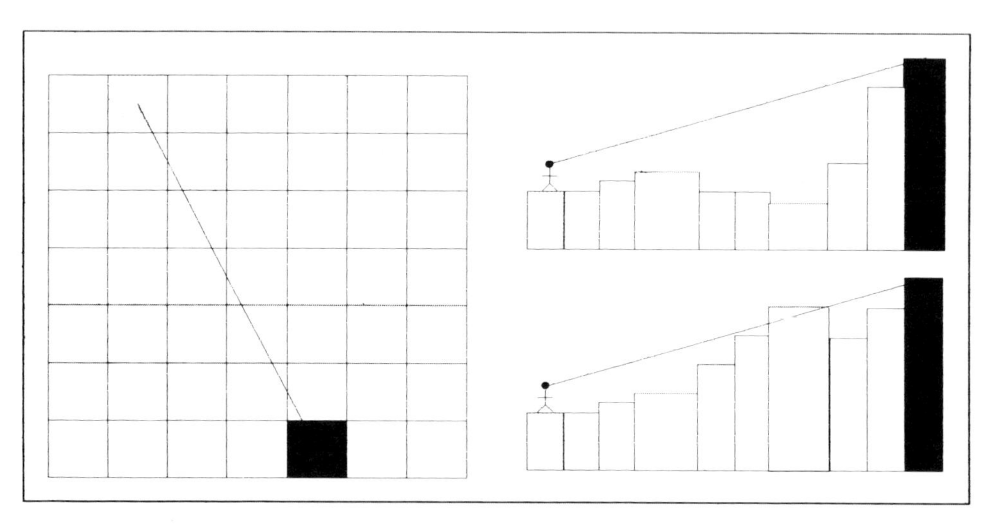

Figure 13.1 Schematic illustration of a line-of-sight calculation. A line is interpolated between two cells of the elevation model (left). If the heights of the intervening cells do not cross this line (top right) then there is a line of sight; if the height of any cell does exceed the height of the line (bottom right) then there is no line of sight.

the result is a binary image, those areas of the landscape which have a direct line of sight from the target cell coded as a 1 and those with no line of sight with 0. Such images are commonly referred to as visibility maps or viewshed maps and have already found some application within archaeology: Gaffney and Stančič, for example, investigating the Greek period of the Island of Hvar, generated a viewshed map for the watchtower at Maslonovik, Hvar. This then demonstrated that a similar watchtower at Tor would have been visible from Maslinovik, which would in turn have been able to pass warnings to the town of Pharos. Their result supported the assumption that such towers formed 'an integral system connected to the town and Pharos whereby watch was kept for any approaching danger' (Gaffney and Stančič, 1991: 78).

Viewshed maps were also suggested by Ruggles *et al.* (1993) as part of a method for investigating the locations of the short standing stone rows of the island of Mull. This method suggested that viewshed maps for each of the standing stone row sites should be combined to create a binary multiple viewshed map consisting of the logical union of the individual maps. From this, prominent landscape features on which the stones may have been aligned could be identified, and viewshed maps generated from these locations. Finally a count could be made of the number of stone row sites falling within these landscape features, and 'the landscape features which best explain the observed placing of the stone rows are those for which this number is greatest' (Ruggles *et al.*, 1993: 127). An example of this method applied is not provided by the authors.

More interesting, perhaps, is the claim by Lock and Harris (forthcoming) that viewshed maps from long barrows of the Danebury region consistently fail to overlap: 'In almost every case the Barrows appear to have been located so as to exclude other Barrows from its viewshed.' This is interpreted as evidence that the Barrows are territorial markers. If the visual interpretation of Lock and Harris is correct (and the images leave little room for doubt) then there should be a means of testing the existence of such a relationship.

13.2.2 Intervisibility within samples of sites

In situations where the intervisibility within a group of sites is of interest, it is possible to obtain a viewshed map for each site location. These individual maps can then be summed, using simple map algebra techniques, to create one surface. This surface then represents, for each cell within the landscape, the number of sites with a line of sight from that cell. For a sample of n sites, the value of this surface will obviously consist of integers which are constrained to vary between 0 and n. Such a map may be referred to as a cumulative viewshed map for the sample of sites within the particular region of interest. The phrase cumulative viewshed map is used here to distinguish this summed result from the multiple viewshed map referred to by Ruggles *et al.* (1993) which is the logical union of a series of individual viewshed maps.

Having generated a cumulative viewshed surface from a sample of sites, it is then possible to perform a point-select operation on this surface based on the site locations, and thus to obtain, for each site, the number of other sites which are visible from it. An adjustment is necessary to account for the line of sight from the site to itself before a hypothesis can then be constructed to test for intervisibility among the sample of sites: this can be made either by subtracting one from the cumulative viewshed surface before the point select operation, or by subtracting one from the result of each point select.

13.2.3 Testing for intervisibility

Once these operations have been completed, the cumulative viewshed surface can be regarded as a statistical population—the population of the number of visible sites of all the cells within the study region—while the new attribute values of the sites can be regarded as a sample from that population. Given these, it is straightforward to construct a pair of testable hypotheses as follows:

- H_0—that the sites are distributed irrespective of the number of other sites which are visible; and
- H_1—that the sites are not distributed irrespective of the number of other sites which are visible.

Using the GIS capability to summarize the characteristics of the cumulative viewshed map (population) and the site attributes (sample), it is then possible to construct a one-sample hypothesis test, at an appropriate confidence interval, to accept or reject this hypothesis. The ability to perform a one-sample test as opposed to a two-sample test, which would involve generating a second sample of random points from the region and obtaining the values of this random sample on the cumulative viewshed surface is of some importance. Unlike the traditional two-sample approach to regional analysis, the one-sample approach makes use of all the available information, and as Kvamme (1990) has pointed out: 'One-sample statistical tests which compare a site sample against a background standard are conceptually and statistically superior.'

There are a variety of tests which may be appropriate to this situation, but one particularly well-suited test if the Kolmogorov–Smirnov goodness-of-fit test, described by Kvamme (1990: 373). This allows the comparison of an empirical sample distribution against an expected referent distribution. The test is performed by plotting the cumulative distributions of the sample and population respectively and obtaining the difference D between these two curves. The maximum value of D, D_{max} is then compared with a critical value d obtained from the sample size and the required confidence interval. For $\alpha = 0.05$, for example, and in a two-tailed test the value of d can be calculated from equation 13.1.

$$d \approx \frac{1.36}{\sqrt{n}} \tag{13.1}$$

If D_{max} exceeds this critical value d, then the sample distribution can be said to be significantly different from the population distribution and H0 may be rejected.

The selected confidence interval will depend to some extent on the circumstances. Although it may seem good practice to select as high a confidence interval as possible in order to avoid the error of claiming a relationship where there is none—a type I error—this must be contrasted with the (admittedly less serious) error of failing to identify a significant relationship—a type II error (Shennan, 1988: 52). In the specific practice of archaeological analyses, this balance usually results in the adoption of $\alpha = 0.05$, with the implication that a type I error may occur five times in every hundred experiments. However, it should be noted that some situations may dictate a greater level of confidence, and the reader contemplating the application of this technique should consider the context of the study, and the use to which the results will be put.

13.3 The method applied—intervisibility of long barrows

13.3.1 Long Barrows in the Neolithic

As an example of how the application of this technique to archaeological sites may provide new insights into the role of monuments within society, we may consider two groups of archaeological monuments in southern England. Before describing the application in detail, some background is needed, although this is necessarily a brief overview. The sites are earlier Neolithic funerary monuments known as long barrows and are by far the most visible and numerous remains from this period in southern England. As such, they are central to the interpretation of Neolithic regional sequences. One group of these sites occurs on the Marlborough Downs close to the later monuments of Avebury and Silbury Hill while another occurs immediately south of this on Salisbury Plain, around the later site of Stonehenge (Grinsell, 1958; Ashbee, 1984).

The geographical groupings of barrows (and other monuments) such as the two used in this study have frequently been interpreted as the remains of discrete political groupings in the past. Renfrew, for example, interpreted the barrows themselves as family tombs, and used the barrow groups to argue for the existence of five Wessex 'chiefdoms' which he characterized (after Sahlins, 1958) as 'a ranked society, hierarchically arranged, sometimes in the form of a conical clan where the eldest descendent in the male line from the clan founder ranks highest, and the cadet branches are ranked in seniority after the main line' (Renfrew, 1973: 542).

More recent accounts have, probably rightly, been less specific about the precise nature of the social groupings which are represented by the barrows. Thomas and Whittle (1986), 'envisage a group rather smaller than the local community as a whole (whatever the absolute size and geographical distribution), closely bound kinship and other ties'. The assumption that these groupings of monuments represent social entities of some sort within Neolithic societies certainly seems justified by the density of monuments, and by the distinctness of the geographical groups. Each group occurs on a defined area of chalk uplands and cannot be adequately explained solely in terms of differential survival. These groupings are also of a physical scale which would have permitted fairly regular travel within each area, involving distances of the order of 10 km—perhaps a day's walk over the moderate terrain of the chalklands. Even the previously cautious Thomas (1991) concludes that the 'extreme density of settlement throughout the Neolithic in the Avebury area possibly entitles us to consider it as a single political unit ...' (Thomas 1991: 163). For the purposes of this example it is assumed that the groups represent some sort of social or political entity, the precise nature of which is unclear.

13.3.2 Typology of the monuments

As well as these geographical groups, the barrows can be broadly divided (e.g. Ashbee, 1984) into earthen mounds, which generally consist of a rectangular or trapezoidal mound of chalk and turf with flanking ditches, and stone chambered mounds (frequently referred to as 'Cotswold–Severn' tombs). These are similar in that they consist of an elongated mound, but differ from the simpler earthen mounds in that they also contain stone chambers and frequently have impressive stone facades at the larger end. The distributions of these two types of tomb are broadly complementary—the stone tombs occur to the north and west of the Avebury area while only earthen barrows occur to the south—but overlap in the Avebury region, where both types occur. The Salisbury Plain group contains only earthen mounds.

13.3.3 Activities at the long barrows

The first phase of activity at these sites may be defined as the selection of the location and the construction of the mound. This was followed immediately by a phase of use at the monument, frequently involving deposition of human remains before some mounds were then deliberately blocked prior to disuse. A final phase of continued existence after the primary use of the site might also be defined, during which the sites remained as important ritual foci. Much attention has been given to the use and blocking of the tombs, while discussion of the construction has usually centred on the architectural form of the tomb or on the mobilization of labour (e.g. Renfrew, 1973). However, little comment has been passed on the monument's role within Neolithic society after disuse, although clearly the monuments must have remained important places within the landscape, marked as they were by structures which were sometimes very imposing.

In this context, a number of recent discussions of the earlier Neolithic of these two areas (e.g. Devereux 1991; Thomas, 1993) have emphasized the role of monument visibility and intervisibility in the interpretation of the social developments and it was with this in mind that the visibility of the Avebury and Salisbury Plain tombs appeared to be an area worthy of attention. If the tombs can be shown to have been constructed in locations which take account of the visibility of other monuments, then this may provide an insight into not only the functioning of the tombs after disuse but also the context of the construction of the later tombs.

13.4 The analysis

13.4.1 Worked example

To apply cumulative viewshed analysis to the two groups of monuments, the locations of reliably identified long barrows were obtained from existing sites and monuments records, and from archaeological accounts (e.g. Barker, 1985; Richards, 1984, 1990). This provided a group of 27 sites in the Avebury region and 31 in the similar area around Stonehenge. These locations were imported to the format for analysis in the IDRISI GIS package (Figure 13.2).

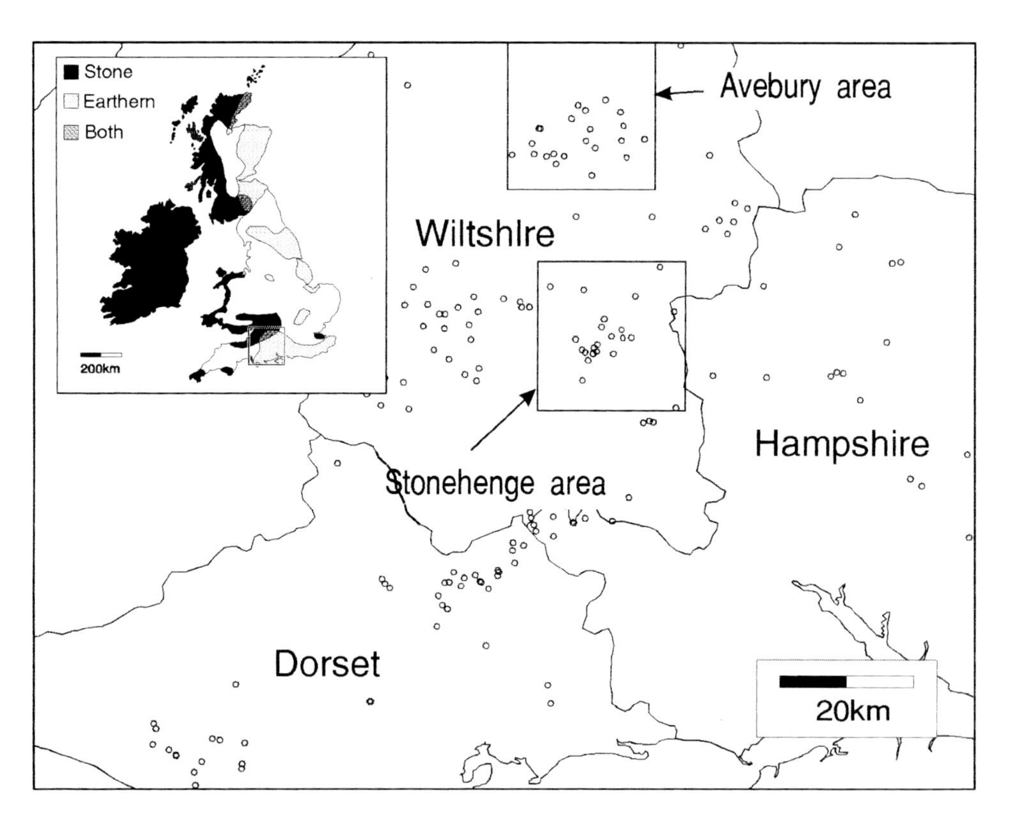

Figure 13.2 The distribution of earthen long mounds and stone chambered tombs in the British Isles (Source: Darvill, 1982), and the locations of the Avebury and Salisbury Plain study areas. Open circles represent long barrows (data for Hampshire, Dorset and Wiltshire only shown).

Elevation models were already available for each of the study areas, having been interpolated from Ordnance Survey 1 : 50 000 digital height data in the form of contours at 10 m vertical interval, with a claimed root mean square error of between 2 m and 3 m. As an acceptable compromise between processing time and resolution, each 20 × 20 km study area was represented as a 250 × 250 raster, giving a cell size of 80 m square.

A viewshed (line-of-sight) map was then generated for each barrow. These maps represent, depending on which way you consider it, either every cell in the landscape which could (theoretically) be seen from the barrow, or every cell in the landscape from which the barrow can be seen. An example viewshed map is given in Figure 13.3). All of these viewshed maps for the barrows in each region were then summed using simple map algebra also available within the IDRISI package. The two resulting maps (shown in Plate 1) are a transformation of the elevation model into a surface in which every cell contains the number of barrows visible from it. The areas of each value in each cumulative viewshed map could then be obtained.

Next, the number of barrows visible from each long barrow in the region was obtained using attribute extraction. These values, of course, include the line of sight from each barrow to itself. The cumulative viewshed map for each area was then incremented by one to account for the visibility of each barrow with itself, and could then be treated as a background population in order to test the null hypothesis that the long barrows in each region are sited with no regard for the number of other barrows with a line of sight from their location. These results, for each area, are

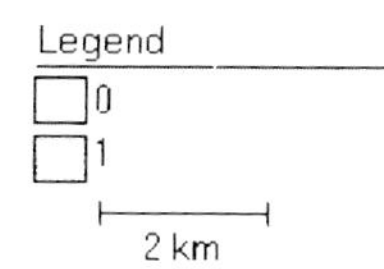

Figure 13.3 Example viewshed map generated from one long barrow of the Salisbury Plain group (shown as a black dot).

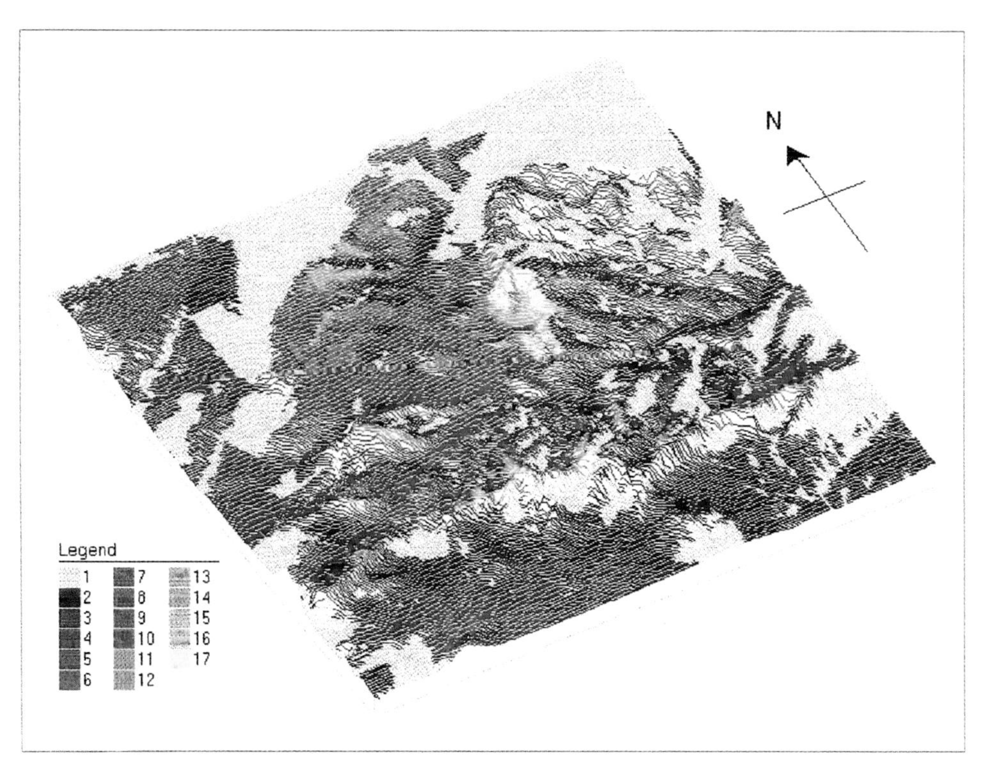

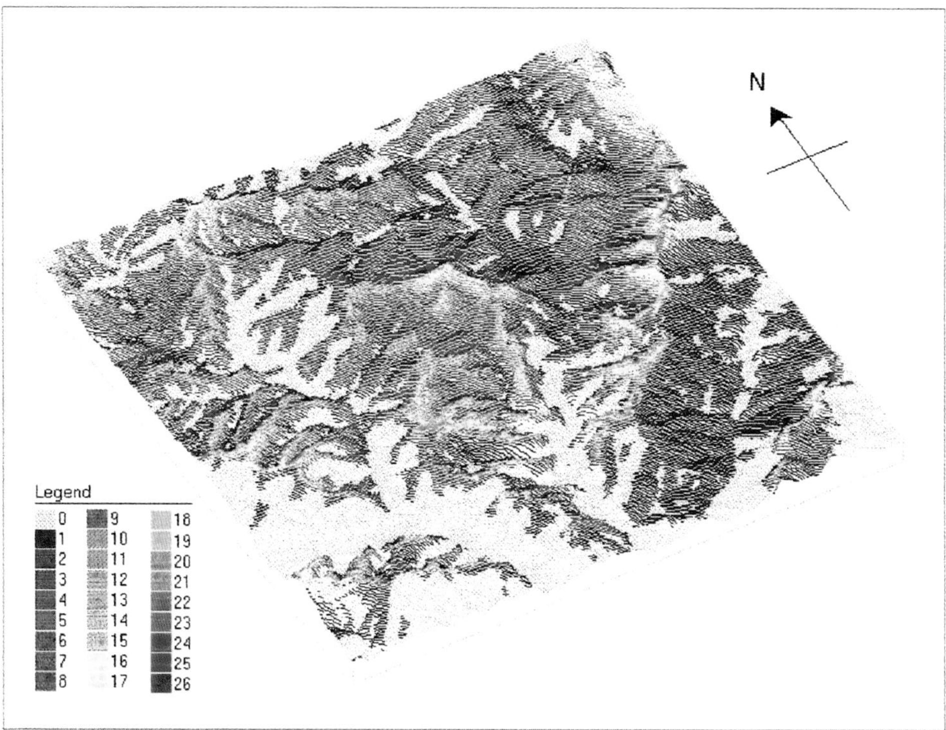

Plate 1. Cumulative viewshed maps superimposed on elevation to show the relationship. Top: Avebury area, Bottom: Salisbury Plain. Both diagrams show the entire 20 km square area which was studied (see Figure 13.2 for location).

shown in Tables 13.1 and 13.2; these are then presented as cumulative percentage graphs in Figure 13.4.

13.4.2 Results

It should be clear from Figure 13.4 that the barrows generally occur in locations with more lines of sight than the background cumulative viewshed population, although this feature is considerably more marked for the Stonehenge group than for the Avebury group. However, it is unclear whether this deviation from the expected distribution of barrows can be explained in terms of pure chance or whether the deviation is statistically significant.

To test this, a one-sample Kolmogorov–Smirnov test was undertaken for each area, adopting a 0.05 confidence level as discussed above. Large sample theory (Kvamme, 1990 citing Thomas, 1986: 506) dictates that the critical value d is approximately $1.36/n$ for the 0.05 significance level, which gives a d of 0.26 for the Avebury region and 0.24 for the Stonehenge group. D_{max} for each group was then obtained from the results in Tables 13.1 and 13.2.

For the Avebury series it can be seen that D_{max} (0.18) does not exceed d (0.26) and the test therefore does not allow the rejection of H0 at the 0.05 level. For the Stonehenge group, however, D_{max} (0.38) does exceed the critical value d (0.24) allowing the rejection of the hypothesis that the deviation may be caused by chance.

13.4.3 Statistical interpretation

The method has revealed, in simple terms, that the long barrows of the Stonehenge group tend to occur in locations from which high numbers of other barrows may be

Table 13.1 Kolmogorov–Smirnov test for Avebury sub-region long barrows

Lines of sight	Population: Area (km^2)	Area %	Cum. %	Sample: cases	Cases %	Cum. %	D
0	0.00	0.00	0.00	0	0.00	0.00	0.00
1	133.08	33.27	33.27	7	25.93	25.93	0.07
2	75.97	18.99	52.26	3	11.11	37.04	0.15
3	70.40	17.60	69.86	4	14.81	51.85	**0.18**
4	45.65	11.41	81.27	4	14.81	66.67	0.15
5	22.17	5.54	86.82	3	11.11	77.78	0.09
6	19.03	4.76	91.57	1	3.70	81.48	0.10
7	10.89	2.72	94.30	3	11.11	92.59	0.02
8	8.31	2.08	96.37	1	3.70	96.30	0.00
9	6.47	1.62	97.99	0	0.00	96.30	0.02
10	3.72	0.93	98.92	0	0.00	96.30	0.03
11	1.86	0.47	99.38	0	0.00	96.30	0.03
12	1.02	0.25	99.64	0	0.00	96.30	0.03
13	0.77	0.19	99.83	1	3.70	100.00	0.00
14	0.42	0.11	99.94	0	0.00	100.00	0.00
15	0.17	0.04	99.98	0	0.00	100.00	0.00
16	0.04	0.01	99.99	0	0.00	100.00	0.00

The test compares the distribution of the population (derived from GIS maps) and the cases (long barrows) with respect to the number of lines-of-sight to other long barrows. D_{max} (**highlighted**) is 0.18, d is 0.26 at the 0.05 level.

Table 13.2 Kolmogorov–Smirnov test for Stonehenge sub-region long barrows

Lines of sight	Population: Area (km^2)	Area %	Cum. %	Sample: Cases	Cases %	Cum. %	D
1	124.54	31.14	31.14	4	12.90	12.90	0.18
2	81.96	20.49	51.64	2	6.45	19.35	0.32
3	34.74	8.69	60.32	1	3.23	22.58	**0.38**
4	31.16	7.79	68.11	4	12.90	35.48	0.33
5	24.56	6.14	74.25	1	3.23	38.71	0.36
6	16.81	4.20	78.46	4	12.90	51.61	0.27
7	14.68	3.67	82.13	2	6.45	58.06	0.24
8	11.01	2.75	84.88	2	6.45	64.52	0.20
9	9.35	2.34	87.22	2	6.45	70.97	0.16
10	8.85	2.21	89.43	0	0.00	70.97	0.18
11	8.01	2.00	91.43	1	3.23	74.19	0.17
12	6.88	1.72	93.15	1	3.23	77.42	0.16
13	5.90	1.47	94.63	3	9.68	87.10	0.08
14	5.01	1.25	95.88	0	0.00	87.10	0.09
15	4.20	1.05	96.93	1	3.23	90.32	0.07
16	3.50	0.87	97.80	2	6.45	96.77	0.01
17	2.71	0.68	98.48	1	3.23	100.00	0.02
18	2.01	0.50	98.98	0	0.00	100.00	0.01
19	1.25	0.31	99.30	0	0.00	100.00	0.01
20	0.88	0.22	99.52	0	0.00	100.00	0.00
21	0.75	0.19	99.70	0	0.00	100.00	0.00
22	0.68	0.17	99.87	0	0.00	100.00	0.00
23	0.22	0.06	99.93	0	0.00	100.00	0.00
24	0.14	0.03	99.96	0	0.00	100.00	0.00
25	0.10	0.03	99.99	0	0.00	100.00	0.00
26	0.04	0.01	100.00	0	0.00	100.00	0.00
27	0.01	0.00	100.00	0	0.00	100.00	0.00

The test compares the distribution of the population (derived from GIS maps) and the cases (long barrows) with respect to the number of lines-of-sight to other long barrows. D_{max} (**highlighted**) is 0.38, d is 0.24 at the 0.05 level.

visible, and that this association is significant at the 0.05 confidence level. The method failed to provide any evidence, however, that the barrows of the Avebury group are anything but randomly distributed with respect to the same variable.

As with all statistical tests of association, however, a note of caution should be inserted. Apart from the obvious scepticism which should apply to any unrepeated statistical test (the LOS maps take some considerable time to calculate, and so time constraints have so far prevented repetition of the experiment), association must be carefully distinguished from causation. Clearly this test might be interpreted as revealing an intention on the constructors' behalf to site long barrows in areas of high visibility. It is equally plausible statistically, however, that another variable (such as proximity to water or elevation) which is itself associated with the area of the viewshed may be the causative factor. On the other hand, if another variable such as elevation, can also be shown to be associated with the barrow locations, then this also cannot be shown to be causative.

It must therefore be left to the archaeologist to interpret the results of such tests, and in this case it seems rather more likely (to the author at least) that the situation of the Stonehenge long barrows on areas of high ground is a side effect of

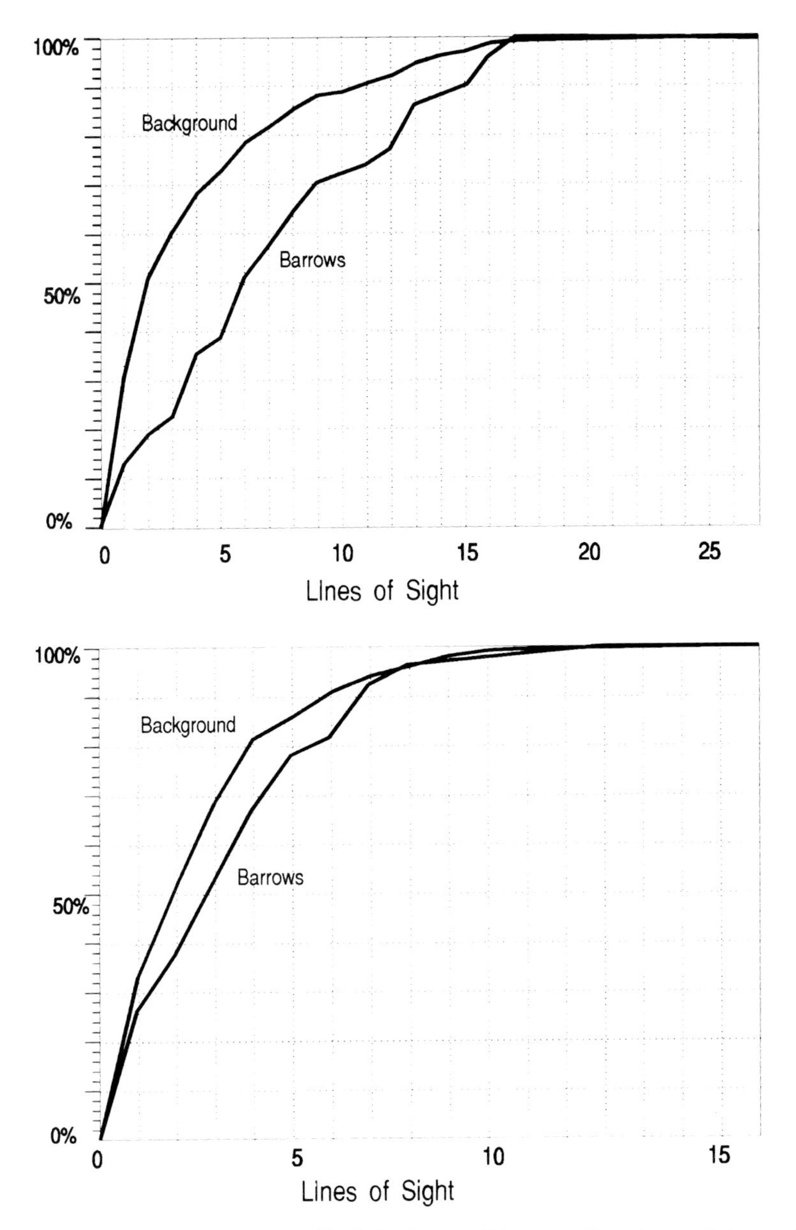

Figure 13.4 Cumulative percentages of both background (expected) and sites (observed) cumulative viewshed values, clearly showing the difference between the Salisbury Plain result (top) and the Avebury result (bottom).

their visibility rather than the alternative, that the visibility of other barrows is a side effect of the selection of high ground.

13.4.4 Error and uncertainty in viewshed maps

Errors are inherent in all data, and geographic databases are no exception to this. The primary sources of quantitative error within the analysis are:

1. elevation values,
2. quantization effects of rasterizing the elevation model,
3. interpolation errors introduced by the isoline to DEM algorithm, and
4. locational displacement of sites.

These errors may then be propagated into secondary errors in

1. individual viewshed maps,
2. cumulative viewshed maps, and
3. the attribute values of sites.

Such errors have been investigated by Fisher (1991) who urges caution over the uncritical use of operations which show locations as either in or out of a viewshed. In a subsequent paper, Fisher (1992) develops the notion of a 'fuzzy viewsheds' which incorporate uncertainty into the calculation of a viewshed map, by registering a measure of certainty for each cell instead of a simple boolean 1 or 0. Such an approach is beyond the scope of this application but clearly offers a method of incorporating uncertainty within the methodology, requiring simply that the boolean viewshed maps from each site location be replaced by fuzzy set maps generated with appropriate parameters derived from the RMS of the elevation model and estimated errors within other parameters.

Another source of information concerning the accuracy of cumulative viewshed analysis may come from sensitivity studies. In these, the height of the viewer can be increased or decreased and the experiment repeated in order to assess the impact of the choices of parameter on the experimental result. It has not been possible to undertake sensitivity analyses in this case but it is encouraging to note Lock and Harris' comment that 'a sensitivity analysis with a Barrow offset height of 4 m showed almost no variation in the patterns so discerned' which compelled them to conclude that the pattern observed was 'unlikely to have occurred through chance but was the result of very careful siting procedures undertaken by native peoples'.

Other than this, however, the only genuinely reliable source of error estimation lies with field observation and, although far from conclusive, the author can report that an afternoon spent identifying features on the horizon from Windmill Hill in the Avebury region provided a very good correlation with the theoretical viewshed map of the same site.

13.4.5 Archaeological interpretation

One particular problem which can be identified in the application of the method to this data is that the landscape was considerably more wooded in this period of prehistory than it is today. For this reason, a line of sight need not infer intervisibility. In mitigation, there is considerable evidence to suggest that barrows tended to be built in cleared areas of the landscape: South Street, for example, was built within a cultivated field (Ashbee *et al.*, 1979), and others such as Horslip, Beckhampton Road and West Kennet have all been shown from molluscan analysis to have been built in open areas.

If we accept the hypothesis that many of the Stonehenge barrows were deliberately located in areas from which other barrows are visible, then some archaeological explanation must be presented for this practice. Furthermore some explanation must also be provided for the fact that the neighbouring, contemporary group of tombs in the Avebury region shows no evidence of the same process.

One clue as to the social function of these locations has already been mentioned: the incorporation of earlier monuments within later ones, as at the Dorset Cursus. Here Barrett *et al.* (1991) have shown that the builders of the cursus included references to earlier monuments in the design of the cursus. This involved not only the incorporation of a long barrow into the cursus bank but also, most interestingly, revealed that the builders deliberately altered the course of the cursus enclose a long barrow so that it 'framed' the midwinter sunset viewed from the cursus terminal. Such physical and visual references to earlier monuments are perhaps an attempt to appropriate the status and associations of an older monument into the new structures, and the same mechanism could be at work in the positioning of the Stonehenge long barrows. In a society in which appeals to past traditions and practices are very apparent in other aspects of the funerary ritual it seems distinctly possible that visual references to other barrows may have constituted part of the mechanism by which social structures were reproduced and renegotiated.

More specifically this could imply that those who directed the building of the monuments felt that the ability to see existing monuments added authority to the new structure through appeals to the historic authority of the existing monuments. Those in control of the new monuments would then gain added legitimacy and be in a better position to retain their own status and authority. This is not to imply that there was no counter to such strategies: other claims to social authority may be evidenced by the use of different locations and types of monument for ritual activities. What may have been revealed by this analysis is just one of many tactics employed by these people to negotiate and transform systems of authority and control.

But if this is the true, then a slightly different mechanism must be at work in the Avebury region. Here, there seems to be no evidence that the barrows are located with reference to the visibility of other barrows. A number of possible reasons might be put forward for this. First, the appearance of Megalithic tombs in the Avebury sequence is different from that of the tombs on Salisbury Plain. In the Avebury sequence, the monuments may have been focused less on the surrounding landscape than on the architecture of the tombs. The creation of the impressive stone facade at West Kennet might be seen to add authority and importance to the tomb as well as focus attention on the activities of those performing at the entrance. In the Avebury region, those who held power were appealing more to monumentality in the structures, and less to their relationship with the past in their attempt to maintain their position.

The difference between the two regions is in itself interesting and might be explained in terms of the resources of the different locations themselves: those who build the monuments on Salisbury Plain may have had no access to large sarsen stones because they are found naturally in the Avebury region. On the other hand, the difference may have been a deliberate assertion of cultural difference: two groups of people constituting themselves in opposition to one another and expressing this through different practices.

In this context, one final archaeological remark may be made concerning the result obtained by Lock and Harris (forthcoming). Although this result was not tested within precisely the same framework as that used here, there seems to be convincing evidence that the group of barrows investigated by these authors were situated for deliberate non-intervisibility from one another, a characteristic which, it is argued, supports the notion that these barrows were used as territorial markers. The simple fact that this result is different from that obtained here need not imply that it is wrong: it has consistently been emphasized here that different mechanisms are at work in the different area. There must be a suspicion therefore that the people responsible for the construction of the barrows of the Danebury region had different motives again to those of the Avebury and Salisbury Plain region in their choice of location.

13.5 Conclusions

This chapter has outlined a new method for investigating the spatial organization of monuments within an archaeological landscape. Through the example of Neolithic long barrows, it has been shown that the method is straightforward to apply using relatively inexpensive GIS technology, and can produce interesting and unexpected results.

The interpretation of the result of the case study must remain extremely tentative: it must be accepted that claims for intervisibility among the barrow groups rest on a particular view of the environmental conditions during the Neolithic, and further evidence regarding the size of the cleared areas around the monuments is needed in order to evaluate the result further. In archaeological terms, such results are not straightforward to interpret because of the need to consider the difference between the statistical association which may be shown by cumulative viewshed analysis (CVA), and causation which can only be inferred from the archaeological context of the particular study.

Whatever the merits of the particular archaeological interpretation of the results of the example application, it is hoped that this chapter has suggested that cumulative viewshed analysis may form a useful tool for the investigation of a variety of different monument types.

Acknowledgements

The author gratefully acknowledges the Royal Commission for Historic Monuments (England) and the Science and Engineering Research Council who have made this research possible and also Dr Stephen Shennan and Dr Arthur ApSimon for their comments and contributions.

References

Ashbee, P., 1984, *The Earthen Long Barrow in Britain*, 2nd Edn, Cambridge: Cambridge University Press.

Ashbee, P., Smith, I. F., and Evans, J. G., 1979, Excavation of Three Long Barrows near Avebury, Wiltshire, *PPS*, **45**, 207–300.

Barker, C. T., 1985, The Long Mounds of the Avebury Region, Wiltshire, *Archaeological and Natural History Magazine*, **79**, 7–38.

Barrett, J. C., Bradley, R. and Green, M., 1991, *Landscape, Monuments and Society, The Prehistory of Cranbourne Chase*, Cambridge: Cambridge University Press.

Darvill, T., 1982, *The Megalithic Chambered Tombs of the Cotswold–Severn Region*, Vorda research Series 5, Highworth: Vorda.

Devereux, P., 1991, Three-dimensional aspects of apparent relationships between selected natural and artificial features within the topography of the Avebury complex, *Antiquity*, **65**, 894–9.

Fisher, P., 1991, First experiments in viewshed uncertainty: the accuracy of the viewshed area, *Photogrammetric Engineering and Remote Sensing*, **57**(10), 1321–7.

Fisher, P., 1992, First experiments in viewshed uncertainty: simulating fuzzy viewsheds, *Photogrammetric Engineering and Remote Sensing*, **58**(3), 345–52.

Gaffney, V. and Stančič, Z., 1991, *GIS Approaches to Regional Analysis: A Case Study of the Island of Hvar*, Ljubljana: Filozofska fakulteta.

Grinsell, L. V. 1958, *The Archaeology of Wessex*, London.

Kvamme, K. L., 1990, One-sample tests in regional archaeological analysis: New possibilities through computer technology, *American Antiquity*, **55**(2), 367–81.

Lock, G. and Harris, T., forthcoming, Danebury revisited: an English Iron Age hillfort in a digital landscape, in Aldenderfer, M. and Maschner, H. (Eds), *The Anthropology of Human Behaviour through Geographic Information and Analysis*, New York: Oxford University Press.

Mercer, R., 1980, *Hambledon Hill: a Neolithic Landscape*, Edinburgh: Edinburgh Unversity Press.

Renfrew, C., 1982, Monuments mobilisation and social organisation in Neolithic Wessex, in Renfrew, C. (Ed.), *The Explanation of Culture Change: Models in Prehistory*, London: Duckworth.

Richards, J., 1990, *The Stonehenge Environs Project*, Southampton: Hobbs.

Ruggles, C. L. N., Medyckyj-Scott, D. J. and Gruffydd, A., 1993, Multiple viewshed analysis using GIS and its archaeological application: a case study in northern Mull, in Andresen, J., Madsen, T. and Scollar, I. (Eds), *Computing the Past: CAA92*, Aarhus: Aarhus University Press.

Sahlins, M. D., 1958, *Social Stratification in Polynesia*, Seatle: University of Washington.

Thomas, J., 1991, *Rethinking the Neolithic*, Cambridge: Cambridge University Press.

Thomas, J., 1993, The politics of vision and the archaeologies of landscape, in Bender, B. (Ed.), *Landscape politics and perspectives*, Providence RI: Berg.

Thomas, J. and Whittle, A., 1986, Anatomy of a tomb—West Kennet revisited, *Oxford Journal of Archaeology*, **5**(2), 129–56.

14

Some Criteria for Modelling Socio-Economic Activities in the Bronze Age of south-east Spain

P. Verhagen, J. McGlade, S. Gili and R. Risch

14.1 Introduction

Questions relating to prehistoric settlement and particularly the spatial and temporal correlates which articulate human/environment interaction, constitute one of the enduring problems in archaeology. The ubiquity of such studies, however, is not necessarily commensurate with understanding; indeed, there are substantial gaps in our knowledge of basic human ecodynamic processes, especially set within a long-term perspective (McGlade, 1993a). The current status of human ecology with its emphasis on stability and equilibrium, actively misrepresents the non-linear interactions which define the evolutionary dynamics of complex socio-natural systems (Terrada, 1987; McGlade and van der Leeuw, 1994). In this chapter, we shall attempt to show how GIS may be able to contribute, if not to a resolution of such problems, at least to helping resituate them within a more productive spatio-temporal framework.

The growing use and increasing sophistication of GIS methods in archaeology have produced a variety of studies which have demonstrated its tremendous potential for interpretation and interrogation of spatially referenced data (Allen *et al.*, 1990). One of the most obvious uses of this technology has been in the realm of regional settlement studies (e.g. Kvamme, 1989; Madry and Crumley, 1990), and particularly with questions relating to the prediction of archaeological sites (e.g. Brandt *et al.*, 1992; Carmichael, 1990).

It has become clear that GIS is much more than a sophisticated cartographic tool; it not only provides archaeologists with a framework within which to explore spatial problems, but, as we shall argue here, it also has the potential to explore less tractable issues such as those represented by socio-ecological questions.

Mapping the socio-natural environment adequately so as to account for its representation as a true ecology of social space, or what one of us has referred to as 'human ecodynamics' (McGlade, 1993b), is of course, non-trivial. Certainly, it cannot be solved by the extant methods of so-called spatial or settlement archaeology since these generate a problematic distinction between the spatial and the social, and effectively privilege the spatial.

What we are suggesting here, is that the full potential of GIS to address such issues has yet to be explored and in many ways can be viewed as a way beyond the simple representation of spatially referenced data. In this vein, the main function of the present contribution is to attempt to demonstrate the utility of GIS, not simply as a methodological tool, but more importantly, as a means of presenting a series of hypothetical scenarios which are relevant to our understanding of human/environment relationships. This will be done principally through the construction of a 'territorial' model designed to articulate a set of semi-autonomous activity spheres which are said to be implicated in the reproduction and organization of a specific archaeological locus, or settlement. Finally, we shall argue for the inclusion of GIS, and especially a time-referenced GIS, as an integral element in theory building within the domain of experimental archaeology.

14.2 GIS and Archaeology

Geographic information systems have, over the last fifteen years, effectively revolutionized the way in which disciplines such as geography, land resource management, ecology and archaeology handle and interpret spatially referenced data sets. Such systems provide a means of manipulating complex, multivariate observations within relatively flexible, interactive computer environments.

Archaeological uses for GIS technology have been expanding since their first appearance in the early 1980s (see Kvamme, Chapter 1), when they began to be used within the context of regional spatial analyses. Despite a number of theoretical issues which have yet to be addressed, one clear advantage which GIS brings to archaeological research is its power of visualization: settlement distributions in two and (pseudo-) three dimensions are by now well known and utilized methods of data representation. In addition, geographical information systems are adept at calculating both distances and surfaces of spatially referenced phenomena in relation to specific geographical features; this fact in conjunction with the possibility of analysing intervisibility by means of digital elevation models (DEMs), has resulted in increasing utilization of GIS methods by archaeologists.

In spite of such developments and the increasing availability of PC and Macintosh compatible user packages which allow easier access, the application of GIS within archaeology has largely been directed at the production of new ways to visualize data, often as a means of verifying environmental determinist models. In short, it has largely been concerned with representation and description (van Leusen, 1993), and for this reason, its use as a potential interpretive tool has yet to be explored (but see Gaffney and Stančič, 1992). For example, while geological or soil characteristics can be represented with some degree of spatial accuracy, the use of geometric polygons is clearly an inappropriate way to deal with spheres of social interaction and their boundaries, since they are characteristically 'fuzzy' (Burrough, 1990; Castleford, 1992). Furthermore, the locational details of archaeological sites are superimposed on present-day geography, and in doing so we create potentially misleading and even spurious correlations. In short, we cannot explain the past by back-projecting the present.

There are, however, more serious problems, when we turn to the theoretical contexts, or the lack thereof, within which GIS systems operate (Wheatley, 1993). The charge that GIS have often assumed a kind of 'theoretical neutrality' in many ways reflects the investment of time and energy in the technological aspects of their development. It is clear that any meaningful discussion on the nature of space (and here we must emphasize that it is not a neutral category, it is socially constructed) demands explicit discussion of the theoretical assumptions within which the particular problem is situated. It is to such issues that GIS methods must now turn their attention, if they are to be more fully integrated into a mature socio-spatial theory.

14.3 Time, Space and GIS

14.3.1 Introduction

As we noted at the outset, understanding regional settlement systems is conventionally perceived as a problem to be set within the context of spatial archaeology. A fundamental, and frequently cited referent in this activity, is said to be the designation of the resource exploitation zone, and this is viewed as constituting the site catchment.

This model, whose underpinnings are borrowed from economic theory, and now widely used in archaeology, presupposes that the activities of food production can be reduced to rational economic behaviour. Thus, human groups are assumed to be interested in 'least cost' energetics. An important corollary states that the closer an area is to a settlement location, then the more intense will be its exploitation. We thus employ terms such as 'optimum resource exploitation zone'.

Perhaps the worse failing of site catchment strategies is their inability to acknowledge adequately the social, cultural and ideological contexts within which human settlements are situated and which structure the day-to-day routine activities such as the procurement, production and processing of food, as well as the array of social and cultural activities involved in the reproduction of society. What we have in fact, is a functional, systemic description in which the social is abruptly disaggregated from the natural. In summary, this model attempts to superimpose an abstract, atemporal Cartesian geometry onto a reality that is fundamentally reflexive, subjective and contingent (Castro *et al.*, 1993; McGlade, 1993b).

14.3.2 Timing space

In addition, this systemic view is articulated within spatial and temporal frameworks which distort the true nature of societal relationships with the environment. This is principally brought about through the separation of time and space, which are often represented as independent entities. In fact, much of the theoretical discourse surrounding archaeology assumes implicitly that time is unproblematic; it is somehow self-evident. Conventionally, it is seen as an abstract container of events supported by chronometric props. Time is presented as sequence and interval; it is objective and quantifiable (Shanks and Tilley, 1987).

Within this abstract framework, the multiple periodicities which make up and define societal existence and reproduction, can, so it would seem, be compressed (smoothed over) and reduced to a sequence of datable events. We can thus make a ladder of history on whose rungs past events are placed. This is the key to understanding, and is rooted in a search for order, for the reduction of difference. What we are arguing is that the seductive logic which underpins this type of archaeological reasoning is both dangerous and pernicious: dangerous, because it privileges a single dimension of time as continuous and linear; and pernicious because it promotes a fictive view of human endeavour and the way in which time is constituted in social praxis (González Marcén, 1991; Picazo, 1993).

The conventional use of space is equally problematic, since it assumes that space as a 'physical' entity can be separated from time and expressed as an absolute value, e.g. as a bounded territory with measurable dimension. In a sense, spatially motivated archaeology and its variants such as 'settlement archaeology' assume that spatial structure is something independent from its social context; there is little appreciation that space is in fact, not a neutral Cartesian concept, but is socially constructed (Lefebre, 1974; Soja, 1980).

The problems posed by the artificial separation of time and space are manifestly obvious, and become especially acute with respect to socio-natural interaction. Here the disaggregation of human ecological processes effects a distortion which severely compromises our ability to interpret the reciprocal dynamic which defines human societal reproduction. If we are to move towards a true ecology of social space, what we have earlier referred to as human ecodynamics, then this must have as a central premise, an emphasis on and an understanding of temporality as a fundamental constituent of the socio-spatial dialectic.

This temporal understanding of space (time–space) demands that we map human societal processes with their individual, engendered, seasonal, generational time-scales onto the array of periodicities and time-scales which define evolutionary environmental phenomena. We thus encounter human ecodynamics as co-evolutionary; humans do not adapt to the environment as is conventionally held in human ecology and archaeology—rather, they are embedded in landscape evolution as a continuous structuring and restructuring of time–space, one that implies no teleological directive. Humans are not passive recipients of environmental information; they are conceived as actively involved in the creation of their environment (Touraine, 1977; McGlade, 1993b).

14.3.3 The territorial map: a time–space description

If we are to come to terms with a true human ecodynamic description of extinct settlement systems, then this requires a conceptual structure upon which analytical and interpretive models can be based: one which is able to address the need for a dynamic ecology of social space. Following Crumley and Marquardt (1990) we shall argue that two types of structures are implicated in the representation of social space: socio-historical structures and biophysical structures.

The first of these structures (socio-historical) is involved with relations of production, inheritance, class, political liasons, defence, trade/exchange and laws, while the second structure (biophysical) includes those natural, geological, pedological, topographical and climatic factors which provide the enabling and constraining features which generate human settlement. Societal organization and its reproduction is thus the product of the reciprocal, mutually reinforcing relations between these two structural spheres. The locus of their time–space intersection we shall refer to as 'territory'.

Thus, we propose that the most appropriate way to view settlement dynamics within a human ecodynamic context is to replace the concept of 'site' or 'settlement' with such a time–space construct; settlement is seen as the interpenetration of multiple coexisting territorial domains (McGlade, 1993b).

Territory here defined is not synonymous with a simple spatial referent; thus the socio-spatial configuration which constitutes for example, the Argaric universe of the Vera Basin, represents the time–space intersection of multiple activity spheres: social, political, ecological and ideological. Collectively, these constitute a web of human–environment interactions in time and space. An additional part of our definition of territory embodies the concept of a field of knowledge and is taken here to mean the locus of particular historical and culturally bound social and natural knowledge-bases, i.e. the intersection of natural phenomena and lived experience. What we have, in effect, are semi-autonomous activity spheres in a specific space–time configuration, implicated in the reproduction of the social group.

As we have previously noted, an important property in presenting such a scheme is the representation of time, or rather temporalities. Time cannot be relegated to the realms of the abstract, nor can there be any single unifying 'time', there is no present moment; rather, time is to be grasped in relation to the particular sets of biological, social, economic, political and ideological processes which articulate societal reproduction. Thus we can say that every social act which takes place within its territorial domain, implicates different temporalities. Moreover, within these territories, we encounter a substantial (as opposed to abstract) time—an arena in which continuous, discontinuous and cyclical periodicities meet.

In order to give these concepts a more concrete realization, a set of fundamental domains has been arrived at, and are regarded as representing a primary set of socio-natural descriptors which may be said to encompass the dynamics of societal reproduction. A more complete elaboration of these ideas is set out elsewhere, particularly with respect to temporal issues (McGlade, 1993c). For our present purposes, we shall present a reduced description of the basic terms of reference:

- The domain of **human reproduction and maintenance activities** is conceived as representing the basic production modes upon which society is constituted and maintained. This territory is structured by the domestic unit, which is defined as the social group or segment that effects control over the day-to-day routinization of household tasks and other activities related to the construction and maintenance of domestic structure.
- The domain of **food production** includes all the fundamental activities connected with gathering, hunting, crop production and animal husbandry. From a temporal perspective, what we have is the superimposition of sets of social temporalities on the natural temporalities of the environment, e.g. the times of soil preparation, of planting, cultivation and harvesting, as well as the imposition of human action on animal reproduction (stock breeding) and behaviour.
- The domain of **material technology production** deals with the extraction, manipulation and transformation of raw materials, and the necessary technological processes which affect these transformations. The temporalities involved here encompass the duration of the different activities involved in the fabrication process.
- The domain of **raw material and artefact transactions** includes all inter- and intra-group transactions, whether equal or unequal. Spatially, this domain is distinguished by its incorporation of other social groups and their territories. This domain is also an important carrier of information across cultural boundaries and thus is involved—either implicitly or explicitly—in the dissemination of myth, ritual, or other specialized knowledge.

- The domain of **political and administrative organization** acts to facilitate the organizational structures, both enabling and constraining, which provide the basis for the perpetuation and legitimation criteria for societal reproduction. Inscribed in this domain are the collection of institutional structures and codified rules and regulations which form the vital underpinning for all social activities and their enforcement. This domain is the locus of control and power.
- The domain of the **ancestors,** by its existence, implies the active participation of the dead in the symbolic reproduction of the living. In this domain are religious and/or ritual practices and the organization of their spaces. It may, in some cases, be identified with a cosmological conception of territory. From a temporal perspective, this domain is unique in that it transcends time as a measurable phenomenon; it is concerned with trans-temporal associations and activities.

Clearly, within this time–space territorial description, not all territorial domains are archaeologically visible (e.g. territories of women differentiated from those of men, territories of ritual, etc.), nor can they be readily accessed via GIS; this however is no reason to ignore or to relegate them as epiphenomena. What we are arguing is that we can make a beginning in terms of addressing some of these issues, and set out some possible interpretive modelling approaches which can be operationalized within a GIS framework. In line with our previous argument, there can, of course, be no privileged model representation of any territory or its domains; what we must work towards is the realization of multiple levels of description, i.e. and assemblage of different interpretive modes, from which 'meaning' can be negotiated. Additionally, as complexes of symbols and meaning structures, the territorial realization that we refer to as a specific archaeological site, incorporated, not simply a set of physical and material criteria, but reflected sets of values as to how space should be perceived and experienced.

14.4 The Vera Basin: the socio-natural context

14.4.1 The natural environment

In order to contribute towards the task of resituating GIS within a more productive archaeological context, one which recognizes the need to generate new ways of articulating human/environment relationships, we shall now focus on an interpretive approach to model building within the context of the Vera basin.

The south-east Iberian peninsula, in which the Vera basin is situated, is the most arid region of Europe. From a climatic perspective, it is characterized by high temperatures and extremely low rainfall (*c.* 250–300 mm per annum), though it is not so much the paucity of rain that is significant here, but rather its intensity and irregularity and the way in which these events are involved in structuring the landscape. The region is broadly defined by the river plains of the Almanzora, Antas and Aguas rivers, and is bordered by the mountain ridges of the Sierra Cabreras to the south, the Sierra de Bédar to the west and the Sierra de Almagro and Sierra Almagrera to the north (Figure 14.1). These mountain ranges are largely composed of metamorphic rock, while the river plains are covered by sedimentary and volcanic deposits from Tertiary and Quaternary contexts. The soils in the mountainous areas tend to be shallow and prone to erosion, and while the river plains tend to have

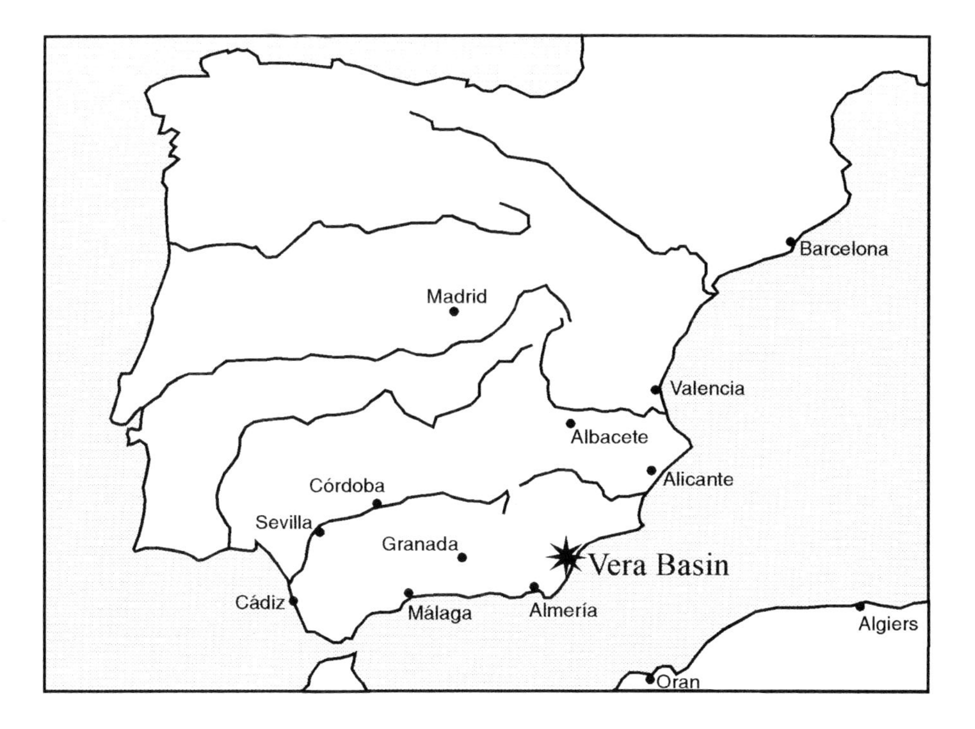

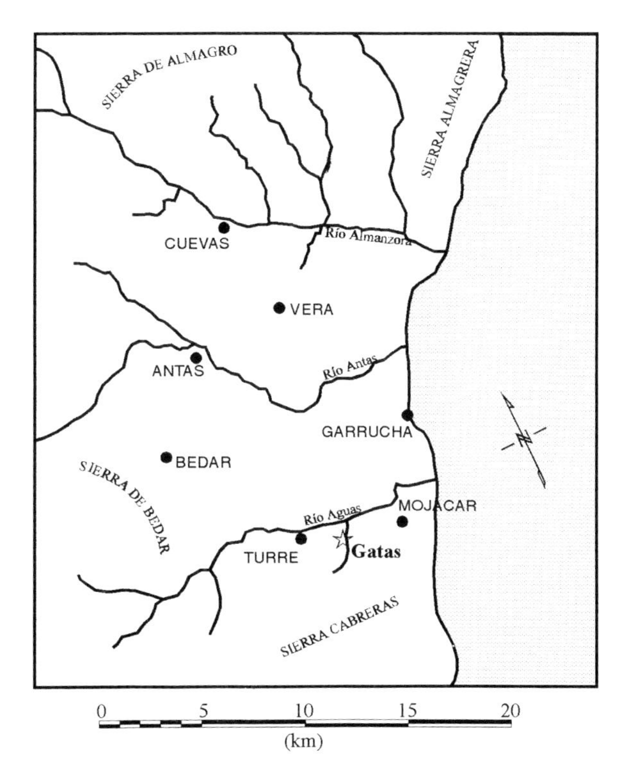

Figure 14.1 The location of the Vera Basin

better soils for agricultural exploitation, they are also prone to lateral erosion by the river channels as well as to salinization (Proyecto Lucdeme, 1988–91).

Although the vegetation cover we observe today is that of species adapted to semi-arid conditions, dominated for example by *Artemisia* or *Lygeum spartum*, it is not clear to what extent these are indigenous species, since in more remote areas which have had minimal human interference, we see examples of light forest cover of *Pinus halepenis* and *Pistacia* (Ruiz *et al.*, 1992). One of the primary aims of the EC-funded ARCHAEOMEDES project is to address such questions within the context of the Holocene period, so as to provide a more rigorous approach to environmental reconstruction focused especially on the role of social and cultural factors and the way in which they have co-evolved to generate the historical trajectory of the Vera Basin.

Preliminary results of the ARCHAEOMEDES investigations have shown that immediately prior to the arrival of humans, the area seems to suggest a less dynamic geomorphological regime than today; a landscape characterized by a smoother topography than we see currently, along with a slow infilling of the low lying regions and the presence of lagoons and marshes both near the coastline and inland (Gili *et al.*, 1993).

14.4.2 The archaeological context

While the specific sets of dynamic processes structuring the palaeo-environments of the Vera Basin are difficult to specify with any degree of precision, what is clear is that the area has been continuously inhabited since *c.*4000 BC. The first Neolithic exploitation of the basin is archaeologically translatable as a pattern of low intensity and high diversity and is consistent with what appears to be undifferentiated socio-economic organization. Geomorphological change during this period resulted in the entrenchment of river beds and there are signs of the beginnings of erosion as is indicated by coastline studies (Hoffman, 1988). Speculations on the causes of such change point to tectonic activity, rather than to climatic factors.

With the ensuing Chalcolithic period (3000–2300 BC), human settlement in the basin shows a significant increase and we have higher levels of productivity and population density especially in the riverine areas. Significantly, the archaeological settlement pattern data shows differentiation based on size and on craft specialization, and this is reinforced by burial data which demonstrate unequal access to material goods.

Around 2300 BC, a radical change in the pattern of human occupation occurred with the beginning of the Argaric period, and is accompanied by a shift from the lowland areas to higher ground, as well as large population concentrations in a few large centres. The radical nature of this change lies both in the rapidity with which it took place, and in the drastic modification in the subsistence base which accompanied it. Additionally, the social discontinuity is clearly reflected in burial practices which now are located inside the settlements and are moreover focused on individual interment. A marked differentiation in grave goods is now evidenced, and this has led to the establishment of a hierarchical model of social stratification with elite control and a possible state level organization (Lull, 1983; Lull and Estevez, 1986). What the available archaeological data demonstrate unequivocally, is that there is no continuity between the preceding Chalcolithic and the ensuing Argaric.

Speculation has arisen as to the possible role of climatic factors in inducing such social and cultural discontinuity or alternatively the role of soil exhaustion as a result of the cumulative effects of long-term cultivation practices in the lowlands. What is interesting, however, is that initially, the change in social organization is not reflected in subsistence practice; it is only after *c.*1800 BC with the full development of Argaric society that we see a transition to barley monoculture. It seems that this change is coeval with a general trend to more arid conditions and the exhaustion of lowland riverine soils, something corroborated by the decline of hydrophytic species in the charcoal and pollen evidence. The available archaeological evidence suggests that this exhaustion seems to have been accompanied by population increase, and this may have forced a shift towards intensive monoculture so as to maintain the existing social and political organization.

This situation seems to have lasted for about 200 years, after which the Argaric system collapsed with the abandonment of the larger settlements and a general contraction of population throughout the region. The resultant dismantling of the close knit Argaric exchange system allowed, for the first time, access to materials from other parts of the Iberian peninsula and from a wider Mediterranean context generally. In addition, the relaxation of hierarchical control seems to have produced a more diversified subsistence economy, in many ways a reversion to the earlier Neolithic exploitation strategies.

The precise causal nature of this system collapse, or rather transformation, has provoked a variety of hypothetical explanatory models (Chapman, 1978; 1990; Lull, 1983; Mathers, 1984; Gilman and Thornes, 1985), which range from the primacy of environmental factors to the determining role exerted by political centralization and social stratification.

Debate on such questions continues apace, and their resolution is thought to reside in the results of future excavations. While it is true that the availability of further data can help to provide better corroboration for particular hypotheses, what we are arguing here, is that:

(a) there is yet a great deal of latent information resident in existing data, and
(b) it may be possible with the use of GIS methods to generate alternative ways of interrogating these data sets.

14.5 The role of GIS in the Vera Basin project

14.5.1 Introduction

The primary aim of the Vera Basin project with respect to GIS, is to provide a means of combining and cross-referencing archaeological, historical and environmental data so as to understand better the way in which socio-natural dynamics have evolved over the long term. Of particular importance is the role of differing socio-political strategies in generating the conditions within which human communities persisted during the period of the later Holocene. What we are attempting to define are the temporal, spatial and social criteria contributing to the array of human ecodynamic relationships which have structured the prehistoric and historic landscape.

The GIS developed for the project is based on a 1 : 100 000 scale. Topography, geology and soil maps were manually digitized, as was a land-use map dating from 1978. More recent land-use data were obtained from the CORINE database (Commission of the European Communities, 1992). In addition, the locations of archaeological sites and present-day mining activities as well as farms were digitized. A DEM with a resolution of 200 × 200 m was obtained from the Spanish National Geographic Institute. A further selection of statistically treated data on sites was added so as to create a working geographical database in GRASS and INFORMIX. Ecological and historical data are currently being added.

From both theoretical and methodological perspectives, an important goal of the research is to go beyond the purely statistical treatment of data and provide a more useful frame of reference within which previously constructed hypotheses might be addressed, reassessed and reconstructed, as well as advancing alternatives.

14.5.2 An integrative, multiscalar framework

In order to work towards such a goal, we have devised an iterative model structure which links the divergent data sets and the GIS with a hierarchical interpretive framework (Figure 14.2). From the figure, we can see that there are three principal levels of 'knowledge' which can be accessed, forming three levels of data transformation. These represent different scales of inquiry or interrogation and range from a mode of representation at the first level, to a mode of description at the second level of transformation to an interpretive/analytical level at the final level of transformation. It is important to point out here that the framework is not simply goal directed to the third level of enquiry, but rather exists as a continuum in which different questions and problem sets can be tackled at a particular level or combination of levels; the feedback loops in the diagram illustrate this point by demonstrating the reciprocal relationship between data and any chosen level of representation or transformation.

14.5.3 The representational level

At this first level of transformation we have conventional statistical treatments of the various databases, i.e. climate, geology, soils, vegetation, etc., as well as historical documentation and the archaeological data from both surveyed and excavated sites spanning prehistory to early historic times. The difficulty of generating accurate spatial representations of the archaeological data lies in the fact that although artefacts are geo-referenced, the spatial dimensions of the organization of their production, distribution and acquisition are unknown. Similar problems beset the environmental data: while the location of pollen samples is known, the precise spatial distribution of the pollen producing species remains obscure. Our third category of information, consisting of historical documents, is equally problematic. Although they provide valuable data on the agricultural properties and their subsistence base, the exact location of the territories of production remains unknown.

With these caveats in mind, the manipulation and cross-referencing of this information can nevertheless generate useful spatial representations at a coarse-

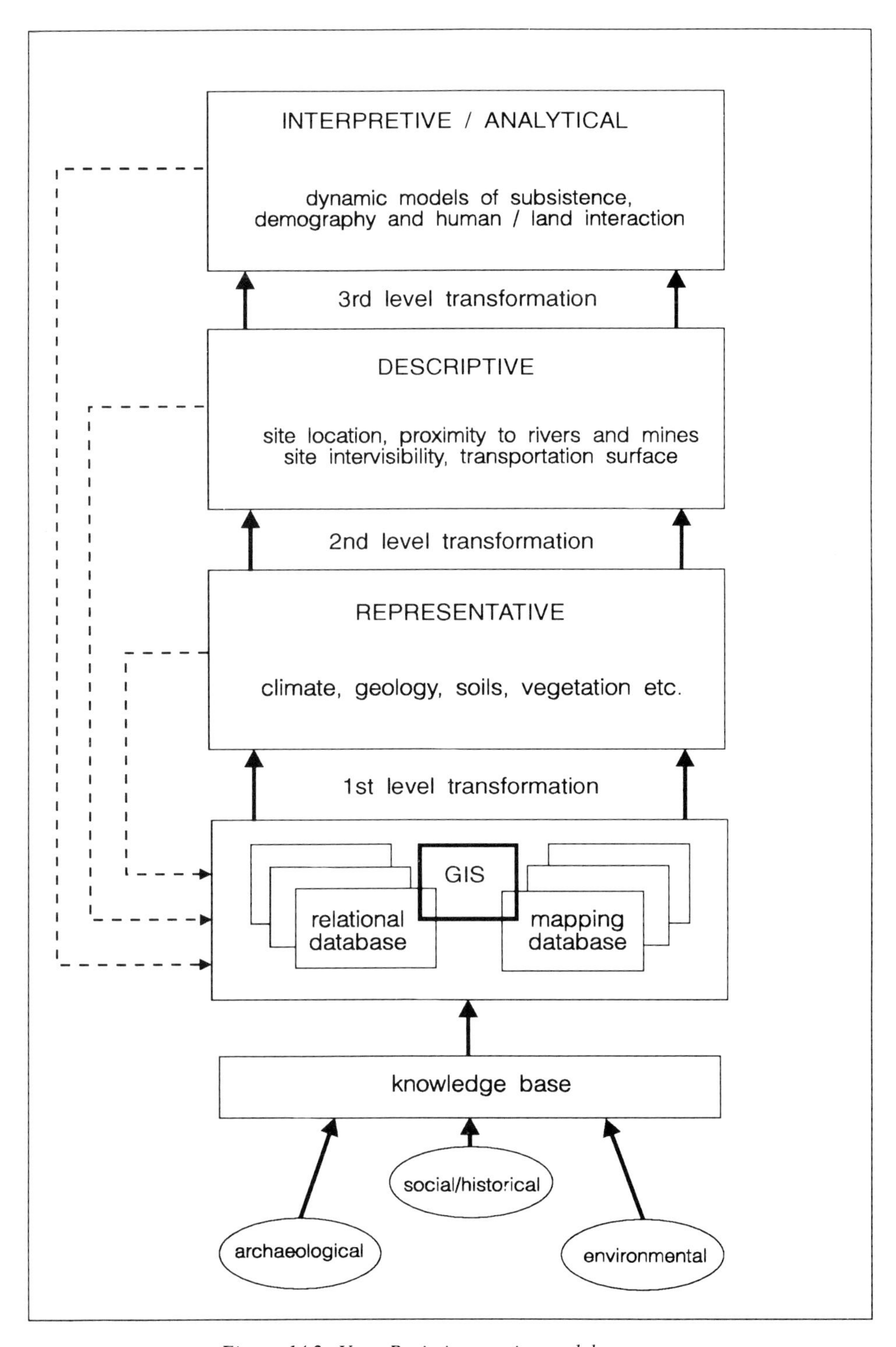

Figure 14.2. Vera Basin integrative model structure.

grained level. With this knowledge, we can move to the second level of transformation—the descriptive.

14.5.4 The descriptive level

At this level, the geological, topographical and archaeological characteristics of site locations have been subjected to univariate statistical treatments in an effort to isolate different site distribution patterns. It should be pointed out here that these patterns are superimposed on present-day ecological conditions and thus caution should be exercised in any proposed correlation with prehistoric conditions. Following this evaluation of sites at the locational level, the next step is to seek non-random associations at the inter-site and regional level, based on an analysis of the immediate surroundings of the sites. To this end, three separate types of analysis have been undertaken:

1. **Distance analysis** For all settlements distances were calculated to *ramblas* (dry river beds), to the coastline, to the nearest neighbouring contemporaneous site, and to the nearest present-day mine with copper, iron or lead (the latter being considered an indication for the presence of silver). The distances calculated are straight line distances, and the actual distances may have been much larger in rugged terrain. To account for the change of coastline over the last 6000 years, the palaeogeographic reconstructions of Hoffman (1988) were used, although the accuracy of these reconstructions remains a matter of debate. Not all of these calculations were useful: the availability of metals cannot be very relevant for the Neolithic period, nor will iron have been of much importance to the Chalcolithic and Argaric cultures.
2. **Analysis of the surroundings of each site** For all settlements the geological, pedological and land-use data were quantified within a 2 km radius from the settlement. This radius of 2 km is a rather arbitrary one and is not to be confused with site catchment analysis, critiqued above. Rather, it serves as a first description of the area around a settlement, and can be used for comparison of the conventional resource environment with the territorial model suggested above. The same technique can be applied for palaeo-botanical and palaeo-faunal data, especially since the locus of deposition and sampling does not correspond to the spatial extent of vegetation or animal foraging areas.
3. **Visibility analysis** As it is assumed that the Argaric society was highly organized and had a state-like character, the analysis of intervisibility can be used to test that hypothesis. Questions like, 'were the most important settlements located in positions where they could see all others?' and 'could the other settlements always see the capital settlements?' can be answered by creating a viewshed around each site, and then calculating the number of visible sites in this viewshed.

Even these relatively simple analyses can add to the verification of archaeological hypotheses. To illustrate this, we shall focus on Gatas, one of the most important recently excavated Argaric sites in the Vera Basin (Chapman *et al.*, 1987; Buikstra *et al.*, 1989, 1992; Castro *et al.*, 1989; 1993; Ruiz *et al.*, 1992). The distance calculations suggest that Gatas' nearest neighbour is the site of Barranco de la Ciudad at a distance of roughly 3100 m. From the visibility analysis however it is clear that the two do not have visual contact. In fact, from Gatas one can see almost all of the

Vera Basin, but to the south-east the view is blocked by the Sierra Cabreras. How can we make the distance calculations reflect this lack of direct contact between the two sites? One of the options is to create a cost-surface around Gatas using slope as a friction factor, and see if we can create a route from Gatas to Barranco de la Ciudad by executing at least cost path. The route that this method provides does not go over the Sierra Cabrera, but descends into the valley, following the Río Aguas almost to the second nearest Argaric site of Cabezo de Guevara, and then follows the coastline before reaching Barranco de la Ciudad. The total distance of this route comes to 8600 m. There is of course no proof that such a route existed (cf. Madry and Crumley, 1990), and the path is not a very efficient one because it first descends and then goes up again. A logical step would be to follow the contour lines, by means of a cost-surface based on the difference in elevation between Gatas and the rest of the area. A route from Gatas to Barranco de la Ciudad using this method is only 4400 m.

When using this last procedure from Gatas to Cabezo de Guevara, a junction is found at the point where the path to Barranco de la Ciudad crosses the ridge of the Sierra Cabrera. The route to Cabezo de Guevara descends following the ridge, the one to Barranco de la Ciudad continues to follow the contours on the other side of the sierra (Figure 14.3). It is interesting to observe that the distance to Cabezo de Guevara is not changing very much using either method: the straight line distance is 4800 m, the distance using the slope map is 5400 m and the distance following the ridge is 5000 m. This suggests that the method to be used is very much dependent

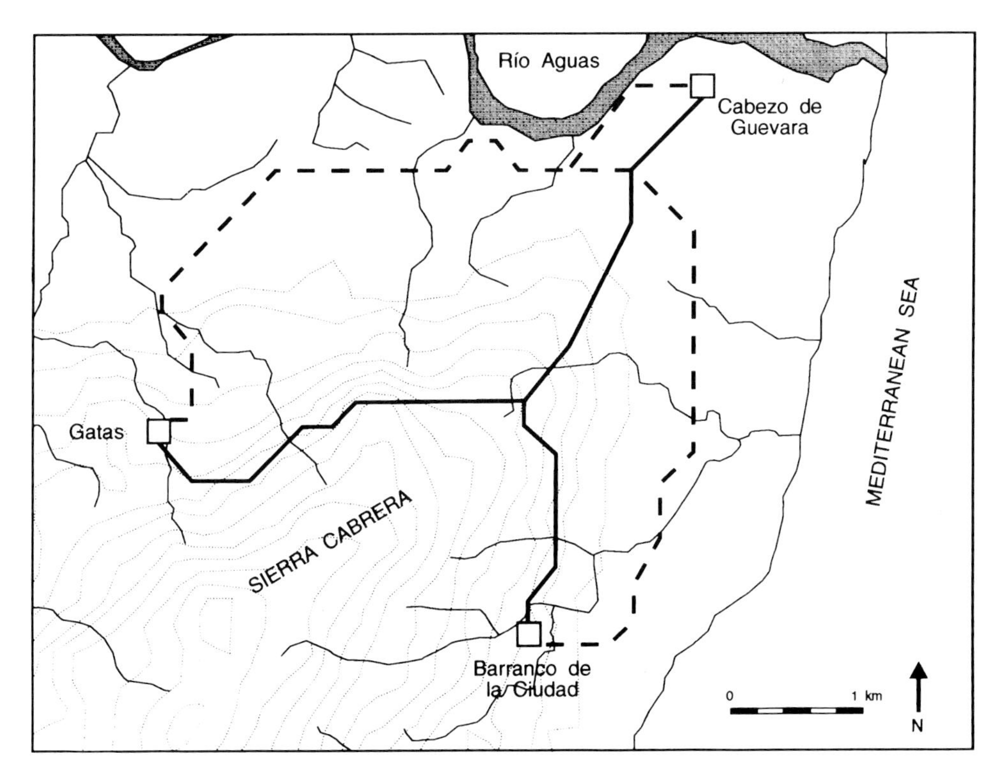

Figure 14.3 Least cost paths from Gatas to Barranco de la Ciudad and Cabezo de Guevara using gradient (dashed line) and elevation difference (black line) as friction factors.

upon the type of terrain and the position of the settlements in this terrain. Both methods however indicate that there was no direct link between Gatas and Barranco de la Ciudad, and that the site of Cabezo de Guevara must have played an important role in their relations.

It seems that the construction of least cost paths can be helpful in finding possible routes of transportation and produces more relevant results for distance calculations, even when used with a relatively crude DEM like the one here. In fact, any factor can be included in the friction map that is used, e.g. vegetation, the presence of water, or for that matter, the location of taboo or proscribed areas. Caution should be applied however with the interpretation of the paths; in fact this little exercise shows that the levels of representation, description and interpretation of GIS derived data are *all* needed in order to tackle our questions.

14.6. An explicative example

14.6.1 Introduction

We shall now attempt to situate some of these ideas with respect to a specific set of issues pertinent to the Vera Basin. Although we feel that it should be possible to apply the methods discussed here in any study of settlement territories, we will be focusing on the possibilities of reconstructing agricultural territories around settlements. One of the most urgent problems to address when reconstructing past land use is the extent of the territories used for farming. Methods like distance buffering and Thiessen polygons have often been used to create settlement territories, but these are not very satisfactory. Of the GIS methods available, cost surfaces made from slope or elevation maps offer the best possibility to obtain non-linear representations of the landscape (Wheatley, 1993). A good example of the application of cost surfaces to create radii of hours' walking is found in Gaffney and Stančič (1992). Although the approach is site-centred, just like the traditional distance buffering, and takes a rather arbitrary cut-off point, it gives a fairly good idea of the accessibility of the terrain around a settlement. We suggest that a different approach can be used by reversing the process and concentrating on the area around the sites first before determining the settlement hinterland. First we will try to find the zones where farming was actually possible, and then make an estimate of what portion was needed to sustain the population of the settlement.

For this approach two issues have to be tackled. First, a comprehensive system for land evaluation is needed from which estimates can be made concerning the productivity of soil types for different crops. Both qualitative and quantitative methods are available and will have to be tested (MacRae and Burnham, 1981) using ecological characteristics of the cultivated crops such as slope, relative humidity and salt tolerance. In south-east Spain, an area which has suffered severe erosion, present-day soil maps can only be used as a guideline for past agricultural potential, since their relationship to the palaeo-environment is exceedingly complex and potentially misleading. Data on soil erosion and a detailed study of palaeosols will be necessary to provide a more accurate picture of past soil types.

Second, demographic estimations are needed, in addition to information on diet, the distribution of resources and the socio-economic organization in general. Clearly, the greater our knowledge on these aspects, the greater will be our ability to generate plausible estimates of the amount of suitable land that is needed for food production for a given community.

14.6.2 Estimating agricultural potential

In order to assess the viability of this approach, we need to make a comparison between the available methods. First, we will consider two of the traditional approaches: distance buffering and walking time. Although the use of cost surfaces will give a more accurate representation of the settlement's surroundings, in level terrain the two methods coincide: a one-hour walk will approximately be the same as a distance buffer of 4 km. The maps of walking time can be created using a slope map in combination with distance buffering, and by applying a rule of thumb that assumes that a healthy person can walk 4 km per hour horizontally, and that each 400 m uphill or 800 m downhill will cost one hour extra. Taking the average of 600 m vertical movement per hour, the maps in Figure 14.4 were created. As we pointed out earlier, these maps cannot account for the effect of walking with the contour lines in mountainous areas. A simple comparison can be made between the 2 km radius from the descriptive analysis and these radii. The soil maps 1:100 000 from the Proyecto Lucdeme (1988–91) provide enough information to roughly assess the agricultural productivity of the area around Gatas, Barranco de la Ciudad and Cabezo de Guevara. Of the soil types mentioned in Table 14.1 and Table 14.2 the *Lithosols* offer very little potential for agricultural use. The best soils

Table 14.1 Distribution of soil types in ha within a two km buffer zone

	Gatas (ha)	Bco. de la Ciudad (ha)	Czo. de Guevara (ha)
Lithosols	405	754	229
Regosols	738	216	217
Vertisols	108	—	149
Xerosols	—	—	28
Solonchaks	—	—	257
Fluvisols	—	60	112
Total	1251	1030	992

Table 14.2 Distribution of soil types in ha within a one-hour walking zone

	Gatas (ha)	Bco. de la Ciudad (ha)	Czo. de Guevara (ha)
Lithosols	209	593	417
Regosols	951	140	598
Vertisols	128	—	264
Xerosols	—	—	454
Solonchaks	—	—	309
Fluvisols	4	68	313
Total	1292	801	2355

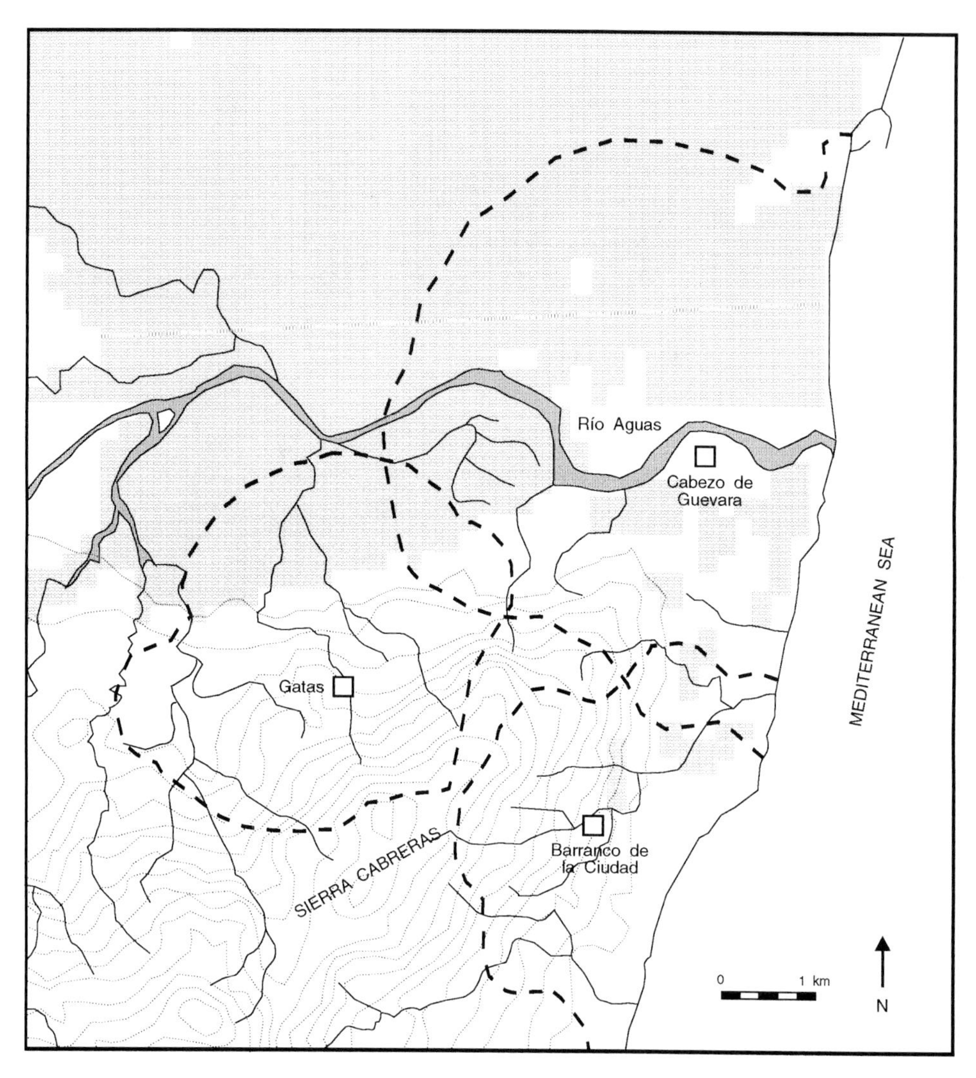

Figure 14.4 Radii of a one-hour walk from Gatas, Barranco de la Ciudad and Cabezo de Guevara. The shaded areas denote soils that have reasonable agricultural potential.

in the region are without any doubt the *Fluvisols* that nowadays are widely used for irrigated horticulture. In fact, irrigation in some areas has been so intense that locally these soils have suffered from salinization and turned into *Solonchaks*. The *Xerosols* are generally well suited for dry farming, while both *Regosols* and *Vertisols* are very much prone to erosion and are only used for terraced agriculture. However, it may well be that these soils have had a much higher agricultural potential in the past.

Both tables show us that the area around each of the three sites has a very different agricultural capability. The mountainous site of Barranco de la Ciudad is clearly dominated by *Lithosols*, so agriculture cannot have been of primary importance to its inhabitants. The site of Cabezo de Guevara shows a completely different picture: a wide diversity of soil types offers many possibilities for both irrigated and dry farming. The proximity of this site to the Río Aguas, however, suggests that the environmental conditions could have changed greatly through the existence of marshes or alternatively, coastline changes as Hoffmann (1988) has suggested for the Antas and Almanzora rivers.

Gatas seems to have an intermediate position with a large proportion of soils that now cannot be considered very well suited for farming, but may have been much better in the past. The main difference between the tables lies in the size of the territory and the relative proportion of the soil types. For Barranco de la Ciudad, the area available is considerably reduced, but the proportion of the soil types stays about the same. For Gatas, however, not much change in the extent of the hinterland is observed, but the amount of *Lithosols* is sharply diminished in favour of the *Regosols*. When we look at Cabezo de Guevara, we see that the catchment becomes 2.5 times as large, with a marked increase in *Xerosols*. The territory of Cabezo de Guevara not surprisingly now approaches a 4 km radius around the site, with the exception of the sea. The comparison shows that the use of this method has a marked effect on the descriptive level of the analysis and over-emphasizes the areas of easy access. It also shows that the nature of this effect is not only dependent upon the methods used, but also on the position of the settlement in the landscape and can give rise to a different interpretation of the site catchments (Wheatley, 1993).

Assuming, for example, that the Argaric settlements were heavily dependent on barley cultivation, at least in the final phase (Ruiz *et al*., 1992), we might consider Cabezo de Guevara to be the most important site, due to its high potential agricultural land, as opposed to Barranco de la Ciudad. Although neither of the two sites has been systematically excavated, if we take into account the total area of artefact deposition and the quality and quantity of archaeological material collected during intensive field survey, the contrary seems to be the case. Moreover, many of the most important Argaric settlements show no clear preference for areas with large amounts of first-class agricultural land.

14.6.3 Reconstructing the domain of food production

Now we shall attempt to make a hypothetical reconstruction of the domain of food production for the Argaric settlement of Gatas. The domain of food production is modelled as a system of mixed agro-silvo-pastoral land use, i.e. a natural type of landscape organization, incorporating cereal cropping, olives, oak groves and animal pasture as a diversified single system. Similar systems of natural land management (*dehesas*) have been identified from palynological remains for the period 4000 BC–AD 1900 for south-west Spain (Stevenson and Harrison, 1992), although as yet there is no reliable evidence for the south-east due to the problems of pollen sampling in semi-arid environments.

Our point of departure involves population estimates. The problems related to the estimation of prehistoric population size and demographic factors generally has been the subject of much debate within archaeology and anthropology (Hassan, 1981; Schacht, 1981). While acknowledging the complexity of this issue, it is not our intention to enter this debate, but rather to use existing population estimates as the basis for calculating a minimum potential area (hectarage) which would be needed to cope with the subsistence requirements for the Argaric settlement of Gatas.

We shall begin with a population estimate of *c*.300, calculated by Chapman (1990) on the basis of burial and settlement area. We can now consider the basic food needs for such a population by utilizing available data on caloric and protein requirements as suggested by World Health Organization statistics (WHO, 1974). Taking an average consumption level of 3000 calories per day, then the human needs for one year become *c*.1 million calories or 1 SNU (standard national unit). One kilogram of wheat or barley will mill down to 900 g of grain, which will produce *c*.3100 calories. This is equivalent to $\frac{1}{3}$ tonne and interestingly, correlates with Roman military rations which allowed each soldier $\frac{1}{3}$ tonne of corn per annum (Mercer, 1981). Using this figure as a minimum requirement per person, the hypothetical Gatas population of 300 would have required 90 tonnes of barley per year.

Yield estimates for prehistoric crops vary enormously depending on soil fertility and climate. In the absence of hard empirical data, most estimates have been based on analogies with classical and medieval sources or from experimental evidence (Gregg, 1988; Mercer, 1981). From these sources it is possible to generate an average yield for barley of between 1.05 and 1.85 tonnes/ha, with a mean of 1.45. Next we must allow for the effects of spillage, disease or rotting and this can reasonably be calculated at around 20 per cent. A similar reduction must be factored in our calculations to account for seed for next year's crop. We thus have reduced the original yield figures by around 40 per cent, giving a new range of 0.63–1.11 tonnes/ha (mean = 0.87). Comparing these estimates with present-day data on land used in the Vera Basin shows that yields for barley are between 0.8 and 0.9 tonnes/ha in favourable years (Ministerio de Agricultura, Pesca y Alimentación, 1982) but on less suited soils the yields may be as low as 0.3–0.5 tonnes/ha. The estimates mentioned before may therefore be too high for south-east Spain. Yields for cereals can however be as high as 1.0–1.5 tonnes/ha when irrigation is applied. We can therefore take two figures for our estimation of the hectarage needed: the original estimate with a mean of 0.87 when irrigation is applied, and the second one with a range between 0.18 and 0.54 tonnes/ha when irrigation is absent. The figure of 0.54 will be more appropriate for well-suited soils like the *Xerosols* and *Fluvisols*, whereas 0.18 tonnes/ha will be a valid estimate for areas with abundant *Regosols*.

Applying these figures to our Gatas example, we can estimate the total hectarage needs, with irrigation, as ranging between 81.1 ha and 142.0 ha with a mean of 103.5 ha. An important assumption we will make is that not all arable land was cultivated. Probably one-third was left fallow, thus we need to increase our hypothetical hectarage at least by this figure, giving new values between 185.6 and 105.4 ha (mean = 134.6). Non-irrigated agricultural estimates give much larger values ranging from 221.7 to 665.0 ha. The subsistence system we are dealing with was not, of course, dependent on cereal crops alone. The available faunal evidence from Argaric sites in the Vera Basin show that both sheep and goats formed a prominent feature of the economy. Our food production domain, therefore, needs to accommodate the additional spatial requirements for these animals. It is known that sheep and goats need *c*.0.7 ha of pasture per animal, thus if we suggest for the settlement of Gatas a mixed flock of 50 sheep and goats, then we need around 35 ha of browse and pasture to accommodate them.

From Table 14.2 it is clear that the amount of arable land within one hour's walking from Gatas is more than sufficient, even when taking the most conservative estimate of 665 ha. This arable land is largely constituted by *Regosols* and *Vertisols*, soil types that are not very well suited for dry farming with yield figures under

0.6 tonnes/ha that do not lend themselves to easy irrigation. When we look at the 1982 land-use map, we see that downhill from Gatas cereals are grown in some places, but only 21 per cent of these areas are actually cultivated and the rest is left fallow. There are, however, some indications that in the past the situation may have been more favourable. Inside the area with *Vertisols* and *Regosols*, patches of *Cambisols* and *Xerosols* are found, soil types that are much more suited for growing cereals. These areas are very much prone to erosion due to the considerable gradient and the unconsolidated material that underlies the soils, so we can assume that the *Cambisols* and *Xerosols* have been much more common in the past, especially in the lower portions of the area. The agricultural potential of the area may have been much larger in the Argaric period. Furthermore, the area around Gatas offers almost unlimited possibilities for goats and sheep. With this knowledge we can construct a map of the hypothetical food production domain of Gatas (Figure 14.5) using the estimate of 221.6 ha for agriculture, located downhill, and 35 ha for livestock uphill. This map can be used as the basis for a dynamic modelling system.

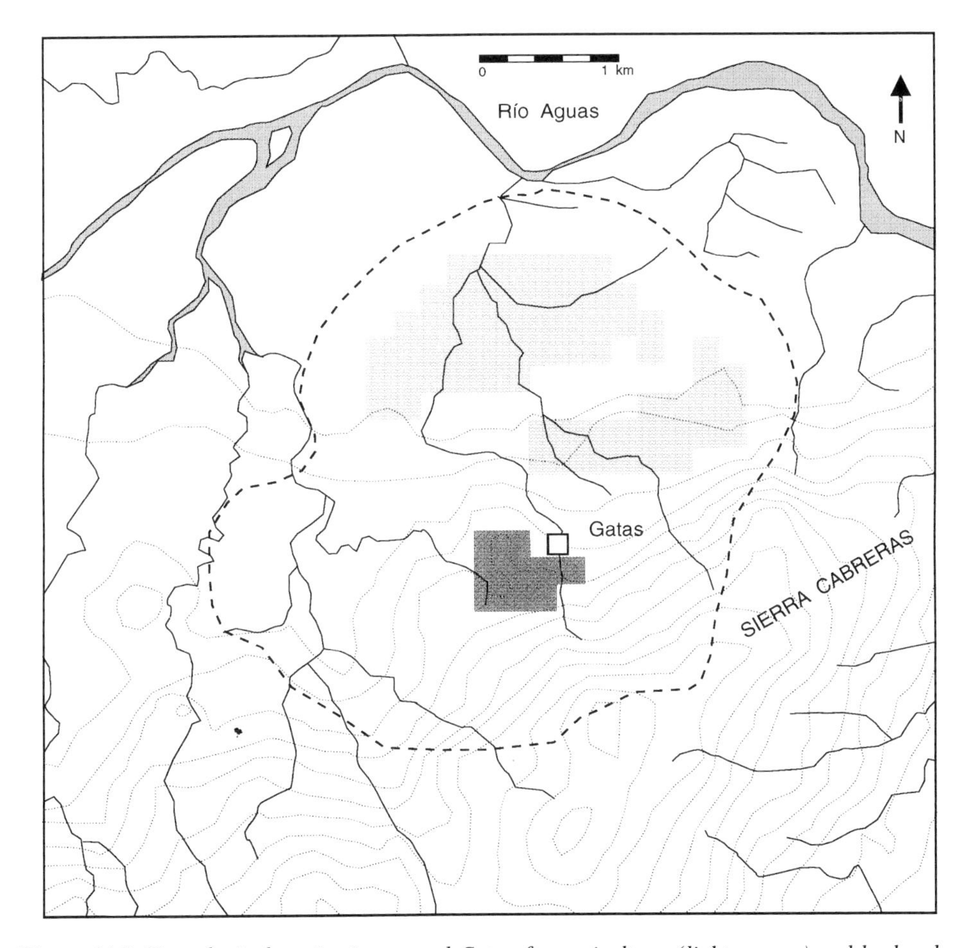

Figure 14.5 Hypothetical territories around Gatas for agriculture (lighter grey) and husbandry (darker grey). The radius of one hour's walking is given for reference.

The GIS offers the possibility to experiment with a number of parametric estimates for our subsistence model, and can provide a representation of possible settlement territories through time. Population changes, erosion or climatic factors can be introduced into the simulation to test some of the hypothetical dynamics concerning the socio-natural evolution of the Vera Basin. The GIS also offers tools for spatializing palynological and other palaeo-ecological data, and the interdisciplinary collaboration which this implies is a prerequisite for a better understanding of spatial interactions. Thus, it is this third level in our original scheme, i.e. the interpretive/analytical one, that can reach further levels of sophistication, demonstrating the fruitful interaction of GIS and dynamic modelling for future archaeological research.

14.7 GIS: towards an integrated model of social space

What the foregoing model building and discussion has shown, is that there is a great deal that GIS methods can contribute by way of demonstrating the utility of an interrogative multiscalar framework. The iterative nature of this model allows formal data to interact with less formal, qualitative knowledge and provides a format within which a variety of social, historical and environmental questions can be posed. A critical dimension in this process is that the model structure is non-hierarchial and pluralistic, such that there is no 'privileged' hierarchical classification of variables. It thus facilitates the construction and reconstruction of a wide variety of possible model, depending on the specific problem or interrogative process desired.

Perhaps one of the most interesting aspects of this scheme, which is still being developed, is the way in which the GIS technology provides statistical and descriptive data which can form the basis for dynamic simulation modelling of prehistoric subsistence practices. While this novel, interactive capacity is still under development, the success of current preliminary models based on data from the ARCHAEOMEDES project in the Vera Basin is encouraging, particularly in the integration of a temporal dimension; it thus suggests a way out of the oft quoted 'static' impasse which GIS is often accused of inhabiting (e.g. Castleford, 1992). Clearly there is much that is yet to be done if we are to move towards a true spatio-temporal GIS: one in which space is not simply a container of events, but the locus of socially defined practices and perceptions.

By way of summary, perhaps the primary implication of the foregoing discussion is the need for a conceptual 'retooling' if we are to realize the potential of GIS not simply as a method of *representation*, but rather as a means of generating sophisticated spatio-temporal *interpretive* models. In this way, archaeology can move to a new level of integrated modelling driven by GIS, and thus its goal of coming to a more complete understanding of the long-term development of human-modified environments, can potentially be realized.

Acknowledgements

The authors would like to acknowledge that part of this work was done in the framework of the ARCHAEOMEDES Project, funded by DG XII of the European

Commission as part of the Environment (Desertification) Programme under contract EV5V-0021. They would also like to acknowledge the co-operation of the team of researchers involved in the Gatas Project (Almería, south-east Spain) as well as colleagues in the ARCHAEOMEDES Project for their help in the preparation of this chapter.

References

Allen, K. M. S., Green, S. W. and Zubrow, E. B. W., 1990, *Interpreting Space: GIS and archaeology*, London: Taylor & Francis.

Brandt, R. W., Groenewoudt, B. J. and Kvamme, K. L., 1992, An experiment in archaeological site location: modelling in the Netherlands using GIS techniques, *World Archaeology*, **24**, 268–82.

Buikstra, J., Castro Martínez, P., Chapman, R. W., González Marcén, P. Lull, V., Picazo, M., Risch, R. and Sanahuja-Yll, M. E. 1989, Proyecto Gatas, Fase II, *Anuario Arqueológico de Andalucía*, **1**, 214–18.

Buikstra, J., Castro Martínez, P., Chapman, R. W., González Marcén, P., Hoshower, L., Lull, V., Mico, R., Picazo, M., Risch, R., Ruiz, M. and Sanahuja-Yll, M. E., 1992, Approaches to class inequalities in the late prehistory of South-East Iberia: The Gatas Project, in Lillios, K. L. (Ed.) *The Origin of Complex Societies in Late Prehistoric Iberia*, International Monographs in Prehistory, University of Michigan.

Burrough, P. A., 1990, Methods of spatial analysis, *International Journal of Geographic Information Systems*, **4**(3), 221–3.

Carmichael, D. L., 1990, GIS predictive modelling and archaeological site location: a case study in the Midwest, in Allen, K. M. S., Green, S. W. and Zubrow, E. B. W. (Eds), *Interpreting Space: GIS and Archaeology*, pp. 73–9, London: Taylor & Francis.

Castleford, J., 1992, Archaeology, GIS and the time dimension: an overview, in Lock, G. and Moffet, J. (Eds), *Computer Applications and Quantitative Methods in Archaeology 1991*, pp. 95–106, BAR International Series S577, Oxford.

Castro Martínez, P., Chapman, R. W., González Marcén, P., Lull, V., Picazo, M., Risch, R. and Sanahuja-Yll, M. E., 1989, Informe preliminar de la tercera campaña de excavaciones en el yacimiento de Gatas (Turre, Almería), *Anuario Arqueológico de Andalucía*, **2**, 219–26.

Castro Martínez, P., Colomer, E., Gili, S., González Marcén, P., Lull, V., Mico, R., Montón, S., Picazo, M., Rihuete, C., Risch, R., Ruiz Parra, M., Sanahuja-Yll, M. E. and Tenas, M., 1993, Proyecto Gatas: sociedad y economía en el sudeste de España *c.* 2500–800 a. n. e., in *Investigaciones Arqueológicas en Andalucía, 1985–1992 Proyectos*, pp. 401–16, Huelva, Consejeria de Cultura y Medio Ambiente de la Junta de Andalucía.

Chapman, R. W., 1978, The evidence for prehistoric water control in south-east Spain, *Journal of Arid Environments*, **1**, 263–74.

Chapman, R. W., 1990, *The Emergence of Complexity: the later prehistory of south-east Spain, Iberia and the west Mediterranean*, Cambridge: Cambridge University Press.

Chapman, R. W., Lull, V., Picazo, M. and Sanahuja, M. E., 1987, *Proyecto Gatas:*

sociedad y economía en el sudeste de España c. 2500–800 a. n. e., BAR International Series 348, Oxford.

Commission of the European Communities, 1992, *CORINE soil erosion risk and important land resources in the southern regions of the European Community*, Luxembourg: Office for Official Publications of the European Communities.

Crumley, C. L. and Marquardt, W. H., 1990, Landscape: a unifying concept in regional analysis, in Allen, K. M. S., Green, S. W. and Zubrow, E. B. W. (Eds), *Interpreting Space: GIS and Archaeology*, pp. 73–9, London: Taylor & Francis.

Gaffney, V. and Stančič, Z., 1992, Diodorus Siculus and the island of Hvar, Dalmatia: testing the text with GIS, in Lock, G. and Moffet, J. (Eds), *Computer Applications and Quantitative Methods in Archaeology 1991*, pp. 113–26, BAR International Series S5776, Oxford.

Gili, S., Lull, V. and Risch, R., 1993, The archaeology of the Vera Basin from the early Holocene to the present: a first approach to the socio-natural dynamics, in Van der Leeuw, S. E.(Ed.), *ARCHAEOMEDES First Year Report*, Preliminary report presented to the Directorate General XII of the Commission of the European Communities, Cambridge.

Gilman, A. and Thornes, J. B., 1985, *Land Use and Prehistory in South-East Spain*, London: Allen and Unwin.

González Marcén, P., 1991, *Cronología del Grupo Argárico: Ensaye de falsificación radiométrica a partir de curvas de calibración de alta precisión*, Tesis doctoral, Bellaterra: Universitat Autònoma de Barcelona.

Gregg, S. A., 1988, *Foragers and Farmers*, Chicago: University of Chicago Press.

Hassan, F. A., 1981, *Demographic archaeology*, New York: Academic Press.

Hoffman, G., 1988, *Holozänstratigraphie und Küstenverlagerung an der Andalusischen Mittelmeerküste*, Bremen: Berichte aus dem Fachbereich Geowissenschaften der Universität Bremen.

Kvamme, K. L., 1989, Geographic information systems in regional archaeological research and data management, in Schiffer, M. B. (Ed.), *Archaeological Method and Theory, 1*, pp. 139–203, Tucson: University of Arizona Press.

Lefebre, H., 1974, *La production de l'Espace*, Paris.

Lull, V., 1983, *La 'cultura' de El Argar: un modelo para el estudio de las formaciones económico-sociales prehistóricas*, Madrid: Akal.

Lull, V. and Estevez, J., 1986, Propuesta metodológica para el estudio de las necrópolis argáricas, in *Homenaje a Luís Siret 1934–1984*, pp. 441–52, Seville.

MacRae, S. G. and Burnham, C. P., 1981, *Land Evaluation*, Oxford: Clarendon Press.

Madry, S. L. H. and Crumley, C. L., 1990, An application of remote sensing and GIS in a regional archaeological settlement pattern analysis: the Arroux River valley, Burgundy, France, in Allen, K. M. S., Green, S. W. and Zubrow, E. B. W. (Eds), *Interpreting Space: GIS and Archaeology*, pp. 365–80, London: Taylor & Francis.

Mathers, C., 1984, Beyond the gave: the context and wider implications of morturary practice in south-eastern Spain, in Blagg, T. F. C., Jones, R. F. J. and Keay, S. J. (Eds), *Papers in Iberian Archaeology*, pp. 13–46, BAR International Series 193(1), Oxford.

McGlade, J., 1993a, Modelling structural change in the later prehistory of the Aisne Valley: evolutionary dynamics and human–environment interaction, in *ERA 12, CNRS, ATP Grands Projets Régionaux*, Paris.

McGlade, J., 1993b, The dynamics of change in the human-modified environments of the Vera Basin, south-east Spain, in Van der Leeuw, S. E. (Ed.), *ARCHAEOMEDES First Year Report*, Preliminary report presented to the Directorate General XII of the Commission of the European Communities, Cambridge.

McGlade. J., 1993c, 'Archeology and human ecodynamics: the interpretation of culturally modified lanscapes', Presentation at the Theoretical Archaeology Group (TAG) Conference, Durham, December 1993.

McGlade, J. and van der Leeuw, S., 1994, Archeology and non-linear dynamics: new approaches to long-term change, in Van der Leeuw, S. E. and McGlade, J. (Eds), *Dynamical modelling and the Archaeology of Change*, London: Routledge.

Mercer, R., 1981, *Farming Practice in British Prehistory*, Edinburgh: Edinburgh University Press.

Ministerio de Agricultura, Pesca y Alimentación, 1982, *Mapa de cultivos y aprovechamientos*, Vera, Garrucha, Sorbas/Mojácar, Madrid: Dirección General de la Producción Agraria.

Picazo, M., 1993, 'Hearth and home: the domain of maintenance activities', presentation at the Theoretical Archeology Group (TAG) Conference, Durham, December 1993.

Proyecto Lucdeme, 1988–1991, *Mapa de Suelos, escala 1:100 000*, Sorbas-1031, Vera-1014, Garrucha-Mojácar, pp. 1015–1032. Madrid: ICONA–Ministerio de Agricultura, Pesca y Alimentación.

Ruiz, M., Risch, R., González Marcén, P., Lull, V. and Chapman, R., 1992, Environmental exploitation and social structure in prehistoric southeast Spain, *Journal of Mediterranean Archaeology* **5**(1), 3–38.

Schacht, R. M., 1981, Estimating past population trends, *Annual Review of Anthropology*, **10**, 119–40.

Shanks, M. and Tilley C., 1987, *Social Theory and Archaeology*, Cambridge: Cambridge University Press.

Soja, E., 1980, The socio-spatial dialectic, *Annals of the Association of American Geographers*, **70**, 207–25.

Stevenson, A. C. and Harrison, R. J., Ancient forests in Spain: a model for land use and dry forest management in south-west Spain from 4000 BC to AD 1900, *Proceedings of the Prehistoric Society*, **58**, 227–47.

Terradas, J., 1987, La resposta dels ecosistemes a les pertorbacions en el context de la teoria ecologica, *Quaderns d'Ecologia Aplicada*, **10**, 11–26.

Touraine, 1977, *The self-production of society*, Chicago: University of Chicago Press.

van Leusen, P. M., 1993, Cartographic modelling in a cell-based GIS, in Andresen, J., Madsen, T. and Scollar, I. (Eds), *Computing the Past, Computer Applications and Quantitative Methods in Archaeology, CAA92*, pp. 105–124, Aarhus: Aarhus University Press.

Wheatley, D., 1993, Going over old ground: GIS, archaeology and the theory of perception, in Andresen, J., Madsen, T. and Scollar, I. (Eds), *Computing the Past, Computer Applications and Quantitative Methods in Archaeology, CAA92*, pp. 133–7, Aarhus: Aarhus University Press.

WHO, 1974, *Handbook on human nutritional requirements*, WHO Monograph Series 61, Geneva: World Health Organization.

15

The impact of GIS on archaeology: a personal perspective

V. Gaffney, Z. Stančič and H. Watson

15.1 The archaeological context of GIS

GIS has already become a standard tool for handling spatial archaeological data (Kvamme, 1989). The allure of the technique is not hard to fathom; archaeological data are often most conveniently stored in map format and the frequent recourse to mapped displays to explain and interpret archaeological material makes GIS one of the more flexible and comprehensible analytical tools available to the archaeologist. Moreover, as more archaeologists use GIS they begin to understand that the technique allows us to handle vast amounts of complex data in a manner that was previously unimaginable. However, in establishing such a profitable relationship there are certain dangers for the unwary. At the simplest level, we should note the alluring visual products of GIS and suggest that there is a danger that archaeologists may be distracted by such outputs to the extent that aesthetics may dominate interpretation. The ultimate aim of the research or the quality of the data for such analysis may be forgotten. These are, however, the pitfalls of a novel technique.

More significant than this is the manner in which the most attractive qualities of GIS technology may mould archaeological thought and practice in a less than desirable manner. If the strength of GIS lies in its ability to analyse mapped data, the archaeological potential of GIS is related to the types of mapped data available, or thought suitable for such purposes. Topographic, environmental and palaeoenvironmental data have to date proved particularly amenable to archaeological GIS analysis. Such variables can be conveniently mapped and measured and have been assessed against archaeological data to provide information pertinent to a variety of problems or models (Allen *et al.*, 1990). However, the aptitude of GIS to analyse such data and the relative sophistication of the results do not legitimize such pursuits archaeologically. Indeed, there are good reasons to suggest that the application of GIS techniques in such a way could ultimately prove to be restrictive to the general development of archaeological thought. In its least harmful form, the indiscriminate use of GIS solely in conjunction with mapped physical data may result in the slick, but repetitious, confirmation of otherwise obvious relationships. In the worst case, it might involve the unwitting exposition of an environmentally or functionally determinist analytical viewpoint of a type which has largely been rejected by the archaeological community (Wheatley, 1993). There is a need, therefore, to explore the archaeological context for GIS applications. Initial inspection suggests that the context of GIS applications within archaeological research has, on the whole, been related to the physical analysis of landscape settlement traces. This is a well-established analytical approach (Clarke, 1968). Following trends largely instigated within geography, archaeology has held a consensus since the 1960s that the archaeological landscape is a structured phenomenon. However, the quantifiable data central to such analysis during the 1960s and 1970s largely resulted in an emphasis on abstract pattern location and economic analysis. The works of the Cambridge palaeo-economists (Higgs, 1972) and publications including Hodder and Orton's (1976) *Spatial Analysis in Archaeology* suggest themselves as good examples of such trends.

However, the acknowledgement of the relative sterility of an abstract or purely quantitative approach to the study of human landscapes occurred early within geography (Taffe, 1974) and an emphasis on the need to incorporate a cultural perspective within analytical approaches rapidly followed (Butzer, 1977). Archaeology has followed this trend with the acknowledgement that 'there is always a cultural geography' (Binford, 1982:7). Cultural and belief systems are increasingly interpreted as being able to order the physical environment within absolute limits (Drennan, 1976; Fletcher, 1977). The need to develop 'cognized models' which incorporate the belief systems and perceptions of past societies has become an imperative (Renfrew, 1982).

In attempting such models we are encouraged by Renfrew's (1982:11) suggestion that 'if people's actions are systematically patterned by their beliefs, the patterning (if not their beliefs, as such) can become embodied in the archaeological record'. The suggestion that cognitive information on the way communities perceive and interpret their environment should be patterned, indicates that such qualities will be measurable and potentially mappable. Today, work on the symbolic properties of material culture, within both archaeology and anthropology, allows some degree of faith in the use of archaeological data for such purposes (Douglas, 1966; Hodder, 1982a; 1982b).

Despite this, some difficulties remain. It is clear that most studies still view artefacts or the relationships between artefacts as the primary channels of symbolic communication. Yet this is only a partial solution. Communities and cognitive systems operate at a number of levels and the need to order the larger environment presents situations in which perception and belief can transcend material culture and incorporate the natural. Archaeology with its reliance upon man-made objects has not necessarily been well placed to investigate such situations. From sacred groves to 'high places', natural features may have an imbued value without needing any concomitant material association, whilst the relationship of paths, tracks and natural markers in the environment suggest that natural and cultural order are not only hard to distinguish (Renfrew, 1982:21), but that the distinction may be artificial and misleading. There is a danger, therefore, in accepting the works of man alone as an adequate symbolic metaphor.

This observation is particularly clear when we consider the context of specific types of monuments. Prehistoric barrows, for instance, may originally have commanded a greater visual impact on the landscape by virtue of their striking appearance, or placed to dominate natural routes or viewpoints (Babić 1984:38–9). However, the difficulties of incorporating such observations coherently into the analytical process are considerable. Evans (1985:83) has suggested that most traditional forms of mapping are unsuitable for such investigations. They involve the reduction of place and space to location and distance, modern abstractions which lose their cultural and cognitive context.

However, most GIS have the capability to overcome some of these difficulties. According to the availability of data, GIS provides access to all the landscape, irrespective of the presence or absence of archaeological sites. Consequently, not only are we presented with the ability to investigate the positive and negative characteristics of archaeological distributions, if we wish, we can explore the larger relationship of archaeological and natural space. There is a more important point. Although earlier discussion has tended to accredit GIS with a powerful ability to analyse space, it it also correct to say that GIS may manipulate space according to variable, imposed values. We can assign value to space which can then be interactively analysed with other forms of mapped data. The simplest example of such a situation is the use of GIS to explore access and movement via least-cost modules. In using such modules, ease of movement across a surface becomes a measurable quantity. However, the use of such techniques in the context of a variety of landscapes or by adjustment to reflect different transportation systems—pantechnicon or pack horses—suggests that distance can be viewed through GIS as a relative, not a constant, variable. GIS is not, under such circumstances, to be considered as an objective observer of patterns implicit within spatial data; rather, it is a tool to create spatial relationships according to values we regard as important.

Some of this potential can be illustrated through two studies carried out by the authors of this chapter. The first relates to a group of settlement and stone cairn data from the island of Hvar in Dalmatia (Gaffney and Stančič, 1991). When published, only a relatively small number of economic models and databases were utilized and the archaeological context of the data was not fully explored. It is therefore worth reconsidering the results within a more explicitly archaeological context in order to explore the implications for further GIS research. The second study is intimately connected with GIS approaches to cognitive analysis and is an interim report on a study of prehistoric rock art carried out to complement recent analysis and field work in mid Argyll, southern Scotland.

15.2 GIS and archaeology on Hvar

The island of Hvar lies off the coast of central Dalmatia, Croatia (Figure 15.1). Its peculiar elongated shape—it is about 68 km long and nowhere more than 15 km wide—results from its origin as an anticlinal peak. The east of the island is a bevelled upland plain containing a series of intermittent fertile basins. The north central section of the island is dominated by a secondary trough which forms the Stari Grad plain, the largest fertile area on the island. Important geological deposits include the Quaternary deposits which occur within the Stari Grad plain and the intermittent outcrops of fertile Eocene flysch which occurs along the south coast of the island, both of which are relatively fertile and well-watered zones.

The period under consideration here is the Bronze Age/early Iron Age (*c*.2200–400 BC). During the early Bronze Age the evidence for human activity is limited and is primarily linked to the appearance of a series of cist burials under tumuli which have been compared with graves from the Cetina culture (Marović, 1985; Marović and Čović, 1983). During the later Bronze Age, inhumation burial under a mound

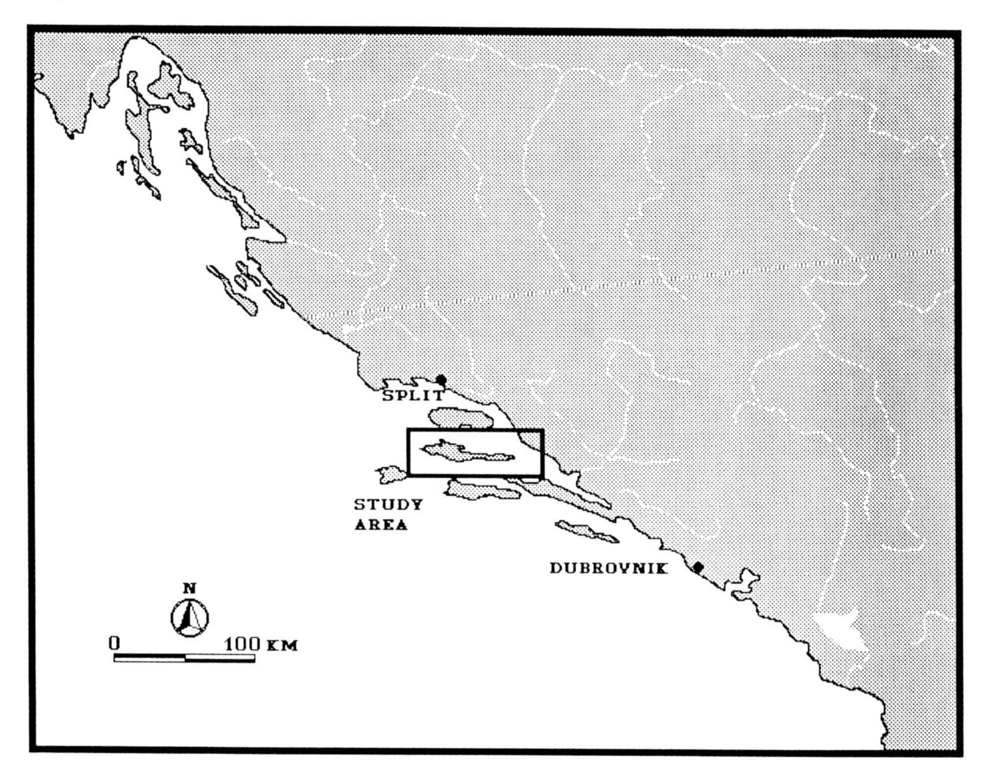

Figure 15.1 Location map of the island of Hvar.

continues in the area. However, there is increasing evidence for permanent settlement in the form of a series of defended enclosures. These vary in size and probably in function. Some are very large and, if interpreted as settlements, others are smaller and are often represented by little more than a small rampart sealing off a mountainous spur. Other habitation evidence is rare, although a series of pottery scatters may represent the remains of unenclosed settlement (Figure 15.2).

The qualitative evidence for Iron Age occupation on Hvar is essentially an extension of that of the late Bronze Age. Defended sites are the principal settlement form, whilst burials continue to occur within tumuli. However, there are reasons to suggest that at least one large site on the island, Hvar Castle, was incorporated within a series of long-distance exchange networks. The finds of eighth century BC Apullian pottery from this site indicate contact with Italy, while the discovery of amber within nearby tumuli excavated during the nineteenth century indicate further flung trade contacts (Petrić, 1979; 1980; 1986).

The initial published analysis of the Bronze/Iron Age data was carried out primarily as a technical exercise in order to illustrate how GIS could be used to establish the relationship between the principal prehistoric sites on the island and their economic basis. In order to achieve this, an investigation of the economic catchments of the largest defended sites on the island was undertaken on the assumption that these sites represented the apex of the settlement hierarchy. Cost surface catchments were constructed for each hillfort using a timed and measured

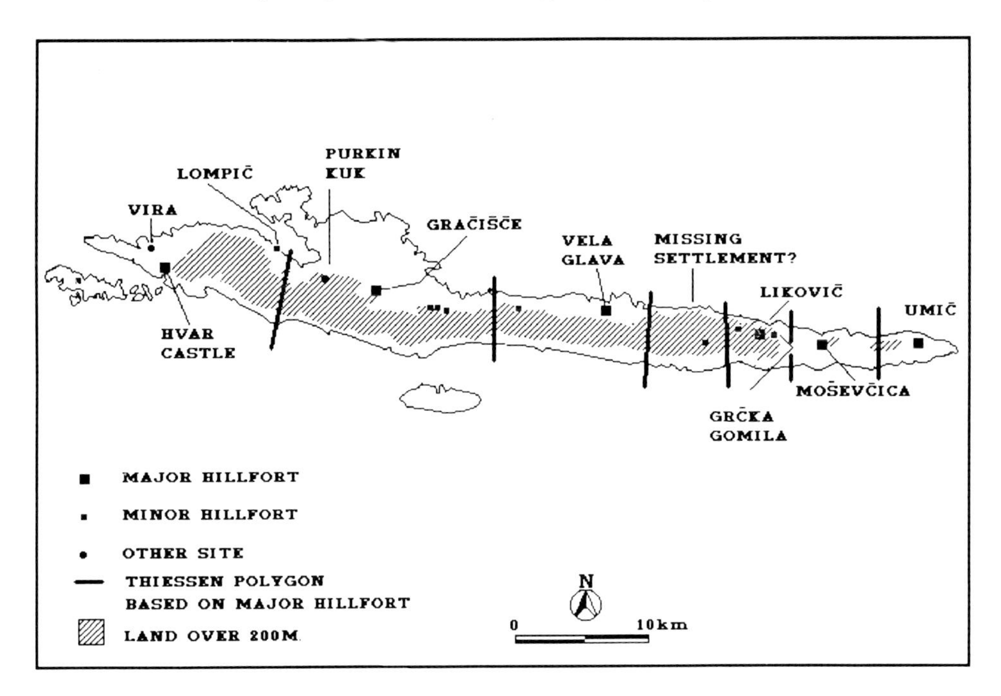

Figure 15.2 The principal Bronze Age and Iron Age sites on Hvar.

journey across the Stari Grad plain as a calibration factor. The details of this process are published elsewhere and need not be repeated here (Gaffney and Stančič, 1991). However, the results suggested that these sites might be interpreted as the central places of small prehistoric communities and that they were situated in order to control large expanses of fertile land (Figure 15.3).

A second supporting analysis was carried out to correlate the distribution of stone tumuli on the island with fertile soils. The rationale behind this lay in the observation that the stone cairns frequently associated with burial could also be interpreted as agricultural features. Excavation of such tumuli on the Dalmatian karst often provides no evidence for burial (Chapman *et al.*, 1987), whilst the nature of the karst is such that stone clearance is a normal by-product of arable agriculture (Gams, 1987). As such, the positive correlation of the prehistoric cairn distribution with fertile soils illustrated by the analysis was interpreted as further evidence for the agricultural basis of the prehistoric communities on Hvar. This supported the interpretation of the economic rationale behind the siting of the major enclosures and cairns on the island (Figure 15.4).

Within the terms of the original publication, the results as summarized above were satisfactory. They demonstrated the utility of GIS analytical techniques and, at a technical level, they suggested that cost surface catchments were a more useful guide to land associated with past settlements than catchments produced in a more traditional manner. The emphasis on the role of arable agriculture within such communities was a welcome counter-balance to the frequent assertion, deriving from later classical sources, that pastoralism was the dominant economic mode during these periods. The isolation of a hierarchy of settlements topped by a series of larger

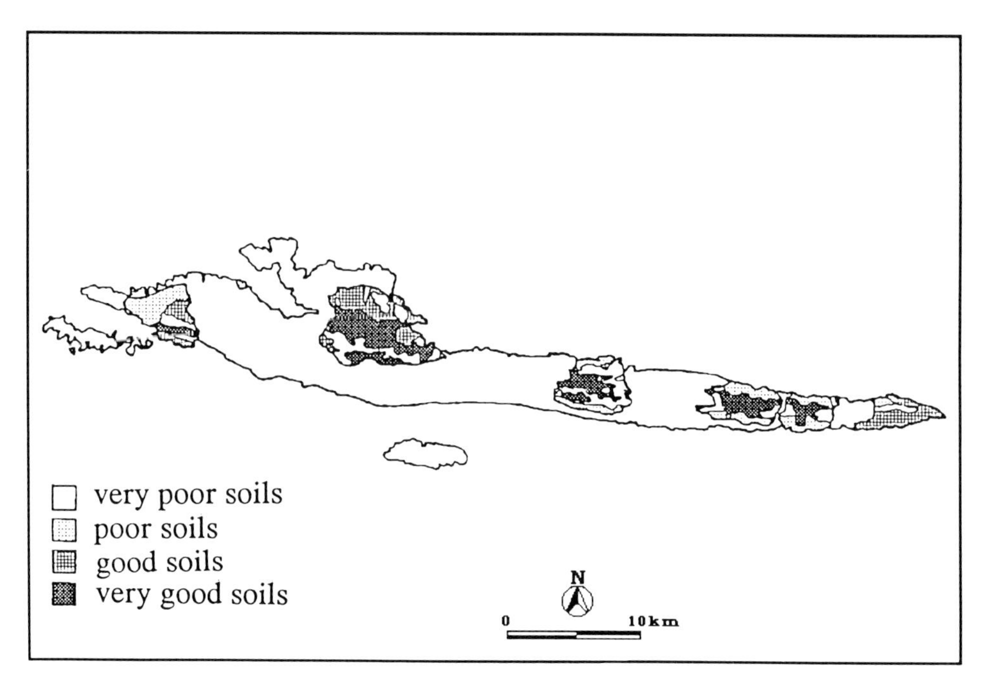

Figure 15.3 Soils within major hillfort catchments.

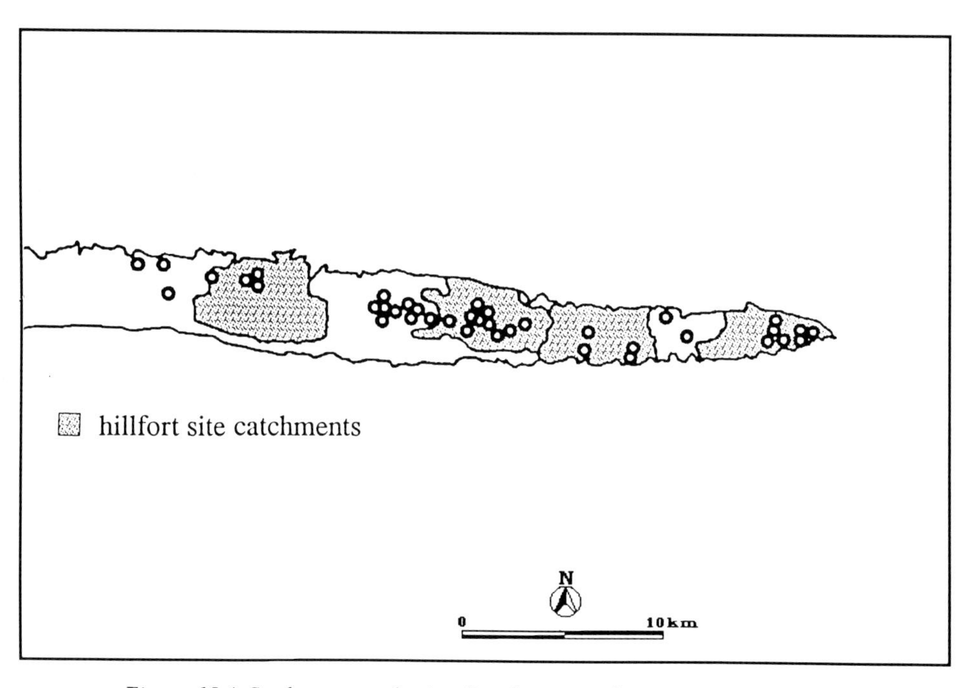

Figure 15.4 Settlements and cairn distribution in the eastern part of Hvar.

enclosures sited to dominate strategic portions of fertile land is also a useful observation.

Although it may be suggested that such results are descriptive rather than explanatory, it is possible to rationalize the emergence of such hierarchies from an economic viewpoint using this evidence. The karst is a fragile environment. It is prone to serious soil erosion when its delicate equilibrium is disturbed by agriculture. Recent work in northern Dalmatia suggests that significant erosion, almost certainly the result of the onset of agriculture, was underway from the late Bronze Age at the latest (Chapman *et al.*, 1987; Shiel and Chapman, 1988). Within such a context, it is no great step to suggest that the threat to critical agricultural resources posed by erosion was met by an increased emphasis on the possession of land and that this is reflected in the construction of defensive enclosures sited to dominate fertile areas.

Presented in this manner, the results represent an essentially environmentally deterministic approach to the data—a straightforward relationship between settlement, economy and society. Although pleasingly supported by GIS analysis, this interpretation is not satisfactory. In suggesting that pressure on land led to the emergence of a hierarchical society, we are side-stepping the problem of isolating the belief systems that permitted such social divisions to develop and be maintained. An alternative explanation can be suggested using the evidence included in the original GIS analysis and other information that we have on belief systems within the prehistoric communities on Hvar.

Key to such an approach is the relationship between agriculture and the dead. It has already been noted that stone cairns, although normally interpreted as funereal monuments, frequently provide no trace of burial and have been interpreted as agricultural clearance cairns. However, it is significant that there is no clear distinction between cairns of apparently different functions. Cairns with burials frequently occur within fertile agricultural zones whilst mounds with no trace of any burial also occur within groups of cairns which clearly are not related to agriculture and are more overtly 'ritual'. The most striking example of the later is at Vira where a cairn group containing 22 mounds is centred on a small peninsular. The placement of the mounds in this group has no agricultural rationale. Indeed, it can be shown that they are sited in order to be intervisible with a single small mound on the central Vira peninsular. Despite this, excavation on the mounds since the nineteenth century has indicated that the majority of the cairns contained no evidence of grave construction. Several of the mounds which did provide evidence for grave cists were otherwise empty, with no traces of bodies being recovered (Petrić, 1979:70). The frequency of anomalous mounds on Hvar and in Dalmatia in general, either as graves devoid of bodies or empty mounds in situations where funereal evidence could have been expected, suggests that in some cases the mound itself was important and the physical presence of a body might be of secondary consequence.

A second related observation is concerned with the evidence of ritual associated with these mounds. Although most published emphasis is placed on the evidence for formal grave goods, the most frequent finds during excavation are simple pottery sherds which may be incorporated within the body of the cairns, or deposited within grave cists. This practice occurs throughout the Bronze and Iron Ages (Marović, 1985; Novak, 1959). The quantities of sherds within mounds may vary from a few individual pieces to large numbers representing more than one hundred vessels (Batović and Kukoč, 1987). Shells, gravel and scraped up earth are also noted as occasional deposits within cists. In many cases, the deposits suggest themselves as domestic refuse. The incorporation of such material within mounds suggests the presence of rituals concerned with cleansing and purity associated with monuments related to the dead. The close relationship between death and purification and the maintenance of fertility through rituals associated with these concepts is often stressed in anthropological studies (Hodder, 1982a). In the case of Hvar, the ambiguous relationship between cairns relating to burial and agricultural clearance, both spatially and in respect to associated rites, makes analogy with such situations particularly attractive.

Treated in isolation, the linkage between mounds, burials, agricultural land and fertility suggests a concern with the legitimation of the control of valuable agricultural resources through the ancestors and the dead. The existence of dummy tumuli and empty graves emphasizes how important such contacts may have been considered. There is, however, a broader context for such an interpretation. Nearly all the 'top ranking' enclosures on the island are also associated with stone cairns of exceptional size. The precise situation of these cairns varies. At the hillfort of Vela Glava, a massive cairn measuring 36 m in length and 4.2 m in height is situated within the enclosure. The mound at Purkin Kuk, the largest on the island at approximately 43 m in length and 5 m high, lies within the probable territory of the Graèišèe hillfort and high on a hill overlooking the Stari Grad plain. The enclosure at Moševćica is composed of three ramparts cutting across a spur. The central rampart incorporates a large tumuli about 5 m in height. At Liković, there is a pair of large enclosures, only one of which is interpreted as a settlement. The second non-settlement enclosure, Grčka Gomila, contains a massively enlarged mound, 6 m in height within its defences. Despite the enlargement of the rampart of Grčka Gomila, its construction close to a hill of approximately the same height, suggests that defence was not the principal function of the enclosure. The only exception to this pattern is the site at Hvar Castle. It, however, has been almost totally destroyed as a result of the medieval and post-medieval settlement of the town of Hvar, but it is significant that the site is linked with the Vira barrow cemetery described above, an exceptional ritual monument on the island.

Such mounds are not uncommon elsewhere in Dalmatia and western Bosnia. A number of interpretations have been mooted for their function, from towers to temples for Sylvanic cults (Benac, 1986). Excavations have taken place on these mounds on Hvar. Unfortunately, the size of the monuments and the small scale of past work has been such that the results cannot be taken as definitive. However, it is interesting that no burials have been recorded in any of these mounds so far, although finds of domestic refuse are frequent.

Consequently, although there is a degree of variation between the situations of individual monuments, the cairns large and small form a coherent group, whilst a morphology and associated rituals link cairns and settlements. This suggests a common concern with legitimation and the maintenance of fertility. However, the scale of the larger mounds sets them apart and their siting suggests that some elements of the rituals associated with them may have had a different emphasis. With the exception of the mound at Purkin Kuk, all the mounds, whilst highly visible landscape monuments, are isolated in some way, either within an enclosing wall or associated with some sort of excluding barrier. This is also true for the Vira cemetery which combines the virtues of general visibility of the central peninsular with the separation from the peninsular by two mounds which may have formed a barrier originally or a gateway to the central mound. Although the mound at Purkin Kuk does not follow this pattern so clearly, it does occupy an isolated hilltop position with exceptional visual connections to virtually the whole of the largest fertile plain on the island. The peculiar position of these monuments suggests that access to associated rituals may not have been freely available and may have involved the physical, but not necessarily the visual, exclusion of a portion of the population.

The possibility that there may have been restrictions on access to ritual on Hvar is an important one. The role of ritual and religion as an arcane means of placating, making obeisance or communicating requests to higher authorities is quite clear. In the social sphere, ritual also gains importance in the relationship that individuals or groups are perceived to have with the supernatural.

Within the context of the data from Hvar, the evidence for the origin of rites associated with burials during the early Bronze Age in Dalmatia suggests that there was a perceived concern with the maintenance of fertility from at least this time. The continuation of such practices and the construction of large monuments and perhaps settlements, intimately linked with similar rituals during the late Bronze Age, suggest that power within these communities was being held by parts of the community through control of rituals maintaining land fertility rather than simply through coercive control of the land itself. There can be little doubt that with the onset of noticeable erosion during the late Bronze and Iron Ages, groups with a perceived preferential access to ritual power associated with the maintenance of fertility would achieve status. Such a situation would be enhanced by the increased isolation of ritual within monuments which physically exclude the remainder of the population. The role of ritual and religion to maintain social conventions and emphasize the status of individuals and groups under the guise of divine authority has a logic within societies with weakly developed political structures (Drennan, 1976:346). Such a context seems most appropriate to the evidence from Hvar.

This chapter is not the place to develop this argument further, although the literature suggests that an extended analysis of the Croatian or central Dalmatian late Bronze Age and Iron Age following these lines would be rewarding (Benac, 1986). What is important in the context of the re-analysis of the Hvar data, is the need to avoid the restrictions of a narrow economic viewpoint which might be encouraged by the use of GIS and to emphasize its potential within the context of an explicit theoretical perspective which integrates the belief systems that engender and maintain social relations—a need to be theory-led, if not theory-laden (Renfrew, 1982:12).

5.3 *Rock art, cognition and GIS*

The second part of this chapter relates to the study of prehistoric rock art and ritual monuments in Argyll in south-western Scotland (Figure 15.5). This is a useful contrast with the data from Hvar, in that we are treating objects which unequivocally carry a culturally embedded message both by virtue of the stylistic and symbolic information inherent within their designs and in their relationship to the larger physical landscape in which they operate. The area under investigation is that of Kilmartin, which includes the estuary of the Crinnan, an important area of lowland

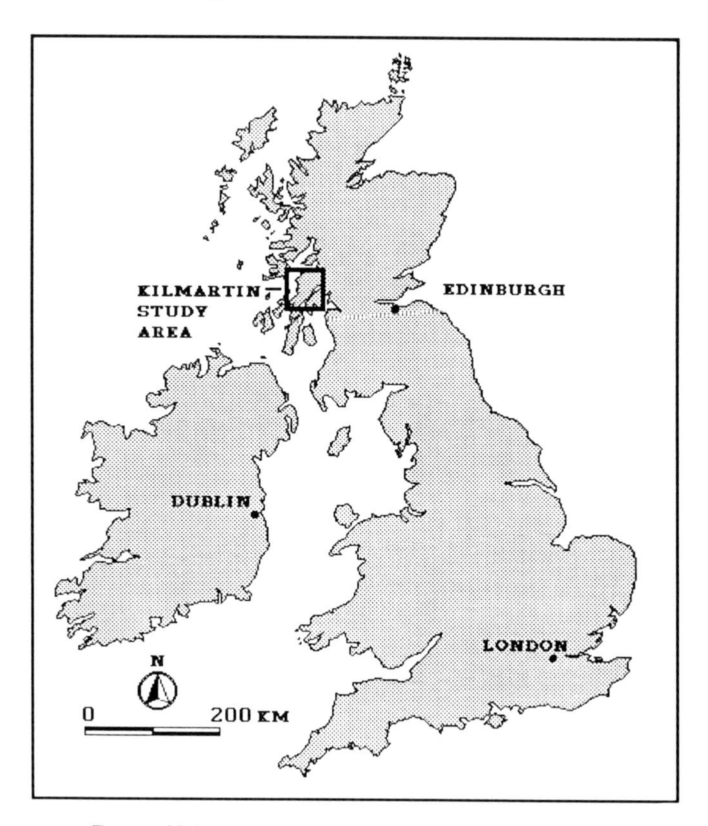

Figure 15.5 Location map of the Kilmartin study area.

north of Kintyre. The Crinnan estuary faces west to the island of Jura and Ireland. It is a significant porterage intersecting with a series of important routes to the north and west. Apart from the agricultural potential that such an area of lowland might have offered to prehistoric communities, the area is also significant in that it contains an impressive concentration of ritual monuments relating to a number of prehistoric periods (RCAHMS, 1988).

During the early Neolithic period, the primary evidence for human occupation is a series of burials within chambered cairns. Settlements of the period are rare and none have been found in the immediate area of study. Interpretation of the evidence has therefore been limited to the burial data and the tentative suggestions that the cairns held a symbolic role and indicated the legitimation of a particular group's rights to land. The cairns seem to have been used over a long period of time and the continued significance of these monuments within the landscape is indicated by the insertion of later cists within the mounds (RCAHMS, 1988:6).

A series of later monuments including an impressive variety of standing stones and stone circles is also concentrated within the study area, although those associated with the timber and later stone circle at Temple Wood are of particular prominence (RCAHMS, 1988:11). During the later Neolithic, at least one henge monument was constructed at Ballymeanoch. Cist burial, frequently associated with cairns and barrows, appears to have been introduced into the area along with Beaker cultural material, and cist graves were constructed throughout the Bronze

Age. There is little evidence for settlement until the Iron Age when a series of settlement types including large hillforts and Duns occur. Specific mention should be made of the fort at Dunadd which stands on a distinctive isolated hill in the centre of the Crinnan plain. The find of a carved Neolithic stone ball in the fort suggests that this peculiar natural feature achieved some degree of inferred importance from an early period (RCAHMS, 1988:156).

The prehistoric rock art associated with this area is one of the largest and most varied groups in the British Isles. Although the majority of the artwork is composed of simple cup or cup and ring marks on natural flat rock faces, more complex designs are also found. These include spirals, starred circles and various grooves and enclosures. The cup and ring marks themselves display significant variation in their composition, association, size and numbers of rings (Bradley, 1991). Rock art is notoriously difficult to date and it is possible that the carvings were produced over millennia. However, as Bradley (1991:79) has noted: 'the ultimate inspiration for this art style ... comes from stone-built tombs in Brittany and Ireland' and there are elements of the Argyll material that link it with the Boyne and Irish Passage grave art (Bradley, 1991, Fig. 14; RCAHMS, 1988:120). The westward orientation of the study area and other evidence for contact with Ireland provide a suitable context for such observations (RCAHMS, 1988:6). It is significant, however, that carved rock panels share motifs with carved stones within standing monuments. This suggests some degree of contemporaneity. When the carving started in the Kilmartin area remains uncertain but carved representations of bronze axes on monuments including the cist graves at Nether Largie north must be late in the sequence (RCAHMS, 1988:68).

The situation of rock art panels in the British Isles is variable, although Bradley (1991) has noted the tendency of principal groups of rock art to be situated both in the vicinity of major monuments and on land interfacing with lowland and highland zones. The integrative quality of the art is also emphasized in the frequent relationship that it demonstrates with dominating viewpoints and important routes. The arcane nature of rock art has laid itself open to a wide variety of interpretations, not all of which sustain rational examination. Bradley, wisely, rejected any attempt at a formal interpretation of the rock art and chose to emphasize the need to treat the art as a channel of information of greater or lesser complexity and has suggested that complexity would be expected where larger and more varied groups might expect to meet and where communication between such groups would be of particular importance. The general relationship of art with public monuments, routes and intermediate land zones suggests that this is a profitable line of enquiry. In the Kilmartin area in particular, there is a relationship between the complexity of rock art and points of entry to the lowland area which contains the major public monuments, prompting Bradley to suggest that the symbolic message of the art may have been directed at those entering the estuary (Bradley, 1991:99).

The complementary study to Bradley's stylistic analysis presented here attempts to use GIS in a manner that allows some quantification of the perception of the monuments in the Kilmartin area. The study incorporates the analysis of viewshed data constructed for all the non-settlement sites listed by the Royal Commission for the Ancient and Historical Monuments of Scotland for the Kilmartin area from the Neolithic to the Bronze Age, a total of 76 sites (RCAHMS, 1988). The decision to use all the pre-Iron Age sites partly springs from the desire to contrast patterns for different classes of site and earlier and later monuments. However, it was also noted that the explicit inclusion of earlier sites into the later cognitive system, and the incorporation of re-used rock art within later cist graves, strongly suggests that the data should be treated as a continuum. Following from this, the monuments were broken down into five basic groups for analytical purposes:

- chambered cairns (6),
- individual or groups of decorated natural rock faces (28),
- standing stones, alignments and circles (15),
- the henge at Ballymeanoch, and
- cists, cairns, burials and barrows (26).

The analysis of these data involved the calculation of the viewshed for each monument, i.e. the area which can be seen from an individual site. This information represents the area within which that monument is likely to communicate visual information. Monument viewsheds can overlap, producing zones in which an observer might be aware of many monuments, all of which may carry information. Presumably, the increasing density of such information can be interpreted in some circumstances as a measure of the importance of a particular area. In the context of this work, it is perhaps better to emphasize the ability of such a procedure to provide a mappable, spatially variable index of perception, which incorporates groups of monuments and plots their visual relationship with the surrounding landscape. Analysis of this data should give an insight into the cognitive landscape within which the monuments operated.

An example of the process is given in Figure 15.6. Here, the viewsheds for the Ballymeanoch Henge and a pair of standing stones at Dunamuck are illustrated

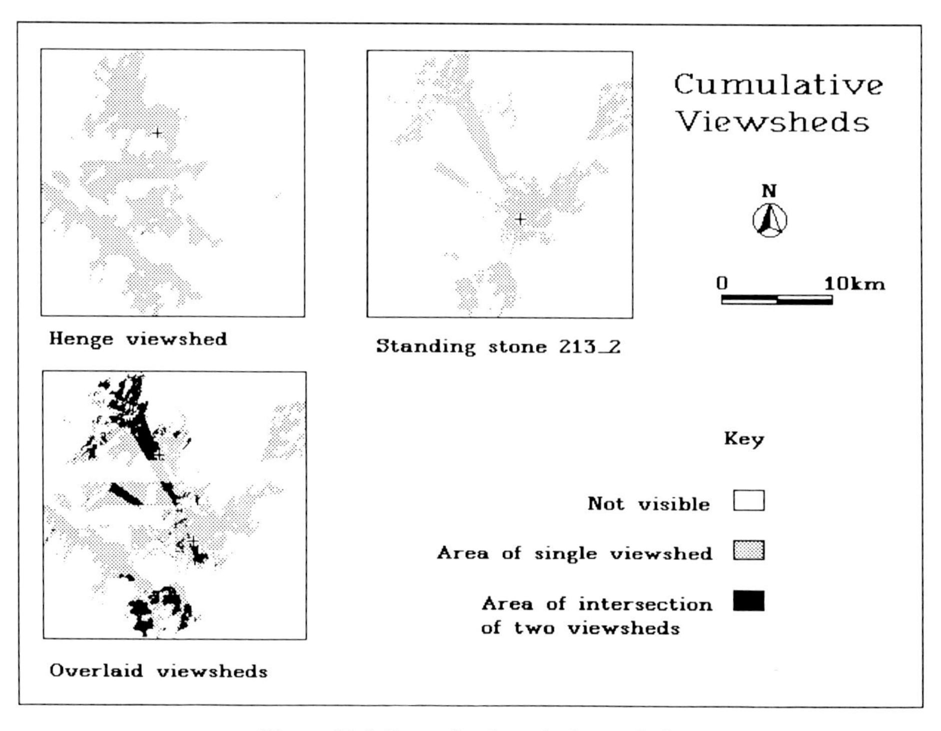

Figure 15.6 Example of overlaid viewsheds.

along with the cumulative viewshed for the two sites. Within the cumulative viewshed the areas which are only visible to one monument have a value of one and the areas which are intervisible have a value of two. The sum of this process for the individual monument groups can be seen in Plate 1.

Analysis of these viewsheds provides a number of interesting patterns. Figure 15.7 illustrates the mean intervisibility values of monument groups. Rock art emerges with the lowest mean value for intervisibility. Chambered cairns also have a relatively low value of intervisibility, whilst there is an increasing level of intervisibility through the groups of standing stones, cists and burials with the Ballymeanoch Henge emerging as the most visible single monument type.

These data can be broken down further to measure the way in which these monuments interact with each other. Table 15.1 illustrates data on intervisibility between monument types. This emphasizes how very little the chambered cairns and rock art interrelate with any group of sites. There is an increasing association between the standing stones and the henge monument, whilst there is a clear positive relationship between the henge and the later cists and barrows.

Several significant points are suggested from results shown in Table 15.1. The first is that chambered cairns, as a group, are not integral with the other monuments even though individual cairns provide evidence for their incorporation into later cognitive systems. It is also, perhaps, surprising that the rock art, despite the relatively large number of such sites, interacts at a very low level with other monuments. However, the henge is clearly a focal point for the standing stones and cists. The strong relatedness of the henge and cists indicates that many of the later burial

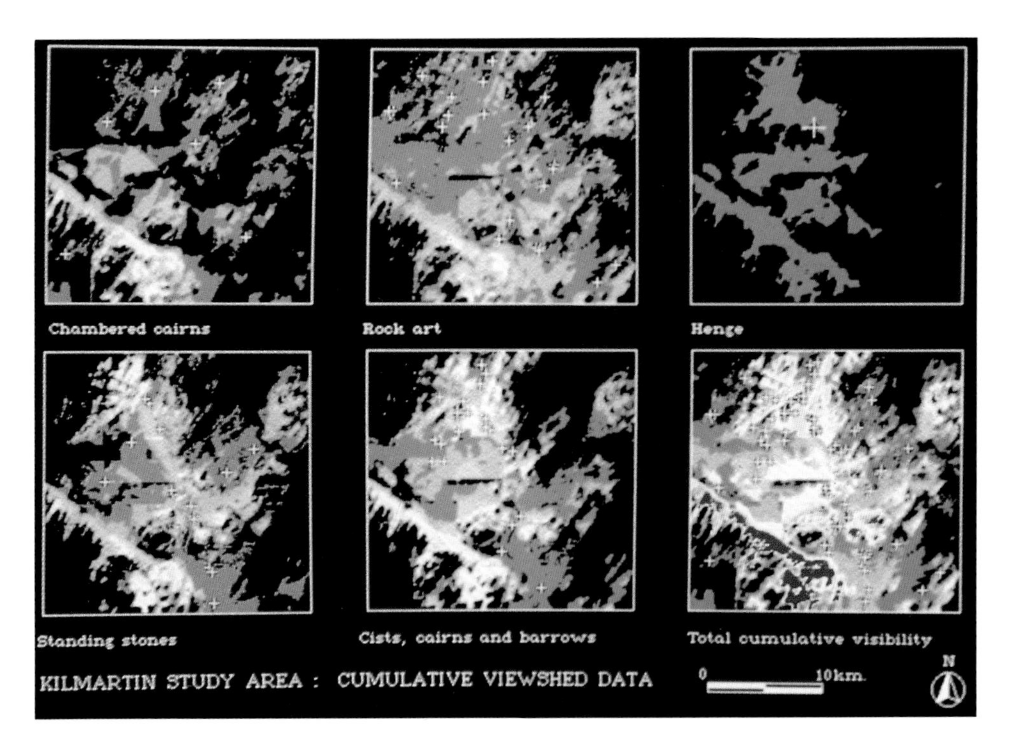

Plate 1. Cumulative viewsheds for individual monument types in the Kilmartin area of Scotland.

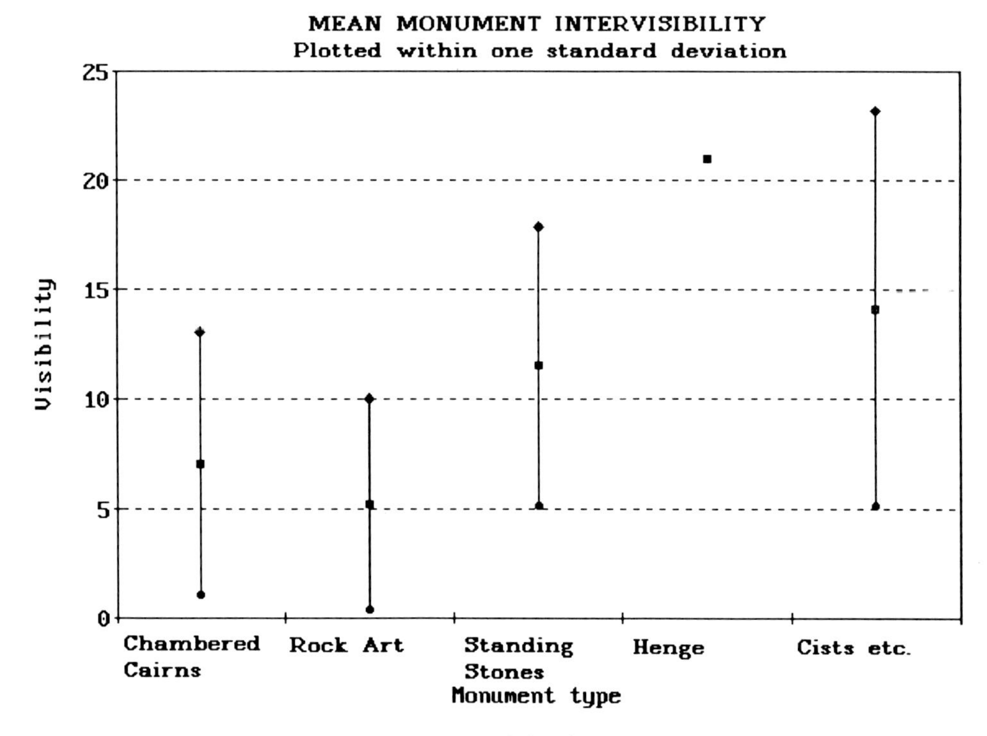

Figure 15.7 Mean intervisibility for monument types.

monuments were deliberately sited to be in visual contact with the henge, emphasizing its special position within the area.

Some insight into the processes leading to this situation can be inferred from the nature of the land contained within monument group viewsheds. Figure 15.8 illustrates the mean proportions of lowland within the viewsheds of each monument group. This shows that although the viewsheds of chambered cairns and rock art may visually incorporate considerable amounts of lowland, there is a trend towards an increasing emphasis on the lowland within the remaining three monument groups, with the Ballymeanoch Henge, again, emerging as the monument type with the greatest mean amount of land, but the viewsheds of the cists and burials also frequently encompass very large amounts of lowland.

Table 15.1 Index of intervisibility between sites (mean number of visible sites divided by the total number of sites of each type)

	Monuments viewed				
Viewpoints	Chambered cairns	Rock art	Standing stones	Henge	Cists, etc.
Chambered cairns	0.19	0.05	0.07	0.00	0.11
Rock art	0.06	0.12	0.11	0.11	0.10
Standing stones	0.07	0.11	0.15	0.33	0.20
Henge	0.00	0.07	0.33	—	0.54
Cists, etc.	0.10	0.09	0.16	0.54	0.31

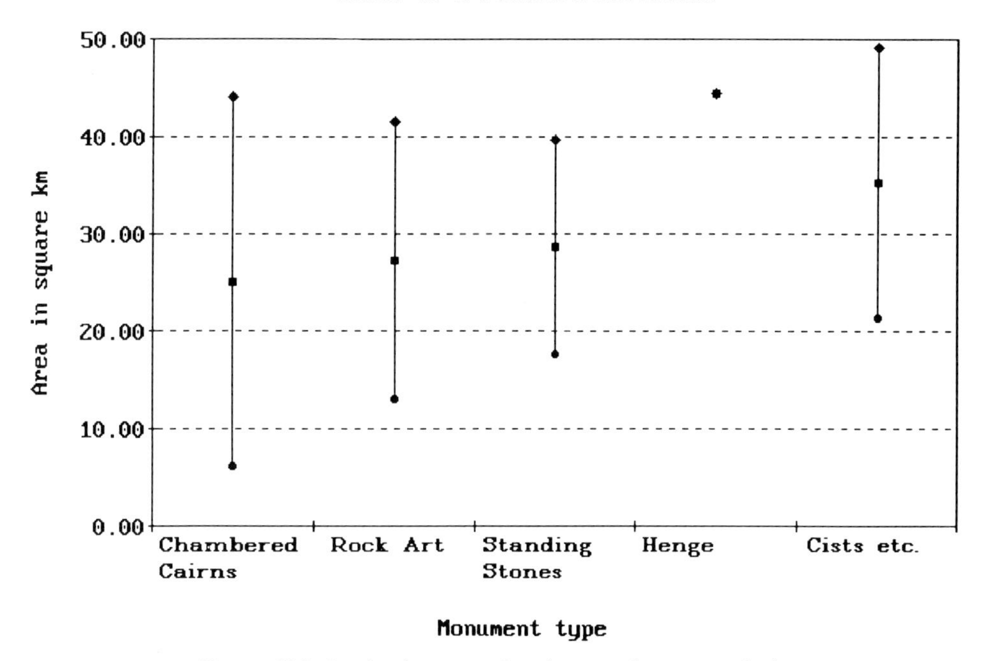

Figure 15.8 Lowland contained within cumulative viewsheds.

In part, this is a chronological pattern. The early Neolithic communities that built the chambered cairns may well have been concerned with the regular exploitation of a wide variety of environments. This is reflected in the variation displayed by the monuments of the period. This may contrast with the later communities associated with the cist graves, who were probably more closely involved in agricultural pursuits. The importance of the Crinnan lowlands to such communities is reflected in the siting of contemporary monuments and visual association with the lowland zone.

How are we to incorporate the other monuments within this scheme? It has already been emphasized that the presence of shared motifs suggests some degree of contemporaneity between rock art panels and monuments including the Temple Wood stone circles. Therefore, we should not seek some simple temporal movement towards the lowland as an explanation for patterning.

What may be more significant is to emphasize the contrasting nature of the decorated rocks and other major public monuments including the standing stones, circles and henge. Bradley (1991) has emphasized the tendency for the most complex monuments, including the henge, stone circle, decorated standing stones and the most complex rock art forms in the Kilmartin area to be associated with the valley entrances to the Crinnan estuary. This may well contrast with the majority of rock art panels whose message is more pervasive and perhaps subtle. Rock art panels occur in a wider variety of settings but essentially integrate upland and lowland. The artwork itself, though a visual phenomenon, is considerably less visible than the other public monuments. The rock faces upon which such carvings occur are frequently almost horizontal or occasionally even in situations which may obstruct views. The information these monuments imparted seems to have been very closely associated with the landscape in a more intimate sense—a point firmly emphasized by the use of natural surfaces for decoration. Its integrative position at the junction of the highland and lowland also emphasizes this role but suggests some degree of ambiguity in its relationship between land zones. Braithwaite (1982:81) has emphasized the role of decoration and ritual to express, communicate, authorize and guide action on boundaries and other ambiguous areas of social interaction. It seems reasonable to view much of the rock art in the Kilmartin area in this light. The large and complex public monuments may be communicating to diverse groups arriving in the area. The less complex rock art panels may be communicating information on areas of ambiguity across the landscape, perhaps to those who utilize these diverse economic zones.

There is clearly a series of potentially conflicting symbolic structures operating within the monuments of the Kilmartin area. These are important and will provide further insights as analysis proceeds. However, it is worth stressing the emergence of a coherent pattern from this disorder over time. It is clear that specific areas achieved prominence as defined by their potential perception and illustrated through the viewshed data. The valley entering the estuary to the north at Nether Largie (Figure 15.9), clearly forms the ritual cognitive focus of the area centring on some of the most important and imposing monuments in the valley. Around this spot are the massive Nether Largie chambered cairn, the Temple Wood stone circles and the Ballymeanoch Henge (RCAHMS, 1988:48–52; 138–42). All three of these major monuments have later cist burials inserted, thus integrating elements of the whole sequence into a single cognitive system, while the cists and burials which

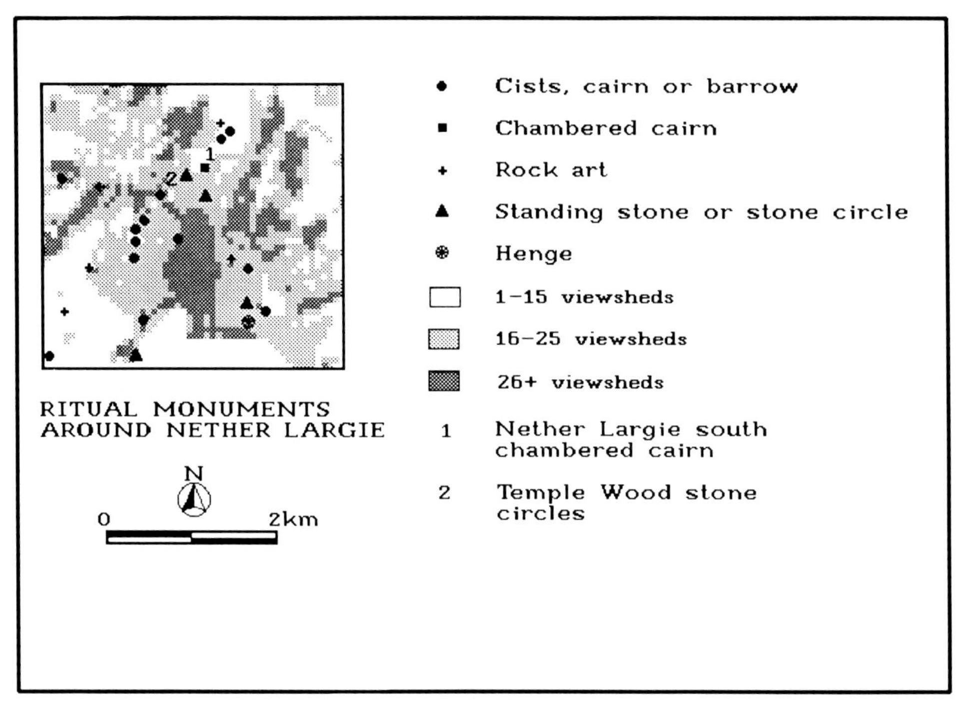

Figure 15.9 Monuments in the Nether Largie area.

cluster around these monuments include several which incorporate earlier carvings, further emphasizing the very real perceived relationship between these temporally separate monuments (RCAHMS, 1988:68–70).

15.4 Conclusions

It does not do to labour a point but it is hoped that the above examples clarify some of the observations made in the introduction to this chapter. Clearly, archaeology has prospered by shedding some of the shackles of functional and economic determinism. We would be well advised to ensure that these do not return in modified and seductive new wrappings. The re-analysis of the Hvar data suggests that GIS, if not used thoughtfully, is open to such criticism. However, archaeology abounds in such cautionary tales and it is more productive to emphasize the new opportunities that GIS offers. The interim analysis of the Argyll ritual monument data clearly illustrates the potential of GIS for the study of large-scale cognitive phenomena and its ability to utilize the full landscape for such purposes. This is new and we are only just beginning to explore the possibilities on offer. In this respect, it is worth emphasizing that the Argyll data were analysed solely through the application of standard GIS modules. While convenient for this exercise, the limitations of modules that were primarily designed for the analysis of physical space will eventually prove restrictive. The future for innovative work in GIS will therefore lie in the development of more sophisticated mathematical modules explicitly for archaeological purposes and within the context of GIS technologies. Archaeology has a reputation as a somewhat eclectic discipline and there is no doubt that in the past this has proved a strength. In this case, however, we have the opportunity to treat GIS as a 'core' technology which can be developed according to our own agenda. Used carefully, archaeology can only benefit from such a situation.

Acknowledgements

We would particularly like to thank Professor Richard Bradley of Reading University, Great Britain, for providing access to important unpublished data and for discussing the results of the analysis of the prehistoric rock art with us. We would like to thank everyone who has commented upon the results presented here. All errors are, of course, our own.

References

Allen, K. M., Green, S. W. and Zubrow, E. B. W. (Eds), 1990, *Interpreting Space; GIS and Archaeology*, London: Taylor & Francis.

Babić, I., 1984, *Prostor izmedju Trogira i Splita*, Kaštel Novi: Zavičajni muzej Kaštela.

Batović, Š. and Kukoč, S., 1986, Podvršje/Matkov Brig, *Arheološki Pregled (1985)*, **27**, 61–3.

Benac, A., 1986, Utvrdjena praistorijska naselja u zapadnom dijelu Jugoslavije,

Materiali XXII (Odbrambeni sistemi u praistoriji i antici na tlu Jugoslavije), Novi Sad, 22–34.

Binford, L. R., 1982, The archaeology of place, *Journal of Anthropological Archaeology*, **3**, 71–6.

Bradley, R., 1991, Rock art and the perception of landscape, *Cambridge Archaeological Journal*, **1**(1), 77–101.

Braithwaite, M., 1982, Decoration as a ritual symbol: a theoretical proposal and an ethnographic study in southern Sudan, in Hodder, I. (Ed.), *Symbolic and Structural Archaeology*, pp. 80–8, Cambridge: Cambridge University Press.

Butzer, K. W., 1977, Cultural perspectives on geographical space, in Butzer, K. W. (Ed.), *Dimensions of Human Geography; Essays on some familiar and neglected themes*, pp. 1–14, Chicago: Cambridge University Press.

Chapman, J. C., Shiel, R. and Batović, Š., 1987, Settlement patterns and land use in neothermal Dalmatia, 1983–4, *Journal of Field Archaeology*, **14**, 124–46.

Clarke, D. L., 1968, *Analytical Archaeology*, London: Methuen.

Douglas, M., 1966, *Purity and Danger: an analysis of concepts of pollution and taboo.* London: Routledge & Kegan Paul.

Drennan, R. D., 1976, Religion and social evolution in formative Mesoamerica, in Flannery, K. (Ed.), *The Early Mesoamerican Village*, pp. 345–68, London: Academic Press.

Evans, C., 1985, Tradition and the cultural landscape: an archaeology of place, *Archaeological Review from Cambridge*, **4**(1), 80–94.

Fletcher, R., 1977, Settlement studies (micro and semi-micro), in Clarke, D. L. (Ed.), *Spatial Archaeology*, pp. 47–162, London: Academic Press.

Gaffney, V. and Stančič, Z., 1991, *GIS approaches to regional analysis: a case study of the island of Hvar*, Ljubljana: Znanstveni inštitut Filozofske fakultete.

Gaffney, V. and Stančič, Z., 1991, Predicting the past: GIS and archaeology, *Geo-Informations-Systeme*, **4**, 27–32.

Gams, I., 1987, Adaption of the karst land for agrarian use in Mediterranean problems of research and conservation (a survey), *Endine*, **3**, 65–70.

Higgs, I., 1972, *Papers in Economic Prehistory*, Cambridge: Cambridge University Press.

Hodder, I., 1982a, *Symbols in Action*, Cambridge: Cambridge University Press.

Hodder, I., 1982b, *Symbolic and Structural Archaeology*, Cambridge: Cambridge University Press.

Hodder, I. and Orton, C., 1976, *Spatial Analysis in Archaeology*, Cambridge: Cambridge University Press.

Kvamme, K. L., 1989, Geographic information systems in regional archaeological research and data management, in Schiffer, M. B. (Ed.), *Archaeological Method and Theory*, pp. 139–203, Tucson: University of Arizona Press.

Marović, I., 1985, Iskopovanja kamenih gomila u Bogomolju na otoku Hvaru, *Vjesnik za arheologiju i historiju dalmatinsku*, **78**, 5–33.

Marović, I. and Čović, B., 1983, Cetinska kultura, in Benac, A., (Ed.), *Praistorija jugoslavenskih zemalja*, IV, Sarajevo, 191–241.

Novak, G., 1959, Prethistorijske gomile na Paklenim otocima, *Arheološki radovi i rasprave*, **1**, 237–44.

Petriæ, M., 1986, Izvještaj o arheološko-konzervatorskim radovima na ruševini Gazaroviæ u Hvaru, *Periodični izvještaj Centra za zaštitu kulturne baštine komune Hvarske*, 7–17.

Petrić, N., 1979, Hvarski tumuli, *Vjesnik za arheologiju i historiju dalmatinsku*, **72–3**, 67–78.

Petrić, N., 1980, Prilozi paznavanju apulske geometrijske keramike na istočnom Jadranu, *Diadora*, **9**, 107–200.

RCAHMS (Royal Commission on the Ancient and Historic Monuments of Scotland), 1988, *Argyll*, Vol. 6, Edinburgh: HMSO.

Renfrew, C., (Ed.) 1982, *The Explanation of Culture Change: Models in Prehistory*, London: Duckworth.

Shiel, R. and Chapman, J. C., 1988, The extent of change in the agricultural landscape of Dalmatia, Yugoslavia, as a result of 7000 years of land management, in Chapman, J. C., Bintliff, J., Gaffney, V. and Slapšak, B. (Ed), *Recent Developments in Yugoslav Archaeology*, pp. 31–43, British Archaeological Reports International Series 431, Oxford.

Taffe, E. J., 1974, The spatial view in context, *Annals of the Association of American Geographers*, **64**(1), 1–16.

Wheatley, D., 1993, Going over old ground: GIS, archaeological theory and the act of perception, in Andresen, J., Madsen, T. and Scollar, I. (Eds), *Computing the Past*, pp. 133–8, Aarhus: Aarhus University Press.

16

Scientific visualization and archaeological landscape: the case study of a terramara, Italy

M. Forte

16.1 Introduction

The principal aim of scientific visualization techniques is to process digital images to produce a simulation model thus increasing the information in the images (Forte, 1992b; 1993a; 1993c). In our experiments, the image processing techniques become fully integrated with visualization techniques. Numerical processing of the image in order to classify it (Forte and Guidazzoli, 1992; Forte, 1992a) combine with visualization to produce an animated simulation model for showing particular features not visible in other ways (Forte, 1992b) resulting in a formal description of the image (Furini *et al.*, 1991; Guidazzoli, 1992; Forte, 1993c). The final objective is to explore and to classify the archaeological sites in the landscape by using texture mapping techniques for investigating the three-dimensional model created by a digital terrain model (DTM) and by image processing to enhance the digital features of satellite and aerial images (Forte, 1993b). This visualization approach is a first step towards the realization of three-dimensional GIS.

Computer simulation, of natural or artificial phenomena, is one of the most attractive and important uses of models or images (Furini *et al.*, 1991). By simulation, we mean any process computed using mathematical models, the results of which can be represented visually (Reilly, 1989; 1991; Forte, 1993c) and then be compared with other information for the furthering of knowledge (Guidazzoli, 1992; Forte, 1993c). For scientific simulation in archaeology, the accuracy of details should be very high, requiring a sophisticated and reliable model. Two major problems exist in the visual simulation of natural ambience and in three-dimensional navigation. The first one is the construction of a reliable DTM (Harris, 1987), the second one is the process of terrain rendering and the perception of colour by the observer. Here the aim is to produce a landscape model, which includes archaeological sites, using a DTM and digital images from aerial photographs. This involves the following steps:

- acquisition of contours from maps at any scale, or the collection of the DTM data on the ground using a total station;
- generation of the DTM (Figure 16.1);
- digital image classification to determine the distribution of pixels for the DTM;
- synchronization of the digital aerial photograph with the DTM;
- texture mapping and generation of the three-dimensional images (Figures 16.2 and 16.3); and

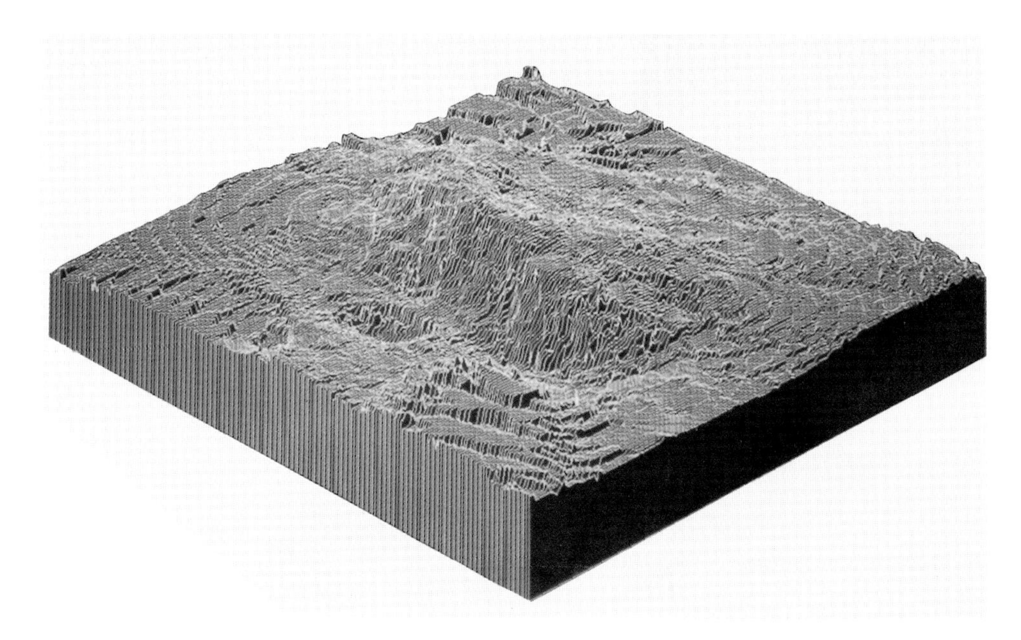

Figure 16.1 The DTM of the terramara.

- interactive three-dimensional model animation, inter-site and intra-site navigation.

The results of this full process produce a digital model through which the researcher can move, navigate and explore like a real landscape. This is especially useful for investigating the evolution of a landscape and its ancient settlements.

Figure 16.2 Digital aerial photograph of the terramara: the crosses are grid sampling points.

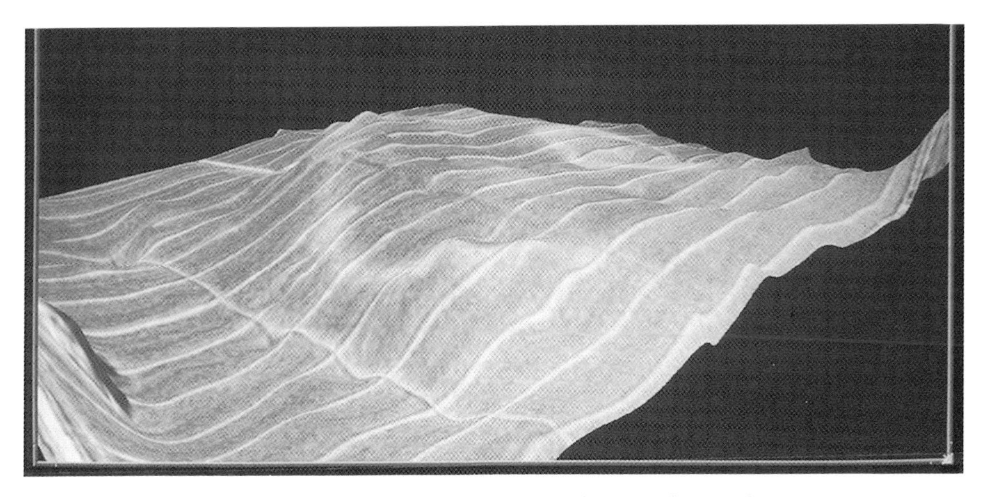

Figure 16.3 Texture mapping of the aerial photograph onto the DTM.

16.2 Visualization of the terramara of S. Rosa, Reggio Emilia: intra-site navigation

S. Rosa terramara, which measures 7 ha, is a middle and recent Bronze Age site which has been excavated since 1984. The terramara was damaged in the nineteenth century by quarry works which are shown in the DTM. Figure 16.2 shows the aerial photograph of the terramara taken after ploughing, and structures are visible on the surface as light soils of the pedogenetic substratum, which are in contrast to the surrounding dark vertisoils. We can distinguish, in the northern region, the small oldest village which is elliptical in shape, and a white U-shaped feature open towards the north. This latter feature represents the enclosure ditch of the terramara and is the larger village of the recent Bronze Age. Situated to the east of this feature there is another, indefinite one which is a farm of the Roman period.

16.2.1 DTM processing and model visualization

Because of the quarry works, the deep recent ploughing and the level ground, the existing elevation data are not detailed enough and have been enhanced by further collection on the ground. A total of 450 points were collected concentrating on the earthwork and other features, which were processed using IDRISI producing some interesting results especially with the enhancement of the z-values (Figure 16.1). In the three-dimensional visualization, the earthwork of the large village is particularly clear-cut while the surroundings of the small village are level and it is low in the centre. In the DTM based on slope values, it is important to emphasize the minimal differences in elevation that are involved, in 7 ha the maximum difference in level is only 93 cm.

The integration of the DTM with the digital aerial images of the terramara produces a visualization of the archaeological landscape which is very realistic (Figures 16.2 and 16.3) and is especially useful for reconstructing the original position of the site before human and natural events changed the surface features. The DTM shows the shallow nature of the ditch of the large village, which is in part filled with deposits since the Bronze Age. The corresponding bank is also negligible;

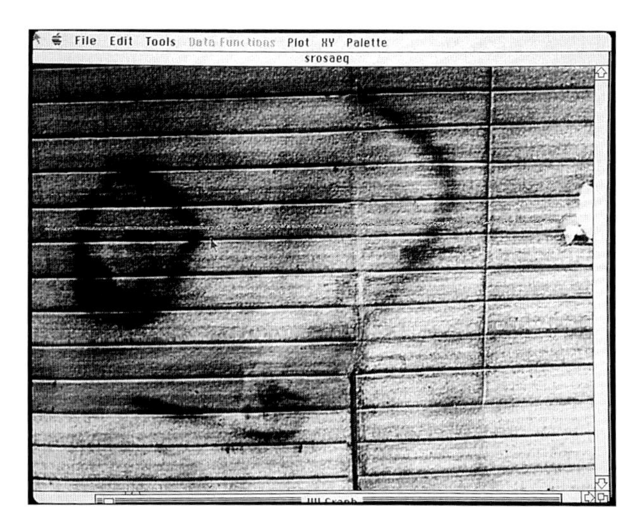

Figure 16.4 Contrast enhancement of the digital aerial photograph.

in fact, after excavation, it was shown to be practically flat due to slippage (Figure 16.3). Moreover, the DTM shows a depression surrounding the small village that divides the site into two different contemporary centres with the highest part not coinciding with the ditches. In this area, there is a large amount of surface archaeological material and recent excavations have shown a complex series of rectangular wooden houses rebuilt many times and superimposed. On the other hand, the small village is flat and slightly concave inside, due to the quarry that has destroyed most of the stratigraphy.

With the completed visualization model, it is possible to simulate the exploration of the site in virtual space by using video-animation for intra-site navigation. For the final rendering, special purpose hardware, an SGI workstation, has been used and for the polygon rendering rather than the usual texture mapping techniques a geometric texturing has been produced based on one polygon per pixel. To produce the video animation, the following steps have been implemented:

1. interactive tools are used to define positions (points) on the image, displayed on a high resolution monitor;
2. selection of the points two by two to analyse the model, the three-dimensional positions of the point of view and the observed point with the relative zenithal and azimuthal angles which produce a set of keyframes stored in different files; and
3. the frames are produced by a process of linear interpolation from the keyframes in order to simulate a fluid movement, producing 3600 frames.

Because of the high model resolution the system produces one image in three seconds—a rate not suitable for real-time animation. To obtain a good animation, a video has been recorded frame by frame at 25 frames per second.[1]

16.2.2 Digital processing of the aerial photograph

The image processing results in the classification of the areas of archaeological interest according to ranges of pixel values. Aerial photographs processed in this way give much more information than the originals. The whole processing includes the following:

- digitization at different resolutions;
- grey-level equalization;
- contrast enhancement in order to increase the chromatic differences and the range of reflectance (Figure 16.4);
- filtering;
- statistical analysis of the reflectance of different materials;
- histograms and digital cross-sections;
- digital sampling and classification (Plate 1);
- processing separate images;
- generation of two different palettes for any image (Plate 1); and
- pseudo-colour processing.

The classification of pixels into significant ranges has identified four main areas each with a particular digital signature (Plate 1). In particular, the digital characteristics of the small village are very different from those of the large and more recent village. The digital classification of these areas identifies

- area 1, later alluvial sediment which hides part of the earthwork;
- area 2, the nucleus of the small village;
- area 3, the ditches of the big village; and
- area 4, a Roman villa.

The final image shows distinctly the formation of natural vertisoils and artificial earthworks, for example in the north-western area the interruption of the ditch indicates burial by flood deposits, it also appears that the two villages are separated by a large ditch. On the basis of these results, it has been possible to reconstruct the

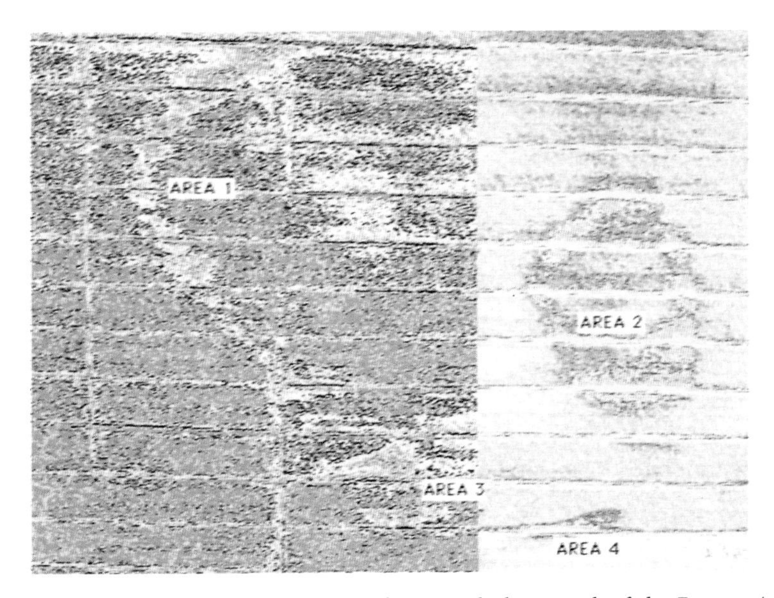

Plate 1. Digital processing and classification of an aerial photograph of the Bronze Age site at S. Rosa, Reggio Emilia, Italy: area 1, alluvial sediment which hides part of the earthwork; area 2, nucleus of the small village; area 3, the ditches of the big village; area 4, Roman villa.

evolution of the site and the environment in different periods from before the Bronze Age to the medieval period.

16.3 GIS and scientific visualization: future directions

The landscape is a very complex information set comprising different types of morphological structures, spatial units, land systems, human installations, and so on. Every structure interacts with the others; therefore, any small artificial or natural modification can change the whole set. The history of the landscape is the history of the evolution of these complicated connections. Moreover, in this context, the importance of the archaeological landscape must be understood in order to survey and to monitor the whole territory, and in order to discover the relationships between the ancient and modern landscapes. The best view of a landscape is a synthetic and significant representation, such as an aerial photograph or a satellite image, because only in this way is information about many aspects available at once, for example the soil, urban centres, roads and, in general, each landscape component or landscape unit. All landscape elements have their own characteristics, such as shape, colour, composition, geomorphological or artificial structures. Because of these features and complex connections knowledge of the landscape depends on an accurate scientific visualization system.

In computer processing, the best view of a landscape is in three dimensions. The aim of scientific visualization in landscape exploration is to acquire in one model the most information, while that of GIS is the efficient and rapid retrieval, manipulation, processing, and display of information. So, if we can process a three-dimensional high-definition model of a landscape which incorporates both these approaches, we can acquire a very important information set. Physical units can be described by the DTM, by slope and aspect and, in particular, the topography of sites can be determined by a visibility analysis (van Leusen, 1993). An intervisibility index determines, for example, which areas are visible for a viewpoint, given the height, the distance, and the presence of morphological structures in the field of view. The study of the DTM is very important at both the micro- and macro-scales; if the DTM is integrated with the texture mapping, a complete landscape representation is produced. In this way, scientific visualization can represent the future of GIS.

The three-dimensional investigation of landscape models which include all the data we need for research, give us a very different methodological approach to GIS by including interactive procedures for exploring the archaeological landscape. When using a landscape model, we navigate inside an information space. A three-dimensional landscape model is the ideal place to implement visualization techniques and at the Scientific Visualization Laboratory of the CINECA, Interuniversity Consortium for Supercomputing, we are developing an interface for the real-time interactive navigation inside three-dimensional models of archaeological landscapes. This could be the first step in experiments in virtual reality, considering the archaeological landscape as a virtual space to explore. At some future time, the researcher will be able to choose all the views and perspectives to visualize data models from the landscape data to excavation data. So, in this way, scientific visualization techniques, fully integrated with GIS, will allow increased knowledge of geographical data, archaeological areas, and so on. Moreover, three-dimensional visualization will be very useful in many fields, in particular for the protection of environmental and cultural resources, or the creation of archaeological parks.

The next phase of GIS must be the creation of three-dimensional GIS, in which it will be possible to visualize landscape data in three-dimensional layers such as ancient sites, roads, rivers, cultivated land, etc. Navigation at the inter-site or intra-site scale with the appropriate topographical information will allow the investigation of settlements and topographical systems or single archaeological sites.

Briefly, the principal possibilities of visualization and computer graphic techniques for the creation of three-dimensional GIS, could be the following:

- the study and analysis of a DTM (cartographic or surveyed);
- the three-dimensional representation in different layers of raster and vector data and images;
- the three-dimensional overlay of all landscape data; and
- archaeological landscape navigation, both inter-site and intra-site, in real time.

16.4 Conclusion

Computer processing integrated with aerial photo-interpretation, shows important new data not obtainable with traditional techniques. In the application described above, the construction of the DTM, texture mapping, and digital image processing have clearly enhanced the structures of the terramara, helping to resolve the problems which emerged in the aerial-photo interpretation. In particular, the interruption of the earthwork of the large village seems connected with burial after the abandonment of the terramara. Also, computer image processing, connected with the DTM and texture mapping of the site, is a portable tool, not only useful for the evaluation of the preservation of Bronze Age deposits in the Po valley. These results are obtained without excavations, using only simple ground surveys.

The methodology described in this Chapter shows the benefits that can be derived from the use of visualization techniques in GIS applications and, particularly, from the analysis of archaeological sites not investigated on the ground. In the past, computing resources have not been available in the humanities to process high-resolution graphic models. Current reductions in hardware costs and increases in computing power allow scientific visualization applications in archaeology with extraordinary results. In the archaeological computing of the future, GIS will be fully three-dimensional resulting in changing the methodology of research with more detailed analysis of the landscape.

Acknowledgements

I am grateful to all the CINECA Scientific Visualization Laboratory staff for the extraordinary co-operation in my research projects, in particular Antonella Guidazzoli, Luigi Calori, Roberto Scardovi, and the CINECA Director Marco Lanzarini. Special thanks are due to Professor Mauro Cremaschi, Department of Geology, Milan University, for the fundamental scientific co-operation, and to the Soprintendenza Archeologica dell'Emilia-Romagna, Dipartimento di Scienze Geologiche dell'Universitá Statale di Milano, AR/S Archeosistemi s.c.r.l. The work was directed by Professor Mauro Cremaschi and Dr Maria Bernabó Brea.

References

Forte, M., 1992a, L'image processing per l'archeologia del paesaggio: sistemi di classificazione del territorio, in *Bollettino d'informazioni del Centro di Ricerche Informatiche per i Beni Culturali*, **II** (1), Scuola Normale Superiore, Pisa: 53–96.

Forte, M., 1992b, Esperimenti di visualizzazione scientifica e paesaggio archaeologico: la navigazione 'virtuale' nel territorio, in *Atti del convegno di ICOGRAPHICS*, Milano: 198–205.

Forte, M., 1993a, Il paesaggio archeologico al calcolatore; *Le Scienze* (Italian edition of Scientific American), June, 46–54.

Forte, M., 1993b, Image processing applications in archaeology: classification systems of archaeological sites in the landscape, in Andresen, J., Madsen, T. and Scollar, I. (Eds), *Computing the Past*, CAA92, *Computer Applications and Quantitative Methods in Archaeology*, Aarhus: Aarhus University Press, 53–61.

Forte, M., 1993c, Un esperimento di visualizzazione scientifica per l'archeologia del paesaggio: la navigazione nel paesaggio 'virtuale', *Archeologia e calcolatori*, **4**, 137–52.

Forte, M. and Guidazzoli, A., 1992, Archeologia e tecniche di eidologia informatica, *Archeologia e calcolatori*, **3**, 37–76.

Furini, P., Righetti, M., Brivio, P. A. and Marini, D., 1991, 'Una metodologia per la simulazione di scene ambientali', *Icographics*, VI Convegno Internazionale e mostra sulle applicazioni della computer graphics nella produzione, progettazione e gestione, Milano: 5–8 March 1991, 405–12.

Guidazzoli, A., 1992, 'Computer graphics videotaping in un ambiente di visualizzazione scientifica', *Icographics*, VII Convegno internazionale sulle applicazioni della computer graphics nella produzione, progettazione e gestione, Milano: 4–7 Feb. 1992, 347–54.

Harris, T. M., 1987, Digital terrain modelling in archaeology and regional planning, in Ruggles, C. L. N. and Rahtz, S. P. Q. (Eds), *Computer and Quantitative Methods in Archaeology*, British Archaeological Reports International Series 393, Oxford: 161–9.

Reilly, P., 1989, Data Visualization in archaeology, *IBM System Journal*, **28** (4), 569–79.

Reilly, P., 1991, Towards a virtual archaeology, in Lockyear, K. and Rahtz, S. (Eds), *Computer Applications and Quantitative Methods in Archaeology*, British Archaeological Reports International Series 565, Oxford: 133–40.

van Leusen, P. M., 1993, Cartographic modelling in a cell-based GIS, in Andresen, J., Madsen, T. and Scollar, I. (Eds), *Computing the Past*, CAA92, *Computer Applications and Quantitative Methods in Archaeology*, Aarhus: Aarhus University Press, pp. 105–23.

Note

[1] The hardware and software used are: workstation IRIS 4D/80 GT, VTR controller LYON Lamb Minivas, videorecorder BVU 950 Umatic SP; device driver written in C to drive the videorecorder; visualization software: C language, GL libraries animation; animation: 180 seconds.

17

Towards a study of ancient Greek landscapes: the Perseus GIS

N. Smith

17.1 Introduction

Archaeologists, who by definition work primarily with the material remains of the past, have recognized the potential of GIS for analysing the spatial patterning of these remains in relation to each other or to natural features. Yet what archaeologists really aim to study is human activity in the past, activity that took place in a culturally constructed and conceptual space, as well as a physical one—in short, activity that took place in a landscape.

Classical Greece is a promising area for a study of past landscapes. Ancient attitudes are represented in the preserved literature of ancient Greece, and literary scholars have become increasingly sensitive to ways in which classical texts project ideologies into the spatial framework of the polis (Loraux, 1993).[1] At the same time, the physical remains have been the object of more than a century of almost uninterrupted archaeological exploration. Recently, regional surveys in particular have made important contributions 'filling in the map', and have provided a basis for reconstructing the rural countryside so rarely described by literary sources.[2] But the very richness of our sources for a study of the ancient Greek landscape can become an obstacle, and attempts to draw on the wide range of available material are not common.[3] This chapter suggests some ways that, in addition to purely archaeological applications, GIS can play an important role in supporting a more broadly based study of ancient landscapes. Spatial relationships link archaeological and non-archaeological material: using a GIS to analyse these relationships can promote the exploration of conceptual areas crossing boundaries of textual and archaeological material.

17.2 The Perseus project and the Perseus GIS

For the past two years, archaeologists working on the Perseus project[4] have assembled an extensive GIS database relating to classical Greece. Initially, our goals for an archaeological GIS closely mirrored the kinds of applications already reported (Allen *et al.*, 1990). We were eager to incorporate information about the present-day environment of Greece into Perseus, and, more generally, we wanted to explore how the experience of using Perseus might be changed by the accessibility of geographic information. These requirements alone demanded the development of a GIS.

We designed the GIS, like the rest of Perseus, with two sometimes conflicting sets of objectives. One of the Perseus project's major goals is to create information resources for the study of ancient Greece that will outlive specific information systems (hardware and software); at the same time, we are unapologetically committed to delivering Perseus in the near term to the widest possible audience on low-end platforms. Using GRASS, we have assembled a variety of resources in our archival (long-term) GIS:

- for an area running from Sicily to western Asia Minor, coverages derived from the Digital Chart of the World (including interpolated digital elevation models);
- for an area roughly corresponding to the modern state of Greece, land-use/land-cover themes derived from eight Landsat MSS scenes; and
- across the Greek world, but with a concentration on the Greek mainland, a database of over 2000 archaeological sites (Crane, 1994).

The development of the GIS mirrored the development of Perseus as a whole in one important respect. After a phase of very intensive work on the GIS and the resolution of such basic issues as what information to include in the database and how to structure this information, we then had to face an entirely new set of problems when we tried to integrate the GIS with other information systems in Perseus. Perseus 1.0 had been published in 1992 (Crane, 1992) using HyperCard on the Macintosh, and many circumstances obligated us to deliver Perseus 2.0 in HyperCard as well. Given the limitations of HyperCard, we decided to create a Perseus 'Atlas tool' emphasizing push-pin mapping on a wide variety of predefined backgrounds. With GRASS's open source code and the excellent support for graphics on the Macintosh, we were able to develop GRASS programs for exporting graphics (both single frames and movie sequences) and convert them to PICTs and QuickTime movies for the Macintosh. Because Perseus is being distributed on CD-ROMs, storage space was not a concern, so Perseus 2.0 includes over 2000 canned background maps, representing a range of themes at various resolutions. The 'Atlas tool' includes some spatial functions—measurement of distance, and spatial queries for objects within a user-defined region, for example—but it is emphatically not intended to be a GIS implemented in HyperCard. The result is a highly visual browser and plotting tool, with support for a limited range of spatial queries (Crane, 1994). The Atlas tool is not a stand-alone module; however, it has to communicate with the other information systems in Perseus. I have previously discussed the challenge of designing a GIS in a complex of intercommunicating information systems (Smith, 1994), but I would like to use three examples to suggest some ways in which a GIS can have an impact on archaeologists' study of ancient landscapes in an environment like Perseus.

17.3 Studying landscapes with Perseus 2.0

17.3.1 Spatial index to Perseus

Perhaps one of the most general and potentially important results of the interaction between geographic information and other information in Perseus is the functionality, for all practical purposes, of a spatial index to Perseus. Perseus's query tools and basic thesaurus-like database of equivalent names make it possible to search through archaeological and textual sources alike for an object like 'Corinth' (whether spelled 'Korinth', 'Korinthos', 'Corinthus'. . .). The results would include artefacts with Corinth as their context, textual and visual documentation about the site of Corinth, and references to Corinth in ancient texts. Users of the simple HyperCard-based Atlas tool could, for example, pose a spatial query for sites by drawing a rectangle on a map. They could then use the results of this query as input to a second query. For example, they could search in all of the ancient literature in Perseus for references to the sites in that region. The possibilities for a study of ancient landscape are, I believe, greatly augmented. Perseus 2.0 includes essentially all Greek literary texts from the beginning until Aristotle, and a number of important later works (including among many others several books of the geographer Strabo, and the entire text of Pausanias the periegete). An archaeologist could now pass directly from a map to a collection of texts referring to sites on the map. The Greek conceptual landscape can be explored from a landscape of the archaeologist's own choice.

In retrospect, this seems very obvious: all we have done is placed a spatial query in front of Perseus's existing capacity to resolve references across different information sources. Yet this was not our original intention in adding a GIS to the collection of information systems in Perseus, and its importance was largely unforeseen. It is a consequence of the synergy of previously discrete kinds of information, made possible by the generality of Perseus's design. A spatial query of ancient literature is radically different from existing tools for working with text corpora, but it raises a familiar problem to anyone interested in ancient landscapes. What spatial entities are represented in Greek literature? The Perseus GIS has a large database of sites, which correspond fairly well to many kinds of object in Greek texts—city-states, sanctuaries, and others. But many other kinds of entities are represented in Greek texts—individual monuments and features, for example. Just as a GIS offers a practical way for archaeologists to support a 'siteless archaeology' approach to fieldwork, GIS should also challenge us to reconsider the scale and nature of what we survey or excavate in Greek texts.

17.3.2 Minting coins in classical Greece

In addition to providing a spatial index of material in Perseus, the spatial perspective of the Atlas tool can sometimes reveal previously unseen dimensions of long familiar information. When the Perseus GIS was under development, one project member asked if it would be possible to analyse where city-states were located in the fifth century. Of course, I replied . . . except that our database of sites still contains very incomplete information about periods of use. We do not yet have the data to identify fifth-century sites systematically. We realized, however, that Perseus includes a large corpus of coins that have good chronological information, and while many city-states did not strike coins, if a city-state struck a coin in the fifth century, there was certainly a city-state there! As a way of quickly taking a large sample of fifth-century city-states, therefore, we searched for classical coins, and plotted the locations of their mints in the Atlas tool against a digital elevation model of Greece. It was immediately apparent that their distribution was anything but random: they were heavily concentrated in the lower elevations. Continuing to work with the Atlas tool, we queried for sites at higher elevations, and recognized many that were of fifth-century date (although this information was not in our databases).

My colleague and I thought we were familiar with the classical city-state and with classical coinage; we had assumed that city-states striking coins were essentially like others, but our assumption had been wrong. City-states that strike coins do not represent an unbiased sample of where fifth-century city-states were located. In a social geography of classical Greece, striking coins is an archaeological marker for a type of city-state that is also geographically differentiated from other city-states. Their social and economic infrastructure in some way preferred a geographic situation at low elevation far more strongly than did city-states without coinage.

It is not my intention here to try to explain this correlation fully.[5] I want to stress instead how this observation affects our thinking about ancient Greek landscapes. While numismatists have long known that some states struck coins and others did not, scholarship has not paid much attention to this distinction.[6] The unexpected correlation between geographic characteristics and the striking of coins suggests that they two groups of city-states must have differed from each other in more fundamental ways than previously proposed. We can recognize a new entity in our archaeological landscape of Greece, although it is apparently not articulated in Greek literary texts, where economic life fades or vanishes altogether from the mental map of the city-state. In the fluid environment of Perseus, a discussion about the polis prompted us to turn to an archaeologically visible set of city-states. The Atlas tool lets us consider these city states spatially and shows that we must disengage the archaeological entity (the minting of coins) from an undifferentiated concept of 'the polis'.

17.3.3 Reading Pausanias 1.1

A final example will illustrate how a GIS can enrich our reading of an ancient text. Pausanias, the periegete, is an author archaeologists frequently dip into but rarely read. This is a pity, for while his text is resolutely geographic and spatial in its organization, his narrative moves at different scales and in different directions at various times, and only a reader with a clear archaeological and spatial orientation will effectively navigate the text. I doubt that any other surviving text from antiquity presents so coherent and detailed a mental map of Greece, and archaeologists could help elucidate the method of this under-appreciated author.

Let us consider just a handful of sentences, the opening chapter of his work (Book 1, Chapter 1, divided into five sections of two or three sentences each.). Some general patterns are evident. As in all ten books of his *Description of Greece*, he begins with a brief geographic orientation from the point of view of a traveller approaching the region.[7] He makes his way more or less directly to the principal polis of the region—Chapter 1.5 has us setting off on the road leading from the coast to Athens, and with Chapter 2 we are already arriving in Athens. Athens is then described, and from this central point, Pausanias sets off on a series of itineraries, but I would like to focus on the transition in Chapter 1.2–1.4, between first approaching Athens by sea and beginning our land route into the city.

The most recent commentator and author of an extraordinary archaeological commentary on Pausanias, Nikolaos Papachatzis (1974), sees in these sections of Chapter 1 a linear progression (Figure 17.1). The points numbered 1 through 9 in

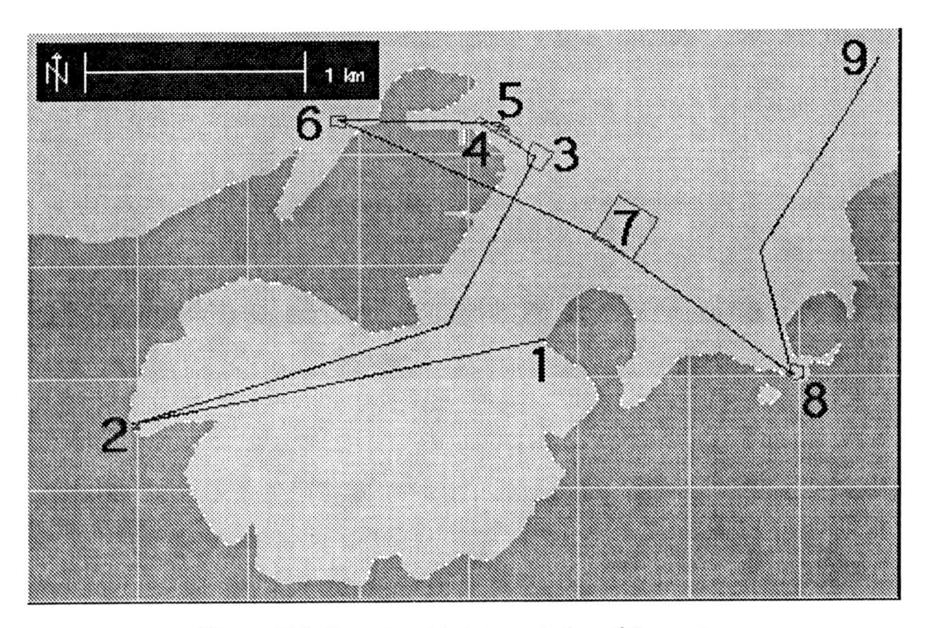

Figure 17.1 Papachatzis's interpretation of Pausanias.

the illustration occur sequentially in order 1 through 9 in Pausanias's text. In Papachatzis's view, Pausanias lands at the harbour of Zea (1 on the plan), detours to the westernmost promontory to see the tomb of Themistocles (2), before doubling back to the 'worthwhile sights' of Piraeus (3–6), and making his way on to Mounychia and Phaleron. For Papachatzis, the linear progression of the text maps directly onto a linear progression in the landscape.

I would read Pausanias's geography differently. After the initial orientation of our approach to Attica and rounding of Cape Sounion, sections 1.2–1.4 represent a transition. These sections are not, it seems to me, a linear itinerary, but an overview of the three ports of Athens: Piraeus, Mounychia and Phaleron (from northwest to southeast, although we had previously been approaching from the southeast). This survey of the three ports provides a change of scale between our orientation to the whole region in 1.1, and the beginning of our land itinerary in 1.5.[8] So 1.2 begins with 'The Piraeus . . .' and 1.4 continues, 'The Athenians have also another harbour, at Munychia, with a temple of Artemis of Munychia, and yet another at Phalerum.' I would see the twisting and winding of Pausanias's text not as a reflecting of his route, but of his mental travels, for as always his text is a description, even a definition, of Greece, culturally as well as geographically. So in 1.2, when he first refers to Piraeus, he provides the historical background to its emergence as the major port of Athens, at the urging of Themistocles. Pausanias notes in passing that Phaleron had previously been the port, and at the end of the section on Themistocles, Pausanias attaches the reference to Themistocles's tomb on the promontory of Piraeus. Appending a reference to a monument after a historical episode is Pausanias's normal practice, and does not need to imply that he is sequentially travelling to Themistocles's tomb. It is also his normal practice to describe the 'worthwhile sights' of areas he describes, and 1.3 then lists the sights of the Piraeus. As we have already noted. 1.4 then takes us to the other two ports, Mounychia and Phalerum.

How can a GIS help us assess these readings of Pausanias (apart from simplifying the preparation of the illustrations!)? Papachatzis's commentary is lavishly illustrated, and includes an original plan of Piraeus, newly surveyed for this book, with 10 m contour lines (*ibid.*, Figure 25). I digitized the plan, and using the thin-plate spline interpolation routine in GRASS (s.surf.tps) interpolated a digital elevation model for the Piraeus. From this DEM, viewsheds were generated from various points on the plan. Comparison of viewsheds from the tomb of Themistocles and the Hippodameian agora is revealing.

The tomb of Themistocles lies on the tip of the westernmost promontory of Piraeus, just outside the defensive walls of the city built by Conon. Even assuming that the original marker on the tomb was several metres high, it is invisible from almost the entire town of Piraeus (Figure 17.2). It is clear that this monument is directed outward, towards the sea, and the approach to the main harbour of Piraeus. This comes as no surprise, since it was normal practice to place tombs in conspicuous locations along the routes approaching a polis.

More surprising, perhaps, is how tidily a viewshed taken from eye-level from the Hippodameian agora encompasses all of the 'worthy sights' of the Piraeus (Plate 1). This group of sites clusters together visually, as it does in Pausanias's text, and is cut off from both Themistocles's tomb and the neighbouring port of Mounychia. If we overlay the Hippodameian street network, however, we see that the one route

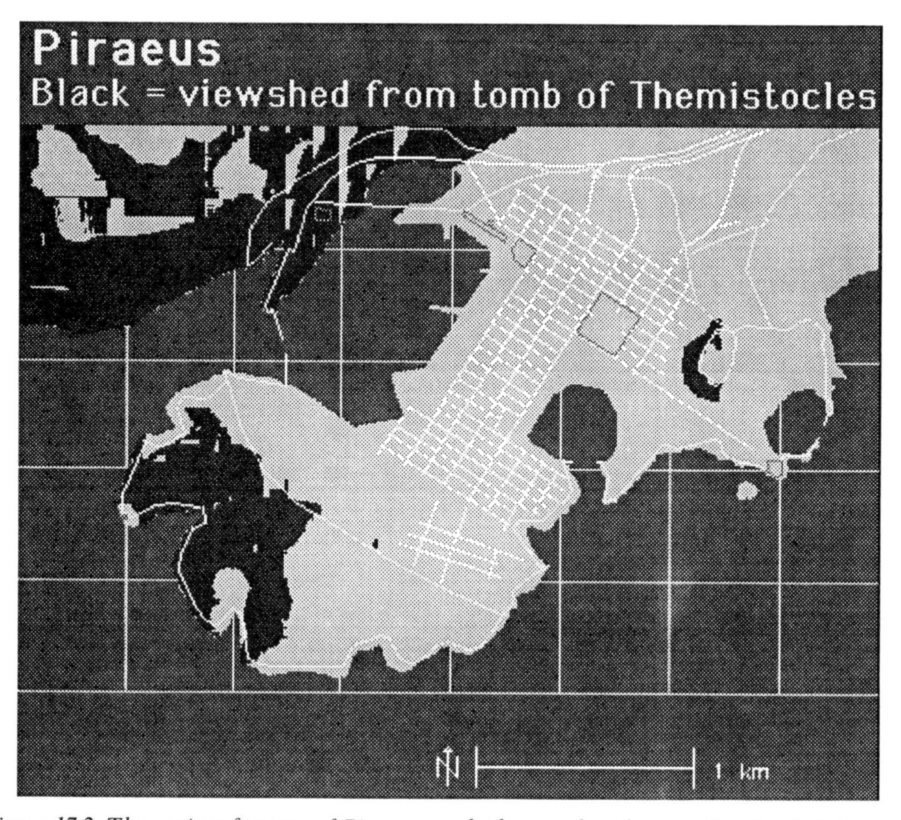

Figure 17.2 The ancient features of Piraeus overlaid on a plan showing the viewshed from the tomb of Themistocles.

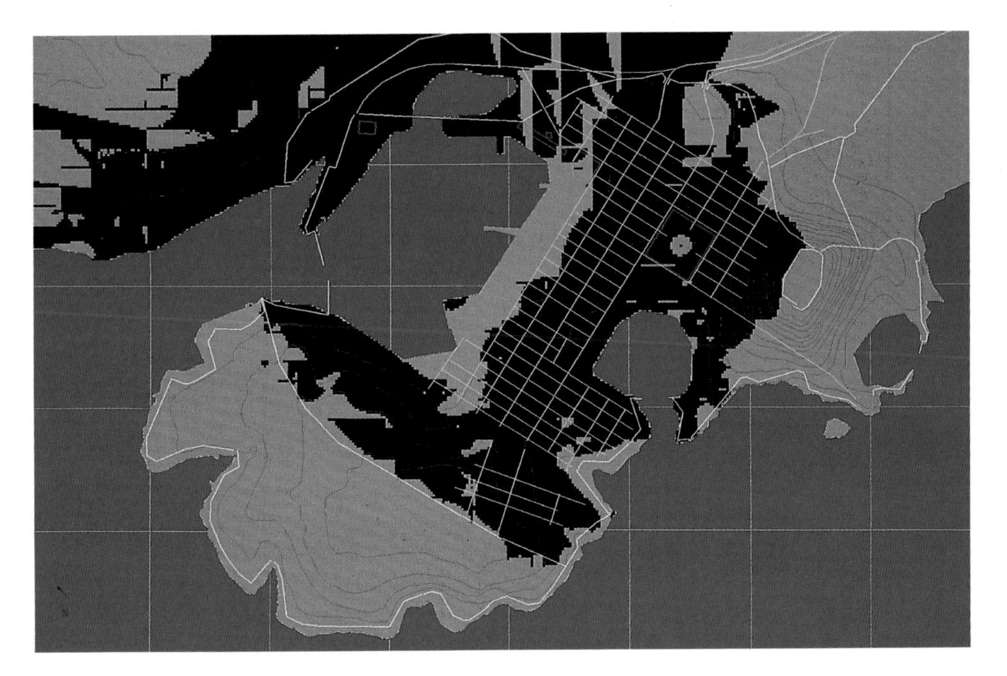

Plate 1. The ancient features of Piraeus, Greece, overlaid on a plan of the agora.

cutting across these viewsheds continues the axis of the Hippodameian agora to the one sight Pausanias mentions in Mounychia, the sanctuary of Artemis.

Thus our model of Piraeus and Mounychia created with a GIS suggests similar clusters of objects and similar spatial relations to the reading of Pausanias proposed above. Piraeus, with its sights, constitutes the main focus (1.2–1.3). They fall within a single, self-contained viewshed, and here Pausanias's text moves through them in order of a linear stroll from the main port. The tomb of Themistocles is isolated from the city spatially, just as Pausanias's reference to it is appended to a historical excursus, rather than part of his itinerary. The port of Mounychia is visually separated by its steep acropolis from Piraeus, but the neighbouring towns are linked together by a single, broad street, as Pausanias lines up the three ports. In spite of his translation of complex spatial structures into the linear form of a textual narrative, Pausanias's conceptual organization finds noteworthy parallels in the physical organization of the Piraeus peninsula.

17.4 Some implications for archaeology

These three examples have focused on how the integration of a GIS with other kinds of information systems enables new approaches to ancient landscapes. As a front-end query tool, GIS can unify and provide access on spatial criteria to information sources like text corpora. Equally important, a GIS can discriminate on spatial criteria objects that might otherwise be lumped together—such as city-states that strike coins, as opposed to other city-states. A GIS can even allow us to assess the relation of conceptual geographies, such as Pausanias's description of the Piraeus peninsula, with physical geographies of natural and human features. In one sense, this is not a particularly novel observation; archaeologists have by now become aware that GIS can be of value for its capacity to integrate diverse kinds of information.[9] The experience of developing the Perseus GIS underscores several more specific points, however.

I think it is significant that the examples described here are really just starting points for further research. In each case, the role of the GIS is not to provide a definitive answer, but to suggest new lines of research. We cannot prove with a GIS how Pausanias conceptualized the polis, but using a GIS we may find suggestive evidence in the physical landscape for the values Pausanias inscribes in his verbal map of Greece. Many archaeological applications of GIS have been directed towards narrowly focused problems—particularly in relation to predictive site modelling and cultural resource management—and we may have grown accustomed to seeing GIS used to provide definitive answers to specific problems. It may even seem unnatural to apply GIS to the kinds of exploratory data analysis described here, but this may become an increasingly significant role for GIS as it becomes just one tool in a large collection of related resources.

One challenge of GIS, to echo a familiar theme, is that GIS creates a tremendous demand for new kinds of data, in frightening quantities. I was able to look at the topography of the Piraeus peninsula only because Papachatzis happened to have published a newly measured topographic plan that included the location of ancient features. Does Pausanias generally explore areas corresponding to viewsheds, and describe in linear itineraries successions of sites following major roads or water routes? No one has the geographic data to test that hypothesis at a meaningful number of sites. We lack both the high-resolution DEMs necessary to calculate a viewshed, and precise locations of archaeological features. Global positioning systems offer a technology for recording data at adequate resolution (after differential correction), but there is an urgent need for geo-referencing familiar, well published archaeological features at high resolution. If Perseus could migrate from a site-oriented archaeological GIS derived from published gazetteers and atlases, to a feature-oriented archaeological GIS based on field work with a GPS, I believe it would enhance our ability to use not only the GIS, but all the databases in Perseus.

Finally, our ability to predict the effects of GIS and other information technologies on archaeology is not impressive. We lack any theory adequate to predict how they might affect our discipline, and need to continue the empirical process of trial and error, with the failures that guarantees. In spite of that, we might hazard that classical archaeology, with its wealth of textual sources as well as archaeological material to draw on, will find that GIS is increasingly important as a technology for unifying approaches to the study of ancient Greek landscapes.

References

Alcock, S. E., 1993, *Graecia Capta: the landscapes of Roman Greece*, Cambridge: Cambridge University Press.

Allen, K. M. S., Green S. W. and Zubrow, E. B. W., 1990, *Interpreting Space: GIS and archaeology*, London: Taylor & Francis.

Crane, G. *et al.*, 1992, *Perseus*, New Haven, CT: Yale University Press.
Crane, G. *et al.*, 1994, Perseus Project Final Report.
Frazer, J. G., 1965, *Pausanias's Description of Greece*, New York: Biblo and Tannen.
Harris, T. M. and Lock, G. R., 1990, The diffusion of a new technology: a perspective on the adoption of geographic information systems within UK archaeology, in *Interpreting Space: GIS and archaeology*, Allen, K. M. S., Green, S. W. and Zubrow, E. B. W., (Eds), pp. 33–53. London: Taylor & Francis.
Loraux, 1993, *Children of Athena.*
Martin, T. R., 1985, *Sovereignty and coinage in classical Greece*, Princeton, New Jersey: Princeton University Press.
Osborne, R., 1987, *Classical landscape with figures: the ancient Greek city and its countryside*, London: G. Philip.
Papachatzis, N. D., 1974, *Pausaniou Hellados Periegesis*, Athens.
Smith, N., 1994, *The Perseus GIS*, Washington, DC.
Snodgrass, A. M., 1987, *An Archaeology of Greece: the Present State and Future Scope of a Discipline*, Sather Classical Lectures, Berkeley: University of California Press.
van Andel, T. H. and Rennels, C., 1987, *Beyond the Acropolis: a Rural Greek Past*, Stanford, CA: Stanford University Press.

Notes

[1] Loraux (1993) gives an excellent illustration of this approach; she defines her work as an attempt to 'inscribe autochthony—the founding myth, the political myth—in the space of the town' (*ibid.*, 43).

[2] The title of van Andel and Rennels (1987) already makes the point. Particularly good overviews of this phenomenon are Snodgrass (1987, Chapter 3) and the introduction to Osborne (1987).

[3] For a notable exception, see Alcock (1993), especially Chapter 5, 'The sacred landscape'.

[4] The Perseus project is a collaborative effort, now headquartered at Tufts University, to build a collection of interrelated hypermedia databases focused on classical Greece. Contributors include archaeologists, historians, scholars of classical literature, and others. The Perseus GIS discussed here was developed by Nicholas Cahill, now of the University of Wisconsin, and myself. I am continuing to work on the GIS as part of the Perseus Atlas Project, centred at Bates College.

[5] I would offer only a few observations. Classical Greek coins in general did not travel far from their mints: they were locally overvalued in relation to their intrinsic worth, while abroad they would be worth only their value as bullion. There cannot then be a facile or direct connection between the preference for lower elevation and trade, since classical coins are generally found concentrated around their mint. In an indirect correlation such as minting coins–low elevation, it is always difficult to identify what relations are primary. Could the correlation with low elevation be a result of a preference for sites near the ocean, for example? I tested this possibility with GRASS, by measuring first the correlation between locations of mints and elevations, then measuring the correlation between locations of mints and distance from the shore. While both were, as one would expect,

strongly correlated (it would be impossible to be closely correlated to one and not the other), the correlation with elevation was much stronger. It seems to be more significant then, and the weaker correlation with distance from the sea only a side-effect of this link.

[6] A recent exception is Martin (1985). One of Martin's main theses is that it is anachronistic to identify the minting of coins as a political right with implications about the political status of classical city-states. The identification of this geographic pattern would support his argument that coinage serves social and economic functions not directly tied to the form of political organization.

[7] Section 1 begins as though approaching Attica by sea from the east: 'On the Greek mainland facing the Cyclades Islands and the Aegean Sea the Sunium promontory stands out from the Attic land. When you have rounded the promontory you see a harbor . . .' English translations of Pausanias are taken from the Perseus edition of W. H. S. Jones's Loeb volume.

[8] This idea seems implicit in Frazer's (1965) commentary, although he is largely silent on the subject of Pausanias's method.

[9] See the perceptive comments of Harris and Lock (1990).

18

The potential of GIS-based studies of Iron Age cultural landscapes in eastern Norway

J. S. Boaz and E. Uleberg

18.1 Introduction

At first glance, the application of the cultural landscape concept in coalition with GIS analysis, provides a number of intriguing possibilities in the study of prehistory. The cultural landscape concept provides an explicit linkage between cultural remains, environment and the humans that created and occupied these environments. On the other hand, GIS analysis provides a means by which large amounts of cultural and geographical data can be studied at varying scales of analysis.

This chapter presents the results of the first stage of a work in progress. The larger goal of this project is the study of the evolution of social inequality during the Iron Age in the Follo region of eastern Norway (Figure 18.1) At this stage, our goal is to evaluate whether or not GIS-based cultural landscape studies provide a useful approach to this complex problem.

In section 18.2, we present a brief overview of the current state of knowledge regarding the Iron Age in eastern Norway. This is followed by a discussion of the cultural landscape concept and how it can be related to the study of prehistory. After this, the GIS methodology that may be appropriate to the study of these concerns is described. In the concluding section, we evaluate the utility of GIS-based cultural landscape studies.

18.2 The Iron Age in eastern Norway

The archaeological record of the Iron Age in eastern Norway largely consists of burial mounds and individual finds of artefacts. Very few houses have been excavated, and while prehistoric field systems are common, there are few that can be reliably dated to the Iron Age. As such, the record of social change in the Iron Age in eastern Norway largely consists of changes in the distribution and contents of burial mounds.

The Iron Age in eastern Norway spans the period between 500 BC and AD 1030. The earliest phase of the Iron Age in eastern Norway, the Celtic or pre-Roman phase (*c*. 500–1 BC), is poorly understood, and is largely characterized by cremation burials with few grave goods, often consisting of only a few needles. The Roman Iron Age (AD 1–400) marks a dramatic change. The number of individual finds and burial mounds increases dramatically; this, combined with the increased utilization

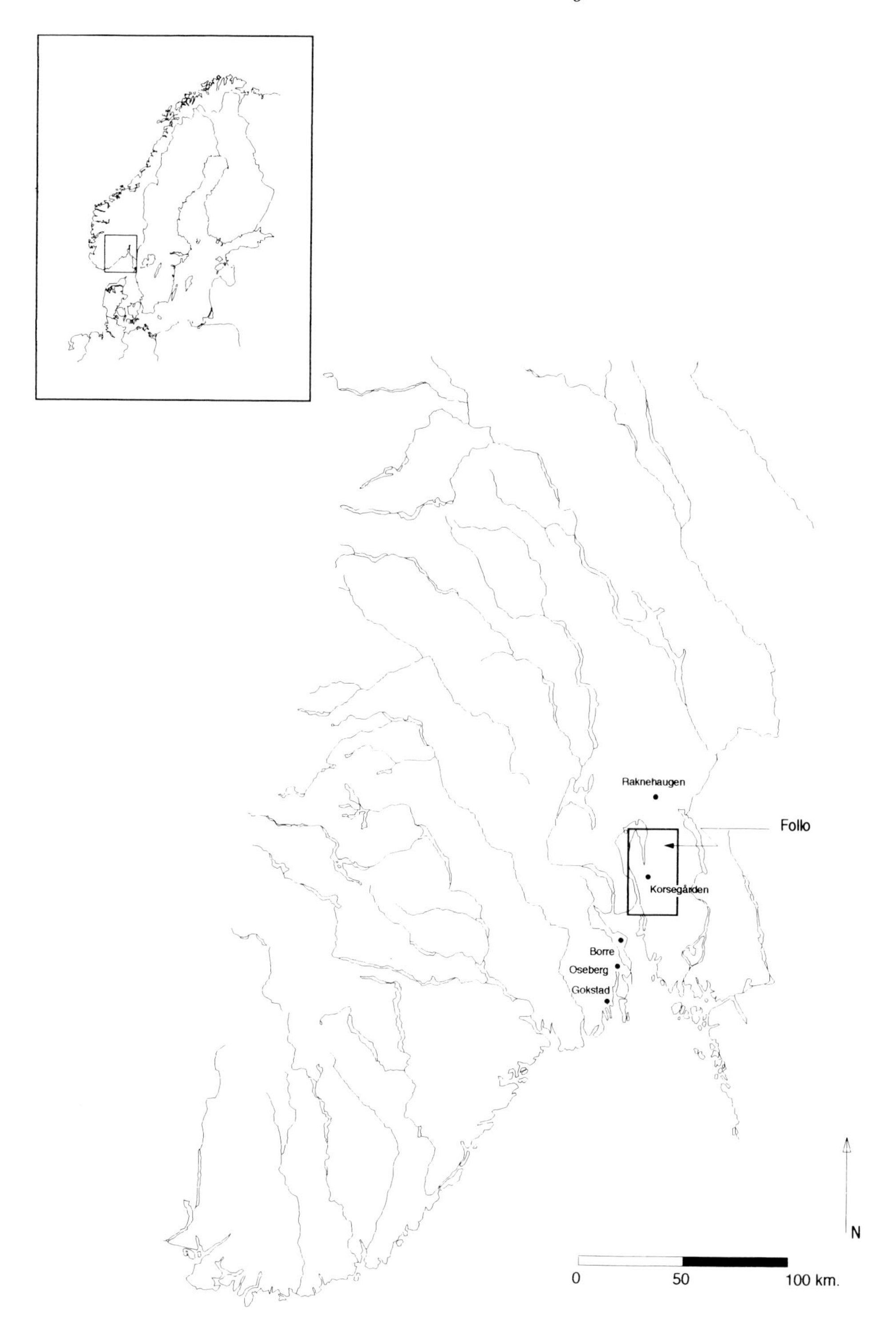

Figure 18.1 Location of study area and sites mentioned in text.

of interior regions, is generally thought to reflect population increases. The burials from this period provide evidence of long-distance trade, as well as social stratification. The last part of the Early Iron Age is referred to as the Migration period (AD 400–550), and is generally thought to mark continuity and gradual expansion from the preceding period. However, there is a great deal of variability in the contents and types of burials from this period, ranging from cremation burials to stone-lined cists, including very large burial mounds, one of which, Raknehaugen, measures *c.* 100 m at its base, and is over 15 m high.

The beginning of the later Iron Age, the Merovinger period (AD 550–800) is marked by a dramatic increase in the number of, and changes in the content and form of the burial mounds. This period is thought to be associated with a population increase as well as a dramatic increase in the number of burial mounds, and increased utilization of inland areas (Myhre, 1992: 168–72). However, grave goods are relatively uncommon. These trends continue into the Viking period (AD 800–1030) in large part marking continuity with the preceding period. Large numbers of burial mounds and other structures are also dated to this period. There is also clear evidence of increased burial elaboration, for example the well-known boat burials of Oseberg and Gokstad.

While in other areas of eastern Norway, such as Borre and Raknehaugen, ceremonial centres are found, no such centres are present in our study area. There are no large burial mounds in Follo, and the majority of the grave mounds are isolated finds, or occur in concentrations of 5–10 burial mounds. The wealth of individual finds of status objects from the Roman period tends to indicate the presence of several small centres. In contrast, the types of finds from the Merovinger period, including a ring sword, indicate that there was no centre in this area but rather that Follo was a subordinate area during the later Iron Age.

18.3 Cultural landscapes in archaeological contexts

When visiting a monumental site, every archaeologist has experienced the conspicuous use of space at these sites. Even the most ardent opponent of post-processual theory would have to agree, at the very least, that at these types of sites someone was utilizing the space to impress something on someone else. This experience of the manipulation of objects in a landscape is what lies at the core of the cultural landscape concept.

In the following sections we first discuss some of the underlying principles of the cultural landscape concept, and then discuss its applicability to the study of prehistory.

18.3.1 Cultural landscape concept

The term 'cultural landscape' has long been in use to denote a landscape that has been modified by human activity. However more recent studies have focused upon the cultural landscape 'as a mediator of the values and ideals produced in society' and 'that perceptions of reality, for example landscape, nature or environment, are produced through an interaction between individual agency and social context' (Jones, 1991: 234).

An important aspect of the cultural landscape concept is its emphasis upon the role of individual experience of the landscape. This focus, by definition, makes the cultural landscape concept a subjective one. To provide an example of this, one can examine the establishment of the distinction between the natural and cultural landscapes. Objectively seen, there is only one physical landscape; however, the determination of where the boundary between the cultural and natural landscape can, or should be, drawn varies greatly among observers. Subjectively, the same landscape can be experienced in a number of different ways, and it is therefore possible to discuss the presence of a number of different landscapes within the same geographical entity. The definition of cultural landscape becomes more complex when one is forced to admit that it is not an objective, describable definition. The description of the cultural landscape changes with, and is dependent upon, the descriptor. The same landscape will have different contents to people with different backgrounds. This brings into and retains a focus upon the role of the prehistoric individual as an active part in nature with their own, culturally dependent perception of the surroundings.

The preceding discussion may give the impression that the cultural landscape concept is solely concerned with idiosyncratic perceptions of the cultural landscape. While this is not the case, the role of perception and experience are critical variables in retaining a focus upon occupants of the area.

A wide variety of factors are involved in both the construction and experience of the cultural landscape. These factors can be described in terms of the socio-historic and physical structures (Crumley and Marquardt, 1990: 74). The socio-historic structures are political, legal and economic. The physical structures include climate, topography and geography, and are largely independent of the occupants of the area. The meaning and importance of these structures are dependent upon the different interpretation of the socio-political structures. The physical and socio-historical structures and their interpretation are 'determinative and mutually definitive of landscape' (Crumley and Marquardt, 1990: 74). Settlement traces, burials and ceremonial sites express a plan or structure, and have therefore reinforced these socio-political structures (Keller, 1993: 60). The landscape is viewed as an ideological expression, and as such, economic change or change in the social relations may reflect changes in the understanding of the meaning of the traits in the physical landscape.

18.3.2 Prehistoric cultural landscapes

At this juncture, it is necessary to digress briefly and view the cultural landscape concept in terms of the archaeological record of our study area. The most critical problem is that the cultural landscape concept is dependent upon the monuments and settlements that are visible or have been excavated. It has been noted by Roberts (1987: 83) that 'visible landscapes are like icebergs: only a small proportion of their real substance lies above the surface'.

In our study area, the lack of excavated settlement sites or studies of Iron Age field systems restrict the potential to include important landscape variables such as: the size, location and structure of field systems, or within settlement structure or the distance between settlements. As such, the elements that would have constituted the majority of the perceived landscape are not available for study.

As was mentioned earlier, the majority of the available data from our area consists of burial mounds and individual finds of Iron Age artefacts. However, the available data for both of these artefact classes, is not complete (Boaz and Uleberg,

1993). Surveys have focused primarily upon agricultural areas, with much more limited effort in the large forested areas. Further, there have been no aerial photographic studies that can allow the assessment of the extent of the destruction of such burial mounds by modern agricultural practices.

Given these limitations, one could clearly question whether or not the cultural landscape concept is applicable in the study of the prehistory of an area with such an incomplete archaeological record. If the cultural landscape needs to be studied as a totality, then this concept is not applicable in this, and in most other, prehistoric situations. However, a number of studies have suggested that the cultural landscape concept can be viewed as consisting of a number of different landscapes. Christophersen (1993: 12–14) has suggested that the cultural landscape should be considered to be a multi-dimensional phenomena in continual change. It is this multi-dimensionality that allows the study of the cultural landscape at a variety of levels of analysis. Just as a landscape can be perceived on a variety of levels, so can it be analysed on a variety of levels.

Keller (1993: 60–1) has recently suggested the use of the landscape room concept as a method for bridging the gap between cultural landscape theory and the real world. A landscape room is a part of the landscape that stands as a bounded entity. These rooms are bounded by topographical variation, and are characterized by homogeneous internal topography. The integration of the structures within and between these landscape rooms can then be viewed as a result of the interplay between socio-historic and economic factors, and reflects a plan or structure of its inhabitants. Therefore, changes in the understanding of these plans and structures may be reflected by changes in the cultural landscape.

18.4 GIS methodology

Wheatley (1993: 133) has noted that archaeological research utilizing GIS methodology has usually adopted an ecological-systems theory approach to explanation. The ecological and systemic emphasis in GIS research, with its emphasis upon the use of environmental data and quantitative methods, is hardly surprising. However, there are aspects of prehistory that can not be studied by such approaches, and the cultural landscape concept is one of these. One method of GIS analysis that we believe has a great potential in the study of cognition and experience is viewshed analysis. In the following discussion, we first present how viewsheds can be utilized in the definition of landscape rooms, and then how these rooms can be compared both temporally and geographically.

18.4.1 Definition of landscape rooms

In the first step of these analyses, we plan to use the considerations advanced by Keller (1993: 64) to define landscape rooms around the individual monuments. These variables are as follows.

1. In which direction is the burial mound oriented? Quite often these mounds are placed on the side of a natural ridge. As a result the view in the direction of the ridge is often blocked.
2. What areas are visible from the mound?
3. From what areas are the monuments visible as mounds?

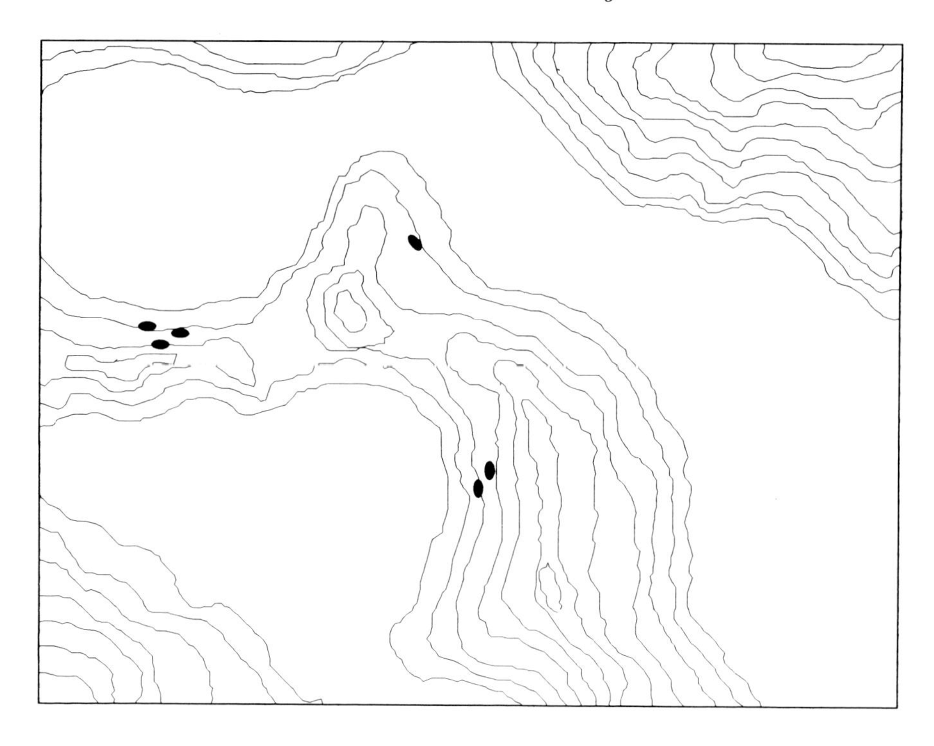

Figure 18.2 Placement of burial mounds for two different phases in a hypothetical landscape.

To provide an example, Figure 18.2 illustrates the placement of burial mounds within a hypothetical landscape for two phases of the Iron Age. While these examples are hypothetical, such locational patterns are typical in Follo. Figure 18.3 shows the landscape rooms that would be defined by the methodology suggested by Keller (1993).

The first of the variables described by Keller (1993)—essentially the slope of the location—can be easily calculated by using a digital elevation model (DEM). The second of these considerations—what areas are visible from the mounds—should be relatively straightforward to determine. However, for these analyses to be effective it will be critical to choose a DEM algorithm that does not eliminate small-scale variation (Kvamme, 1990). In many cases, the difference in the distance between the tops of these mounds and the ridges is usually only a matter of a metre or so. As such, there is the possibility that incorrect location of the mound, or lack of detail in the DEM, could produce a false picture.

The quantification of the third of these variables—the areas from which the monument is visible as a burial mound—is more problematic, from both a methodological and theoretical point of view. The visibility of a given mound is dependent upon an array of considerations, the size of the mound, the presence or absence of visual hindrances, such as trees or buildings and the distance of the observer. It is, of course, impossible to reconstruct the locations of obstacles such as forest cover, and it will be necessary to assume that such features were not present. Keller (1993: 64) suggests the relationship of the monument to the horizon will be important in the definition of the silhouette of the monument. The development of an adequate measure of this variable will, in all probability, require some experimental archaeology which, hopefully, will allow the development of an equation which can then be utilized to reconstruct a zone around the burial mound from which it is visible.

18.4.2 Characterization and comparison of landscape rooms

The second step of this analysis will utilize the capabilities of GIS methodology to describe the characteristics of these areas in terms of topography, types of water bodies and soil types, within these landscape rooms. In the case of the data from Follo, we believe that the difference between the viewshed of the individual burial mound, and the area from which these monuments are visible will provide a critical, and quantifiable, comparison of the variation between landscape rooms. For example, there will be a difference in the ratio of these zones in the landscape rooms shown in Figure 18.3. In the first of these figures, there should be a much larger difference between these two zones, while the difference between these zones in the second figure should be much less, and in some cases non-existent.

The comparison of the sizes and shapes of these zones can be done both visually and quantitatively, and the areas where these zones overlap can then be identified and described. The comparison and contrast of the size and shape of these zones, combined with standard GIS methodology will provide the means for a wide range of quantitative descriptions of the environmental variables that characterize these areas.

One of the more interesting of these is the comparison between the zone which a burial mound has visual control over, and the zone from which these mounds are visible. The difference between these two zones with each landscape room can be taken as a measure of the relative importance of the need to signal the importance

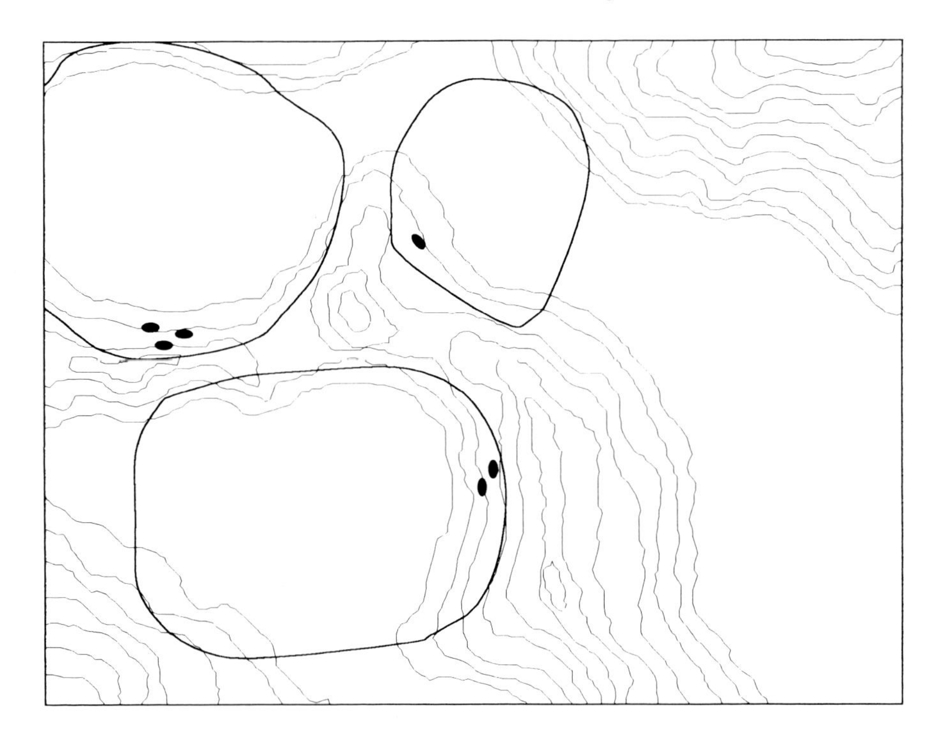

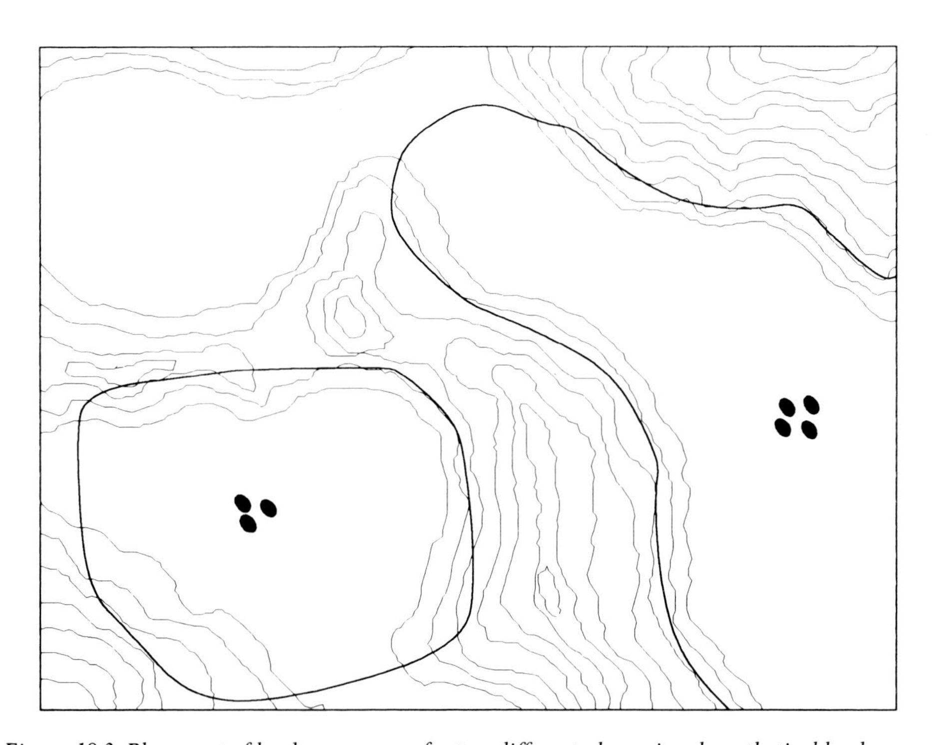

Figure 18.3 Placement of landscape rooms for two different phases in a hypothetical landscape.

of these mounds, and may well be related to control and reinforcement of group membership. As such, the use of this measure, combined with the capabilities of GIS analysis to describe the environmental characteristics of each landscape room, provides the potential for a wide variety of quantitative comparisons.

At this juncture, it will be appropriate to create descriptions of environmental characteristics from randomly generated locations (Kvamme and Jochim, 1986: 6). Another useful, and potentially more interesting, form of comparison could be between similar monuments from other time periods, such as Bronze Age burial mounds, or medieval or modern churches.

18.5 Evaluation

Every 20 years or so, it seems that the old data from the Sites and Monuments Record is trotted out of its stable, and made to jump through the hoops of the latest theory and methodology; this chapter is no exception to this tradition. The monuments that a few years ago were territorial markers, that developed as a result of the economic potential of the soil types found in the area, are now used to designate landscape rooms which may reflect the occupants' perception of the landscape in these areas. It is clear that more primary data are necessary; however, until the beginning of large-scale settlement excavations, these are the only data that are available, and if we are going to study this period, these are the data that we will have to use.

To return to our original goals, we do believe that the combination of the cultural landscape concept and GIS analysis provides a useful beginning from which to study social evolution in the Iron Age in eastern Norway.

The landscape room concept has a great deal of promise at a local scale of analysis, and allows the division of the study area on the basis of considerations that may have had some meaning to the occupants of these areas. However, an important consideration is the development of an understanding of what types of settlement units these landscape rooms represent. As noted by Keller (1993: 65–6) it is tempting to view these rooms as representing settlements or farms, and in our study area this interpretation is supported by the recent excavations at the site of Korsegården (Uleberg, 1990). While it is tempting to describe these areas as representing settlement units, in the absence of more concrete data regarding the distribution of settlements it is most likely more profitable to view them as areas that were bound together by at least a ritual connection (Dillehay, 1990). However, the question of whether these landscape rooms were occupied by a single or several settlement units will require further research.

Both Wheatley (1993: 135–37) and Gaffney and Stančič (1991: 79–81) have suggested the use of cost-surface analysis to define site territories that may have had some meaning to the occupants of the sites. However, the utility of this method can be questioned on both theoretical and methodological grounds. The use of cost-surface analysis presumes that the occupants' perception of the area was based upon the single economic variable of transport costs. While such considerations are an aspect of cognition, focusing on a single economic variable is clearly an over-simplification. Methodologically, cost-surface analysis ignores the importance of unknown or seasonally variable factors affecting transport costs. For example, in areas such as eastern Norway, the presence of snow cover, or the seasonal variation in the density of the vegetation cover can have dramatic effects on transport costs.

While a wave pattern model of travel costs (Tomlin, 1990: 134–40) provides a more realistic description of the costs of moving through a landscape, concepts of distance and travel costs are culturally bound and are more complex than are allowed for in cost-surface analysis.

In contrast, we believe that the use of viewshed analysis in the definition of landscape rooms provides a more powerful analytical tool. The explanation of the meaning of an area defined by such a means is not as precisely defined as that defined by cost-surface analysis. However, it is this ambiguity that allows for the inclusion of a variety of experiences and understanding of the cultural landscape, and as such provides room for a wide variety of understandings, which is likely to be a more accurate representation of prehistory.

To turn from a local scale of analysis—the landscape room—to the regional scale of analysis, again returns us to the concept of perception. It seems clear that an individual's perception of the landscape is, at the very least, influenced by the understanding of how the immediate landscape relates to the other landscape rooms in the immediate area, as well as at a regional scale. For example, the similarities and dissimilarities in the placement of the burial mounds in the landscape rooms across a region, or in the environmental characteristics of these rooms, or in the placement of these rooms relative to transport routes, both within a given phase and over time, provide the possibility for the study of an almost unlimited number of questions, all of which can be quantified. In the case of the excavated grave mounds, it will also be possible to search for patterns related to male and female burials, and the types and amounts of grave goods that are present. The relationship between these landscape rooms could show either hierarchical or heterarchical relationships (Crumley and Marquardt, 1990: 74–5). Similarly these relationships could vary over space both within and between different periods.

In conclusion, we return to our original question, 'Does the combination of the cultural landscape concept and GIS analyses provide a useful means from which to study the nature of social development within our study region?' We believe that the answer is yes. The use of the landscape room concept combined with viewshed analysis provides the potential to divide the study area into units that may have had some meaning to the occupants of the area. The comparison of the viewshed from, and the view back to, the burial mounds provides a powerful means to compare the nature of the placement of these monuments over time—and finally, standard GIS analysis provides the potential to characterize and compare these landscape rooms over both time and space. While a number of practical problems (data quality and accuracy, regional scale and dating) remain, we believe that such analyses will provide a number of new insights into the nature of social evolution in the Iron Age in eastern Norway.

References

Boaz, J. and Uleberg, E., 1993, Gardermoen project—use of a GIS in antiquities registration and research, in Andresen, J., Madsen, T. and Scollar, I. (Eds), *Computing the Past: Computer Applications and Quantitative Methods in Archaeology, CAA92*, pp. 177–182, Aarhus: Aarhus University Press.

Christophersen, A., 1993, Det kultiverte naturlandskapet, *Universitetets Oldsaksamling Årbok 1991/1992*, 7–14.

Crumley, C. L. and Marquardt, W. H., 1990, Landscape: a unifying concept in regional analysis, in Allen, K., Green, S., and Zubrow, E. (Eds), *Interpreting Space: GIS and Archaeology*, pp. 73–79, London: Taylor & Francis.

Dillehay, T., 1990, Mapuche ceremonial landscape, social recruitment and resource rights, *World Archaeology*, **22**, 223–41.

Gaffney, V. and Stančič, Z., 1991, *GIS Approaches to Regional Analysis: A Case Study of the Island of Hvar*, Ljubljana: Filozofska fakulteta.

Jones, M., 1991, The elusive reality of landscape. Concepts and approaches in landscape research, *Norsk Geografisk Tidskrift*, **45**, 229–44.

Keller, C., 1993, Visuelle landskapsanalyser i arkeologien, *Universitetets Oldsaksamling Årbok 1991/1992*, 59–68.

Kvamme, K. L., 1990, GIS algorithms and their effects on regional archaeological analysis, in Allen, K., Green, S., and Zubrow, E. (Eds), *Interpreting Space: GIS and Archaeology*, pp. 112–25, London: Taylor & Francis.

Kvamme, K. L. and Jochim, M. A., 1986, The environmental basis of Mesolithic settlement, in Bonsall C. (Ed.), *The Mesolithic in Europe: Proceedings of the Third International Symposium*, pp. 1–12, Edinburgh: University of Edinburgh, Department of Archaeology.

Myhre, B., 1992, Borre- et Merovingertidssentrum i Øst-Norge, in Økonomiske og politiske sentra i Norden ca 400–1000 e.Kr., *Universitetets Oldsaksamling Skrifter* Ny rekke Nr. 13, pp. 155–79.

Roberts, B. K., 1987, Landscape archaeology, in Wagstaff, J. M. (Ed.), *Landscape and Culture: Geographical and Archaeological Perspectives*, pp. 77–95, Oxford: Basil Blackwell.

Tomlin, C. D., 1990, *Geographic Information Systems and Cartographic Modelling*, Englewood Cliffs: Prentice-Hall.

Uleberg, E., 1990, En gård fra eldre jernalder i Akershus, *Nicolay*, **54**, 48–54.

Wheatley, D., 1993, Going over old ground: GIS, archaeological theory and the act of perception, in Andresen, J., Madsen, T. and Scollar, I. (Eds), *Computing the Past: Computer Applications and Quantitative Methods in Archaeology, CAA92*, pp. 133–138, Aarhus: Aarhus University Press.

19

GIS applications at the Hungarian National Museum, Department of Information

K. T. Biró and I. Sz. Fejes

19.1 Introduction

Within the Department of Information and Conservation of the Hungarian National Museum, regular training of museum staff in computer applications has been practised since 1990. This has concentrated on basic needs, but with the growing availability of more powerful computer systems and the requirements for spatial analysis, especially in the case of large-scale excavations, it was decided in 1992 to widen the scope of studies to include GIS.

The potential and possibilities of a GIS approach in archaeology are recognized within the Museum although the hardware and software requirements as well as computer skills and costs prevent the routine spread of this technique to include most practical archaeological work. In spite of many problems, mainly financial, we have attempted some GIS-based analysis. The equipment available was rudimentary: an AT-386 DX PC with VGA monitor, and a 12 × 12 inch digitizing tablet. Due to the lack of large-scale digitizing possibilities, we concentrated on pilot projects involving the intra-site analysis of archaeological data and spatial distribution studies. Emphasis has been placed on integrating other software, spreadsheet and database programs, with IDRISI to establish a more-or-less direct retrieval system for archaeological features which can be presented in the form of GIS images.

Applications so far are focused on three main areas: first, the interpretation and archaeological recording of fieldwork and find material at the lithic workshop and exploitation sites around Szentgál–Tüzköveshegy, second on the elaboration of a large multi-period excavation at the site of Gyoma–133 and third, the wider possibilities of using GIS as a tool for distribution studies.

19.2 Specific studies

19.2.1 Szentgál–Tüzköveshegy and its environs

Excavations on and around the Tüzköveshegy (Flintstone Mountain) at Szentgál have been in progress since 1983. In the course of the work, several satellite settlements have been found around the lithic material source together with various activity areas at the source itself. GIS techniques were used first to solve very basis

Figure 19.1 Szentgál–Füzilap. The excavations of 1985–88 showing lithic artefact densities.

problems of distribution density within the site (Biró, 1994). The site plan was registered using the coordinates of the trenches and sections and the counts of lithic artefacts, by number, were input via a spreadsheet and text editor. Using area calculations, the density of stone tools within the excavation area was determined which suggested centres of stone-working activity at the site (Figure 19.1).

During the excavations of the 1993 season, some of the data for finds and features were registered in IDRISI (Figure 19. 3). A parallel sondage, consisting of a double line of test pits 10 m apart, was used to delimit the extent of the workshop area. One of the activity areas identified by the sondage included a depot of hammerstones which was excavated in 5 cm layers, each one being registered as a GIS coverage (Figures 19.4a,b). Thus the images can be superimposed to produce an approximate three-dimensional model of the section. The main purpose of this was to create a demonstration for exhibition purposes within the museum where it is planned to show the excavations and environs as a DTM (Figure 19.2) Problems encountered here include the choice of an appropriate resolution and, for the recording of mining shafts, an adequate representation of the original surfaces in three dimensions.

19.2.2 The site of Gyoma–133

Within the framework of a major project of the Archaeological Institute of the Hungarian Academy of Sciences, the Alföld Microregional Survey Project, a multiperiod site was excavated in the vicinity of Gyoma, SE Hungary. The settlement is called Gyoma–133 and dates in parts to the Neolithic, late Bronze Age (Gáva culture) and to several early medieval periods assigned to the Sarmatian and Avar peoples. The complete area excavated comprises 14 600 km^2.

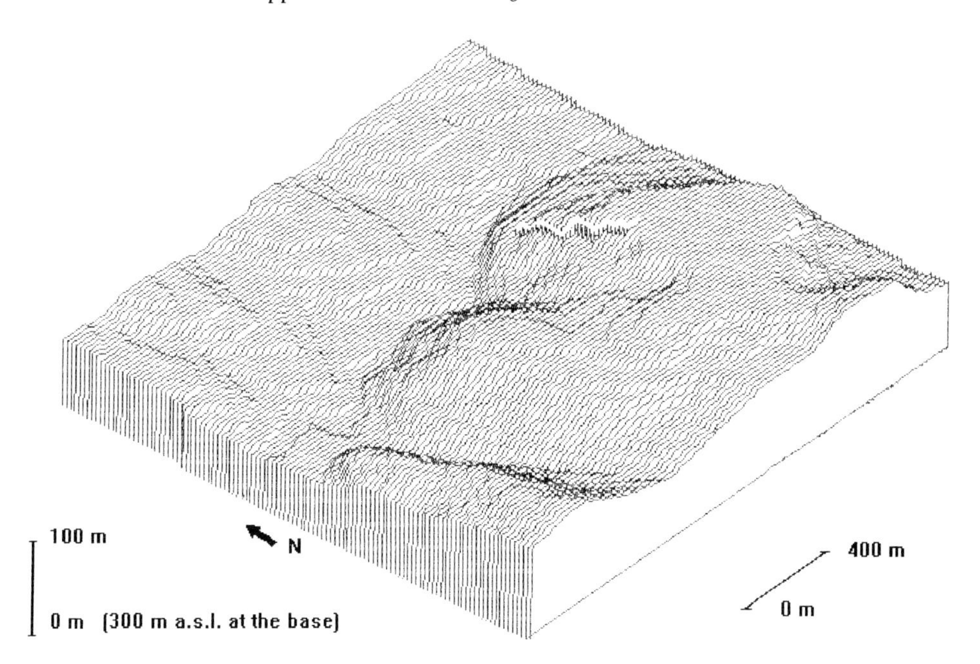

Figure 19.2 The DTM of the Tüzköveshegy with the location of the test pits.

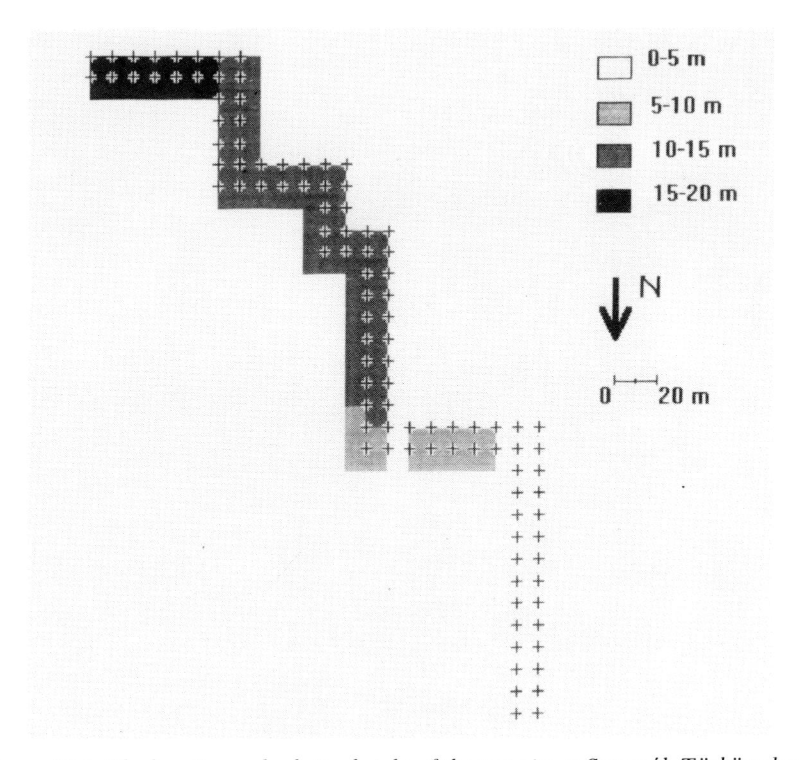

Figure 19.3 The location and relative height of the test pits at Szentgál–Tüzköveshegy.

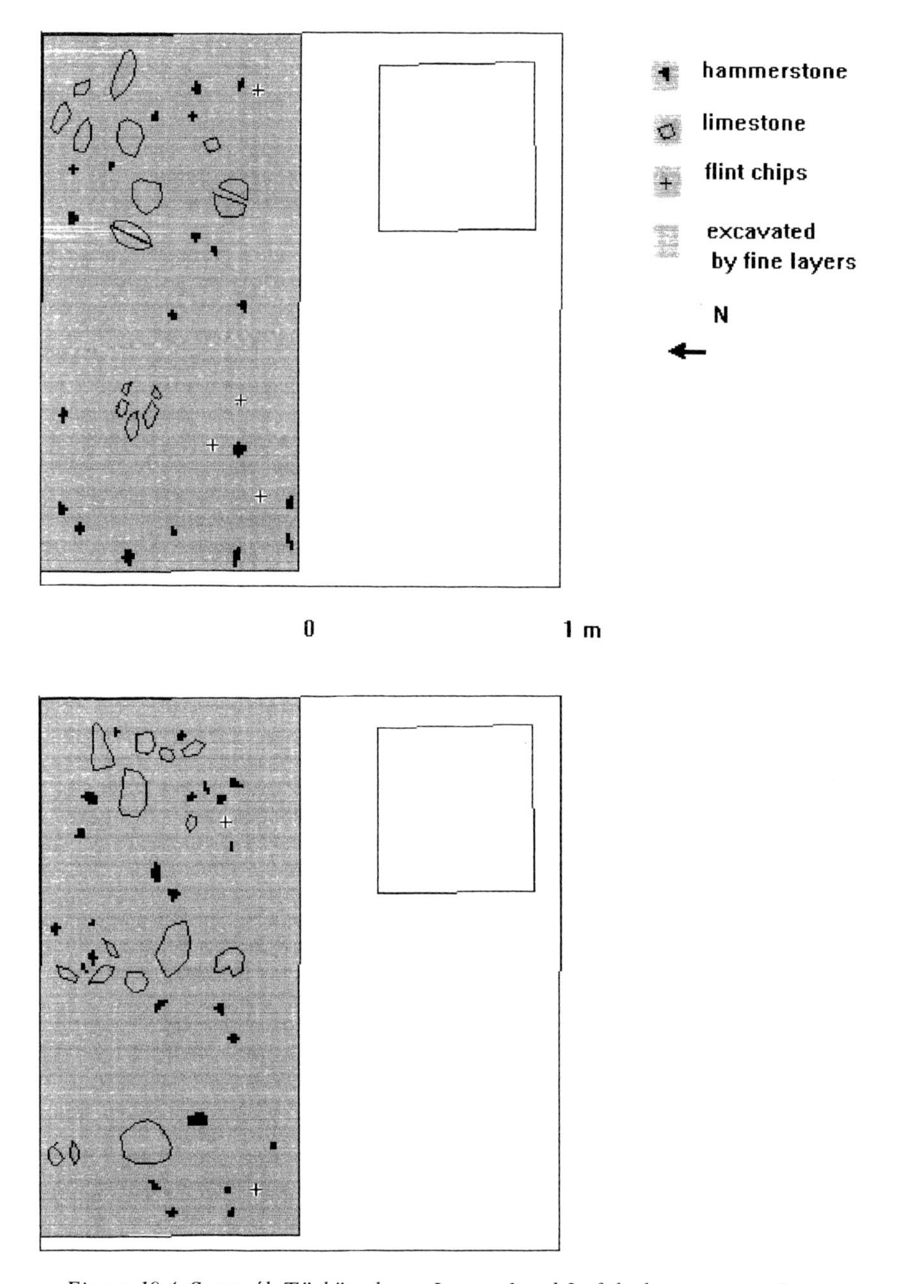

Figure 19.4 Szentgál–Tüzköveshegy. Layers 1 and 2 of the hammerstone depot.

The archaeological interpretation of the site involved a variety of data and the intensive study of refitting pottery allowed a very fine chronology within a large settlement practically devoid of stratigraphy. After a primary analysis of the material by traditional archaeological methods, GIS techniques were applied to the excavation data. The site plan was digitized and linked to a relational database and a GIS allowing the three-dimensional representation of archaeological and natural features and the generation of spatial statistical data (Figure 19.4).

Technical considerations
Starting where traditional archaeological analysis ended had its advantages as well as drawbacks for our GIS-based study. The primary archaeological work resulted in a large amount of data. Information was available by feature number, with a summary of data for each feature consisting of its character, maximum depth, types and numbers of contained finds.

The Gyoma–133 site was planned according to normal archaeological practice at various scales ranging from 1 : 10 for detail to 1 : 1000 for the overview plan of the site. For GIS input we digitized twenty-one A4 sheets at 1:200 scale (10 m equals 5 cm). Consequently, our range of accuracy is on the order of a few dozen centimetres, which is adequate to handle the archaeological features as units with statistically relevant information. The raster size for the GIS analysis was set to a resolution of 0.5 m resulting in archaeological features under the size of 0.5 m^2 being represented by points, i.e. one pixel. Another problem was the three-dimensional representation of the excavated surface because, for most units, we used the maximum depth of the feature based on the site report which does not allow us to give a more realistic model of the original surface.

Several problems were encountered in the course of digitizing: for example we had to name (i.e. number) all features that had no numerical identifiers; also, complex features with several numbers had to be considered as one. Much experimentation and learning of the software resulted in strategies being developed. For example, it was easier to perform vector transformation within a spreadsheet by manipulating the x, y shift of coordinates and then set the scale by modifying the IDRISI.dvc files. Once in IDRISI, the whole image was much easier to manipulate and using 'reclass', it was possible to give new identifiers to the features.

When building the three-dimensional image, however, we hit the limits of our system and it was necessary to return to other software because too many features—over 400—had to be modified at a time. Unfortunately, the limits of the spreadsheet were also encountered so a relational database was used which also offered a solution to attaching statistical information to the image. The database was also used to modify the site plan and its details according to object number and features by

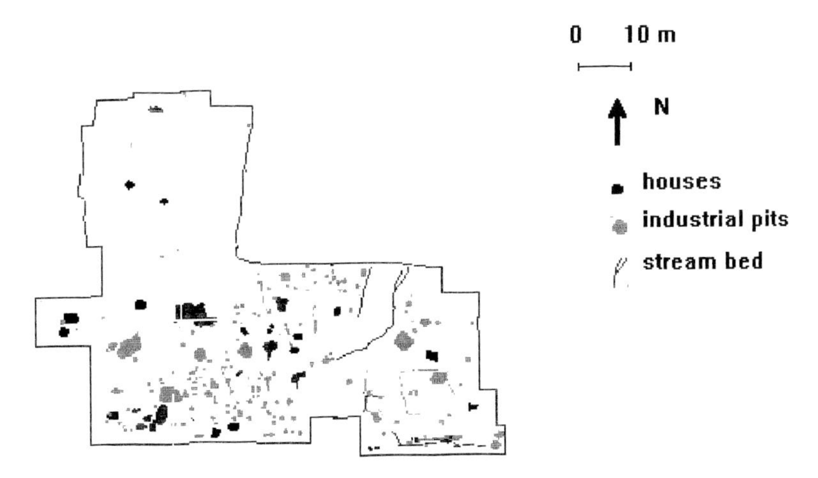

Figure 19.5 Gyoma–133. The excavation area and archaeological features.

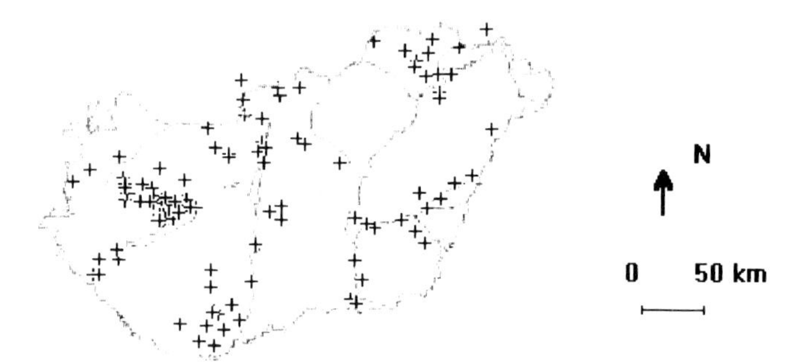

Figure 19.6 The base map of Hungary showing the main hydrographic features and the main archaeological localities for lithic distribution studies.

importing identifier and coordinate information in the format necessary for an IDRISI.vec file. Plans categorized by feature number, feature type (Figure 19.5), the three-dimensional surface, together with the statistical interpretation of finds were prepared.

Our work was not necessarily made easier by the use of GIS. A basic problem was transferring data obtained by traditional archaeological methods to computer, as the original recording was designed for paper records only. This inevitably involved certain compromises mainly regarding resolution and accuracy. Finally though, the Gyoma–133 data are ready for archaeological questions and we hope that interesting new correlations can be demonstrated using these tools.

19.2.3 Distribution studies

Using existing coordinate data, the base map of Hungary, as well as the localities considered for lithic distribution studies (Figure 19.6) were entered into IDRISI.

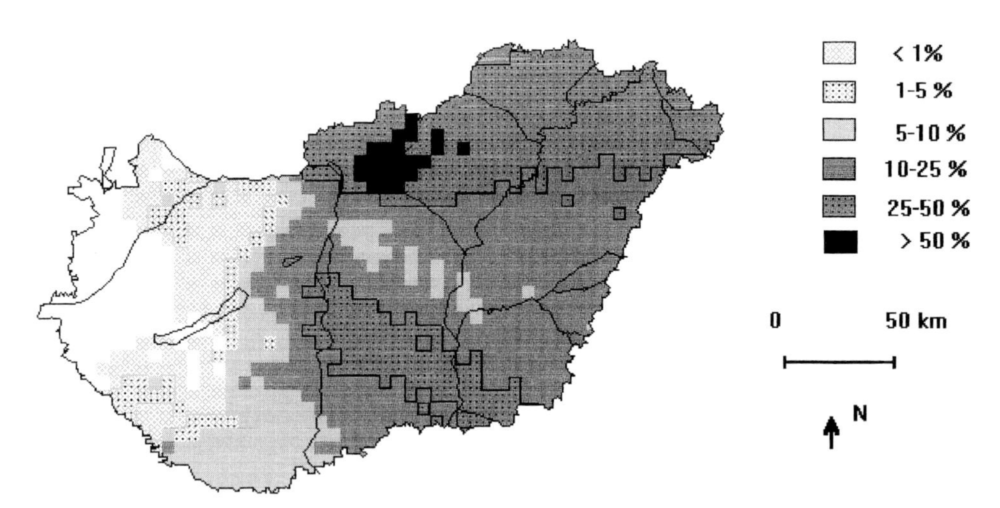

Figure 19.7 Obsidian distribution in Hungary showing the percentages of assemblages. No filters were used on age or size of assemblage.

Following previous spatial statistical analysis on percentage distribution of raw materials using SURFER (Biró, 1991), the 'Interpol' facilities of IDRISI were used to obtain distribution percentage data in the form of a distribution map (Figure 19.7).

19.3 Conclusions

GIS techniques obviously have great potential. Our limited experience in their use gave us only an insight into the advantages we can expect from this type of approach, primarily visualization. Statistical features of GIS, however are also going to be of great importance in our work, in the presentation of spatial and statistical information.

References

Bartosiewicz, L., Berecz, K., Choyke, A., Medzihradszky, Z., Vaday, A., Vicze, M. and Vida, T., in press, Site Gyoma–133 of the Békés County Microregion Research Program in Bökönyi (ed.) Cultural and Landscape Changes in South-East Hungary.

Biró, K. and Regenye, J., 1991 Prehistoric workshop and exploitation site Szentgál–Tüzköveshegy; *Acta Archaeologica Academiae Scientiarum Hungaricae*, **43**, 337–75.

Biró, K., 1991, A kései neolitikum köeszközei és nyersanyagforgalma a magyar Alföldön, unpublished PhD thesis, Budapest, 1–180.

Biró, K., 1993, Computer applications in the study of the Szentgál–Tüzköveshegy mine and workshop complexes, Múltunk jövôje '93, Hungarian National Conference on Computer Applications in Archaeology, Budapest.

Biró, K., 1994, A Szentgál-Füzi-kùti kèsöneolit településs (lithic material of the later Neolithic settlement Szentgàl-Füzi-kùti, *Veszprèm Megyei Mùzeumock Közlemìnyei*, **19/20**, 89–118.

Vaday, A. and Fejes, I. 1993, Data analyses and interpretations based on information content in practical and theoretical archaeology, Hungarian National Conference on Computer Applications in Archaeology, Budapest.

20

GIS and excavation: a cautionary tale from Shepton Mallet, Somerset, England

S. Biswell, L. Cropper, J. Evans, V. Gaffney and P. Leach

20.1 Introduction

The use of GIS within European archaeology is no longer news. The hardware and software costs of systems are so low that virtually every institution, and many individual archaeologists, can have access, if they wish, to some sort of GIS. The results of this situation is that GIS is taking its place along with databases, word processers and desktop publishing systems as a computer tool which will be used on a daily basis by many archaeologists. Along with the increasing availability of the systems themselves, there is also a growing body of literature devoted to GIS; either in specialist journals devoted to the technology or in case studies of archaeological applications (Allen *et al.*, 1990).

On the basis of the available evidence, it might be presumed that arguments for or against GIS applications within archaeology have been won. However, such a belief would be premature. A review of the literature suggests that whatever consensus may be emerging about the essential utility of GIS for archaeologists, current uses have been very restricted in their application. Virtually all published works relate to landscape studies, the analysis of large-scale survey data or the management of cultural resource data at the landscape level (Gaffney and Stančič, 1991). It is worth considering the reasons for such a trend. However, while this pattern can be observed within the North American, as well as the European literature, we wish to consider the reasons for such a situation within the specific context of British archaeology.

An initial and somewhat simplistic observation can be made at the outset. It is very clear that most of the individuals responsible for the dissemination of GIS within Europe have, by and large, held research interests which have been primarily related to large-scale survey and/or some form of landscape analysis (Gaffney and Stančič, 1991; Lock and Harris, 1991; Kvamme 1989). While explaining the initial limited take up of GIS within a relatively narrow community, this observation does not explain why such a group should be more interested in GIS than, for instance, archaeologists who are more active in site-based research. A more fundamental explanation probably lies in the nature of the analytical modules within the GIS software suites available to archaeologists during the early 1990s. The majority of modules tended to have been designed for landscape analysis and to use mappable landscape data. In such a context, it is not surprising that survey archaeologists have been relatively successful in applying GIS technologies to their data. Indeed, it would have been a considerable cause for concern if this were not the case.

However, if we assume that many of the spatial analytical procedures used by archaeologists are as applicable at the site as at the landscape level, some further explanation of the *status quo* is necessary. Here we may make several fundamental observations on some of the differences between much archaeological research at the site and at the landscape level. Without wishing to be condescending, it can be stated that much of the data and the analytical procedures used within many landscape analyses have tended to be relatively simplistic. The processing of much landscape data is also relatively cheap. In this context, we should also note that the rapid development of landscape archaeology and extensive archaeological survey in Europe during the during the late 1970s and 1980s occurred largely because survey techniques were believed to provide a cost-effective alternative to excavation, and in some areas of Europe were perceived as an alternative research route in countries where excavation licences were becoming more difficult to acquire.

This situation contrasts with that associated with excavation. Excavation is neither simple, nor cheap. The procedures relating to good excavation have become increasingly sophisticated and horrendously expensive as our understanding of site-formation processes and recovery practices have improved. Post-excavation costs have also risen dramatically, as more and more specialist procedures are regarded as standard practice.

There is nothing wrong with such a situation—provided that you can fund the work. Indeed, on the surface, it might appear that GIS should add little to the ongoing costs of an excavation. After all, spatial analysis within site-based archaeology is not an innovation linked to GIS, the concepts have been known for years—GIS should be a more efficient manner of carrying out this work. The thoughtful use of databases within GIS and the transfer of data capture from AutoCAD to analysis within ARC/INFO is not such a great conceptual step.

There are, however, reasons to believe that the applications of GIS within site-based archaeology are not quite so straightforward, and we should try to understand some of the reasons behind this situation. At least some of the blame must be attached to recent developments of the structure of British archaeology, particularly with respect to funding of site-based work. Direct funding revenue of field institutions from central authorities has diminished from *c.* 70 per cent in 1978 to *c.* 33 per cent at the start of this decade (Spoerry, 1992). The shortfall in funding during this period has been almost entirely met by an increase in developer funding which now accounts for nearly 50 per cent of all funding within Britain. This trend has been accompanied by a move towards competitive tendering for archaeological contracts and the proliferation of organizations seeking such funding.

How are these observations relevant to GIS and excavation programmes? The move towards developer funding and competitive tendering actually has major implications for archaeological field methodologies and data analysis strategies. These activities can be correlated with GIS data collection and preparation phases, activities which represent the greatest cost in any primary project which does not have ready access to digital data (Kvamme and Kohler, 1988).

The provision and preparation of site-based spatial data (an essential within any GIS project) is therefore extremely expensive, and it is generally accepted—on the basis that the polluter pays—that the developer should provide funding for such provisions. In opposition to this situation is the unfortunate fact that developers will usually pay as little as they possibly can. This situation is made more problematic by the lack of a consensus among archaeological curators about minimal spatial data collection requirements, despite a number of recent publications relating to the preparation and content of archaeological briefs and specifications, (Institute of Field Archaeologists, nd; Association of County Archaeologists, 1993). There is, therefore, a conflict between the curators' desire to investigate any archaeological remains to the highest standard possible and the developers' aim to proceed with the least cost that is acceptable.

We should be clear about this situation. This is not simply a curatorial problem. Although exceptions may be cited (Brereton, nd; Lobb, 1988; Shott, 1987), there is a fundamental lack of an archaeological literature relating to the techniques of archaeological evaluation and the effectiveness of chosen methodologies within archaeology as a whole. The result of such a situation is the frequent implementation of evaluation schemes which all too often have little relevance to archaeology beyond the fact that they cost no more than the developer is prepared to pay. One indication of such a situation is the monotonous reference to the use of a 2 per cent sample within evaluation briefs. This is not indicative of the relevance of such a sample; it simply reflects the lack of any alternative methodologies or techniques of assessing methodological procedures or needs.

In those cases where excavation is more extensive, the situation is slightly different. Here, developer funding and competitive tendering has resulted in a shift in emphasis away from the spatial and towards the more strictly chronological and functional. Function, in this situation, largely being defined by the simple presence or absence of particular finds or structures, rather than a more sophisticated analysis of the spatial configuration of archaeological material. Archaeology is much cheaper this way. Unfortunately, the ultimate result is the development of what can only be described as a budget-based archaeology.

Here it must be emphasized that in making such an assertion we are not trying to suggest that archaeology was better prior to competitive tendering. This would be totally untrue. The introduction of competitive tendering has probably raised standards of field archaeology enormously. However, priorities may well have changed over time.

20.2 Excavations at Shepton Mallet

The potential implications for GIS-based applications of a funding base which can, on occasion, pay only lip service to the needs of spatial analysis are significant. These can be illustrated with reference to a GIS analysis carried out as part of an excavation and post-excavation programme associated with the Roman settlement at Shepton Mallet in Somerset (Figure 20.1).

The Roman site at Shepton Mallet lies adjacent to the Fosse Way, the Roman road which ran south to Exeter and Ilchester and north via Bath and Cirencester. The existence of Roman finds associated with the site were recorded as early as 1864, when a series of pottery kilns were discovered during the construction of the Anglo-Bavarian Brewery (Leach, 1991). More than a century later, in 1988, the discovery, by a metal detector user, of an inhumation burial within a lead coffin and a number of other metal finds also indicated the existence of a site of some significance. In February 1990, plans for the development of the area for domestic housing

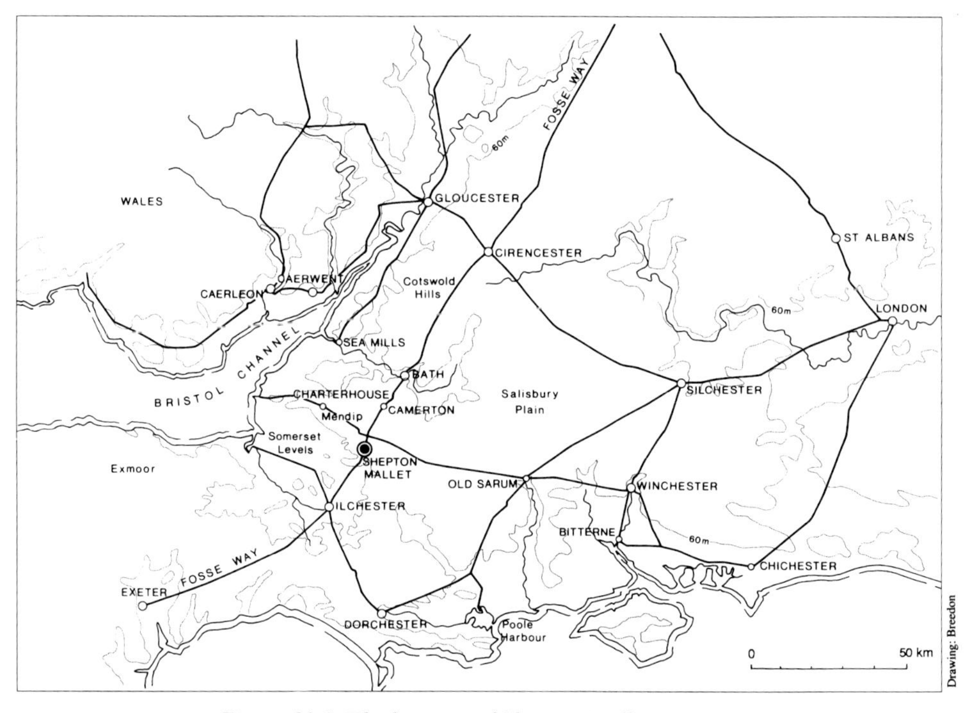

Figure 20.1 The location of Shepton Mallet, Somerset.

and a large warehouse prompted the commissioning of Birmingham University Field Archaeology Unit (BUFAU) to carry out an evaluation of the site. This evaluation exposed part of a Roman cemetery and indicated the presence of substantial surviving archaeological remains. The extent and preservation of these remains demanded extensive excavation.

Over a period of three months, the BUFAU field team investigated an area of *c.* 2.5 ha. This intervention provided evidence for a dense complex of enclosures, buildings and tracks, as well as locating numerous other features including ovens and graves. The evidence suggested that the area under investigation lay within a more extensive Roman settlement, probably a small roadside town. Apart from the evidence for domestic structures, the majority of features represented boundary walls and ditches defining a series of plots, most of which we can assume to be agricultural in character, but some of which had specialist functions including use for burial. Presumably, the area actually fronting the Fosse Way, which could not be excavated, was more heavily developed.

Topsoil was removed by machine within the investigated area. Beneath this, almost the entire area was cleaned and planned photogrammetrically. Remains which could be preserved *in situ* were planned. Where development threatened total destruction, archaeological features were investigated more thoroughly. Finds were located across the entire site and were bagged individually and recorded using an EDM. A total of *c.* 44 000 finds were recorded from 20 000 individual contexts or spatial units.

Despite the relative intensity of the archaeological investigation, it was clear from an early stage that there were a number of fundamental problems with the data. The nature of the limestone bedrock and the intensity of later agricultural practices suggested that timber structures were unlikely to be located. Post-depositional disturbance also meant that the majority of finds occurred outside traditional archaeological contexts and could effectively be interpreted as unstratified material.

Photogrammetric plans of the site were transcribed and digitized within AutoCAD 12. Unexcavated features were phased, retrospectively, through the study of these plans and through the recorded relationship of excavated and unexcavated features. All finds data were held within a FOXPRO database.

By British standards the excavation and post-excavation project was reasonably well funded, costing *c.* £250 000, with this sum breaking down roughly by 1:2 in favour of the post-excavation budget. Use of a GIS for the purposes of spatial analysis was not, however, included within the original post-excavation programme. Initially, any spatial analysis which was deemed necessary would have taken place simply with reference to the visual assessment of point data distribution over AutoCAD plans. However, in 1991, a GIS-based analysis was costed into the post-excavation budget.

Analytical restrictions were imposed on any attempted GIS study by differential preservation of deposits and the selective excavation strategy adopted during fieldwork. Consequently, at the start of the post-excavation programme the principal aims of the GIS-based analysis were:

1. to try to integrate the structural and material databases, held in AutoCAD and FOXPRO respectively, in a meaningful manner;
2. to use the finds distributions to isolate activities areas, or finds concentrations associated with buildings which had either been destroyed during the post-Roman period or were timber buildings which were not located or did not leave any trace; and
3. to retrospectively utilize finds data to confirm phasing derived from the photogrammetric plots.

Although the unit currently runs a variety of GIS platforms, at the time the Shepton Mallet project was carried out IDRISI, the raster PC-based GIS, was the only software available within the Unit for such work. Although IDRISI has many limitations, it proved to be quite an effective vehicle for the Shepton Mallet study. The raster base was particularly suitable for the work and data were readily interchangeable between AutoCAD, FOXPRO and IDRISI via standard ASCII data and script files. Analysis was carried out using a cell resolution of down to 10 cm.

Analysis of the artefact data and the photogrametrically derived plots suggested that the Shepton Mallet data could be broken up into seven principal periods (Figure 20.2).

- Period 1 is dated from the late first to the early second century AD. Evidence for this period is largely represented by a series of field boundaries and tracks with no evidence for built settlement structures. Presumably these existed on the Fosse Way frontage.
- Period 2 is dated from the later second to mid-third centuries. The first of a series of structures appear during Period 2. These include building VII. This structure was destroyed by fire and levelled, to be succeeded by later structural phases. Consequently, the building has provided one of the best stratigraphic sequences on the site.

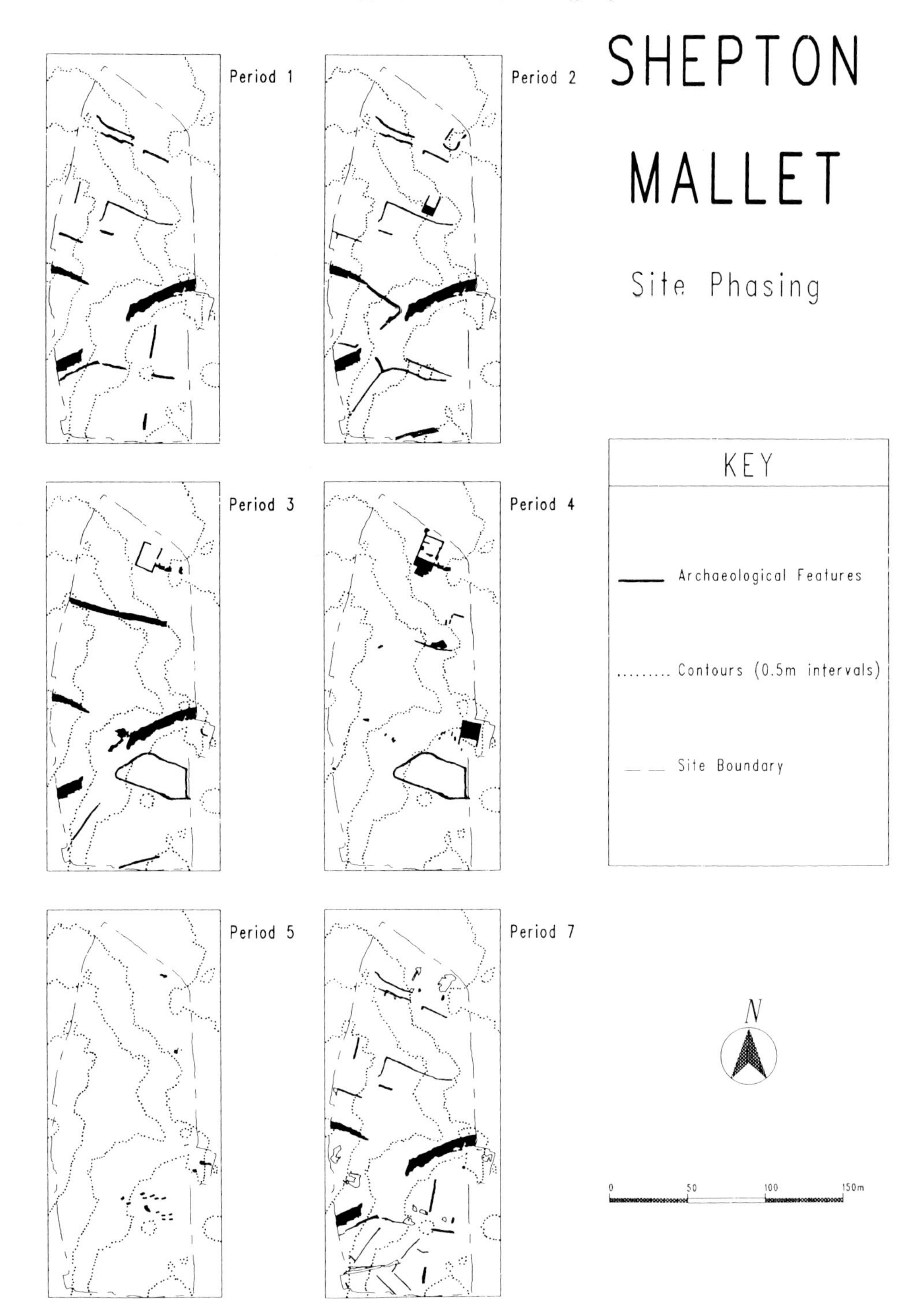

Figure 20.2 Site phasing.

- Period 3 is dated from the later third and early fourth centuries AD. Period 3 is very much an intermediate period. Some of its principal features originate in Period 2, while several of its key features, including building IX, and the large enclosure (Area 3) which is constructed in the south eastern part of the site, have functions which are actually defined better in later periods.
- Period 4 is dated from the fourth century. There is considerable development within Period 4. Perhaps the most notable feature associated with this period is the expansion of building IX—a large house with an aisled hall, and the construction of buildings I and VIII.
- Period 5 is dated from the later fourth and early fifth centuries. While the structural evidence for Period 5 is relatively ephemeral, it contains some of the most important archaeological data from the site. The period is primarily associated with a series of graves, most of which are contained within the enclosure initially constructed during Period 3. Excavation of one of these graves exposed an inhumation burial associated with a silver chi-rho amulet dated to the early fifth century. This amulet and burial is probably one of the most reliable indications of an early Christian burial in Britain and it is possible that the owner of the amulet was a priest. A replica of the Shepton Mallet amulet was worn by the Archbishop of Canterbury at his investiture in 1991. The evidence suggests that there was a small Christian community at Shepton Mallet by the fourth century.
- Period 6 contains all features which are post-Roman in date, while Period 7 contains all features for which there was no available data evidence. Neither were included in subsequent analyses.

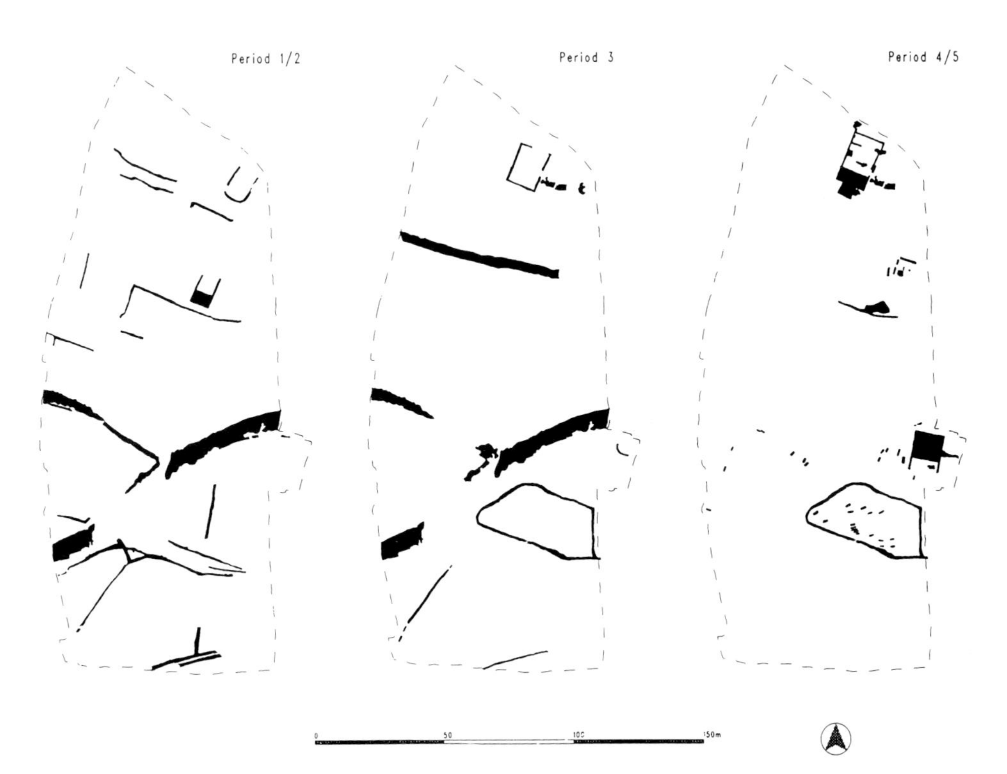

Figure 20.3 Principal periods used during the Shepton Mallet GIS analysis.

Consideration of the archaeological evidence suggested that the data should be divided into three periods for analytical purposes. These are shown in Figure 20.3.

- Period 1/2—the early Roman period dating from the first to mid-third centuries AD.
- Period 3—the late third to early fourth centuries AD.
- Period 4/5—the later Roman period dating from the early fourth to early fifth centuries AD.

Initial analysis of the Shepton Mallet data was primarily distributional. The main thrust of the GIS analysis was to import artefactual data into IDRISI and to use the software's overlay facilities and mathematical operations to form groups of disparate objects for visual comparison with the graphic structural database. Although essentially a simplistic use of such software, this procedure was of some considerable value.

Within Period 1/2, the distribution of pottery related to this period shows a number of clear associations with surviving buildings (Figure 20.4). However, several concentrations are not associated with known structures. One of these (A) may relate to buildings which lie outside the trench, perhaps in association with the original Roman road frontage. However, the relationship of the material with undated cobble spreads suggests that we may be dealing with a timber structure which has not survived.

If this is the case, the association of Samian pottery with built structures and the increased loss of coins in association with these structures suggests that lower-

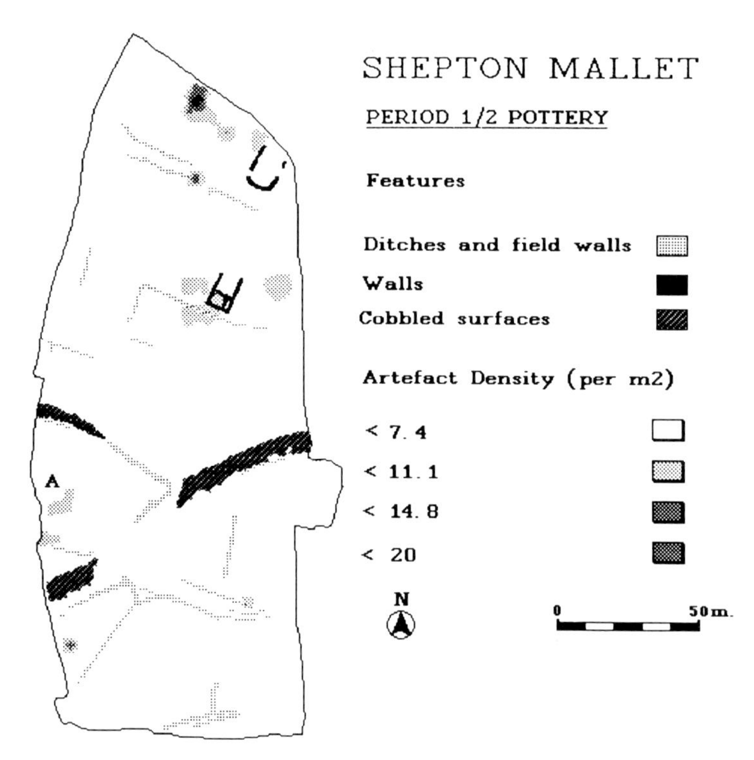

Figure 20.4 Period 1/2: pottery and cobbled surfaces.

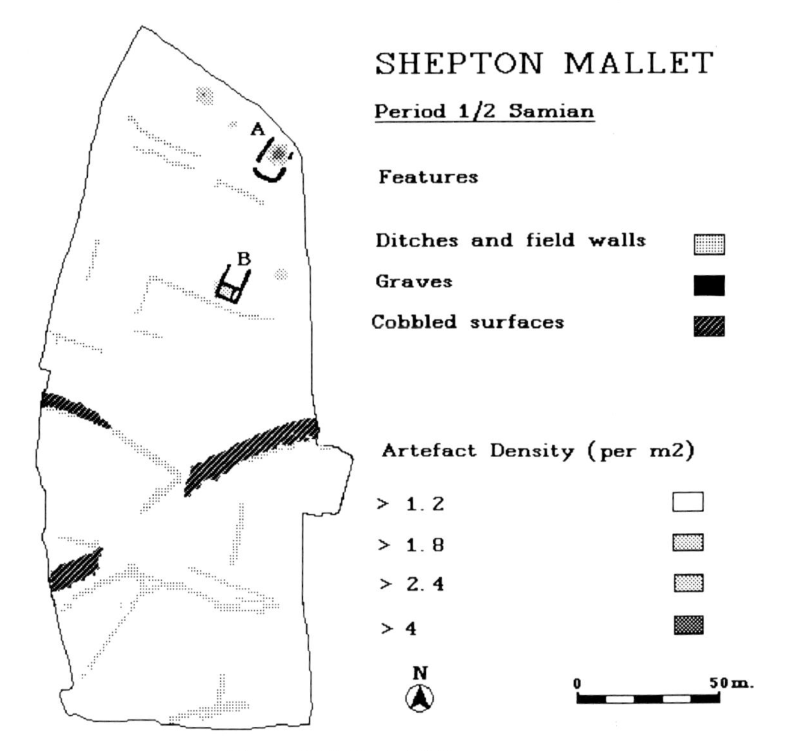

Figure 20.5 Period 1/2: samian.

Figure 20.6 Period 1/2: Coins and later cobbled surfaces.

order pottery foci were of different status or function (Figures 20.5 and 20.6). However, we must be cautious in interpretation. Analysis suggested that many classes of artefacts tended to cluster in areas adjacent to structures. This may indicate that artefact distributions relate in part to differential discard behaviour, perhaps associated with work areas. Certainly pottery types associated with food preparation are associated with most pottery foci, suggesting that whatever the status of these scatters, domestic activities were associated with most structures (Figure 20.7).

As suggested earlier, Period 3 is poorly defined and contains much data that relates it to both earlier and later periods (Figure 20.8). The distribution of artefacts associated with this period also illustrates its equivocal nature. Pottery foci relate strongly to later structures, suggesting either some taphonomic process at work, which we do not understand, or that the initial phasing of the site may not be correct. Once again, there are also a number of artefact foci (C and D) which do not correlate with known structures (A and B), suggesting the presence of timber or other structures which were not located during excavation, or the action of differential disposal practises.

Artefact distributions for Period 4/5 provides similar evidence. Major structures are associated with significant artefact concentrations, but several artefact concentrations hint either at disposal practises or buildings which are no longer extent (Figure 20.9).

Finewares tend to be more ubiquitous in Period 4/5 in comparison to Period 1/2 (Figure 20.5 and 20.9). However, this may reflect the nature of the pottery supply

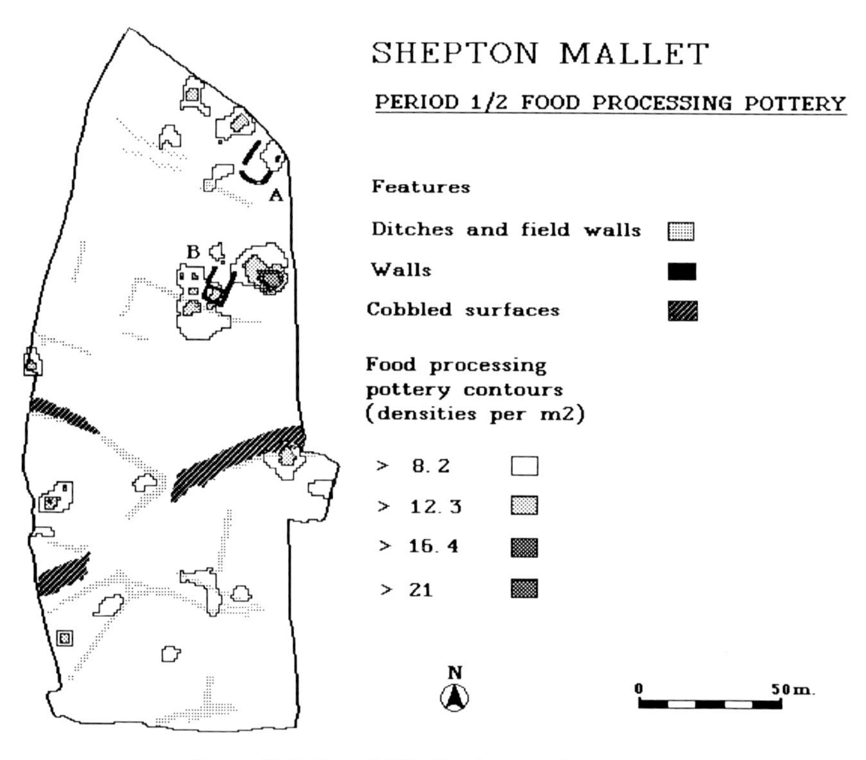

Figure 20.7 Period 1/2: Food processing pottery.

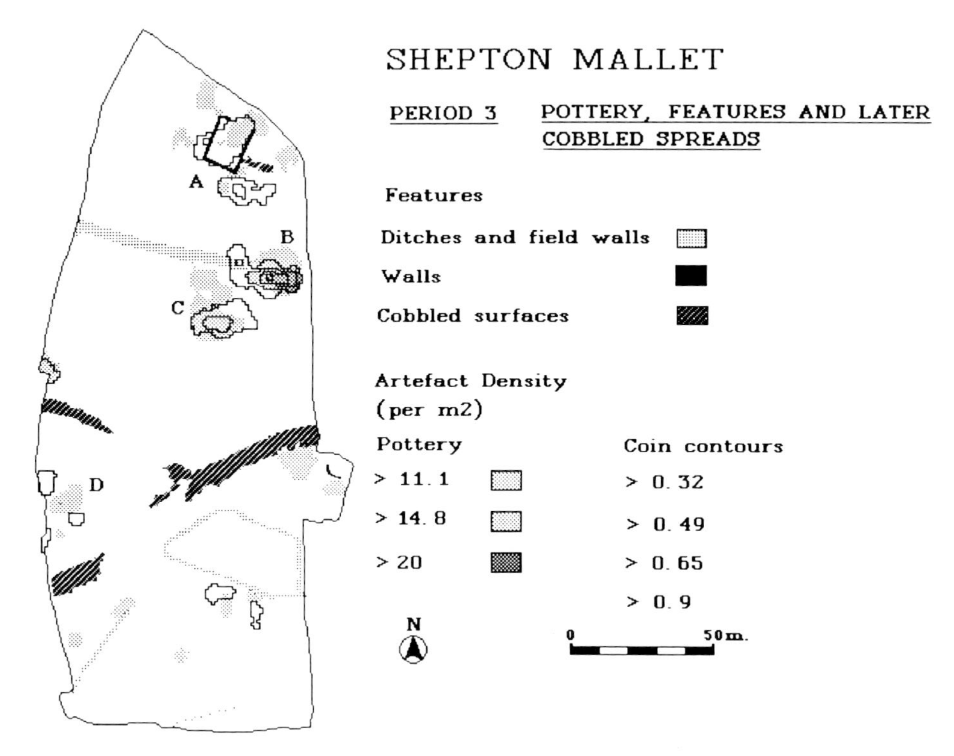

Figure 20.8 Period 3: Pottery, features and later cobbled spreads.

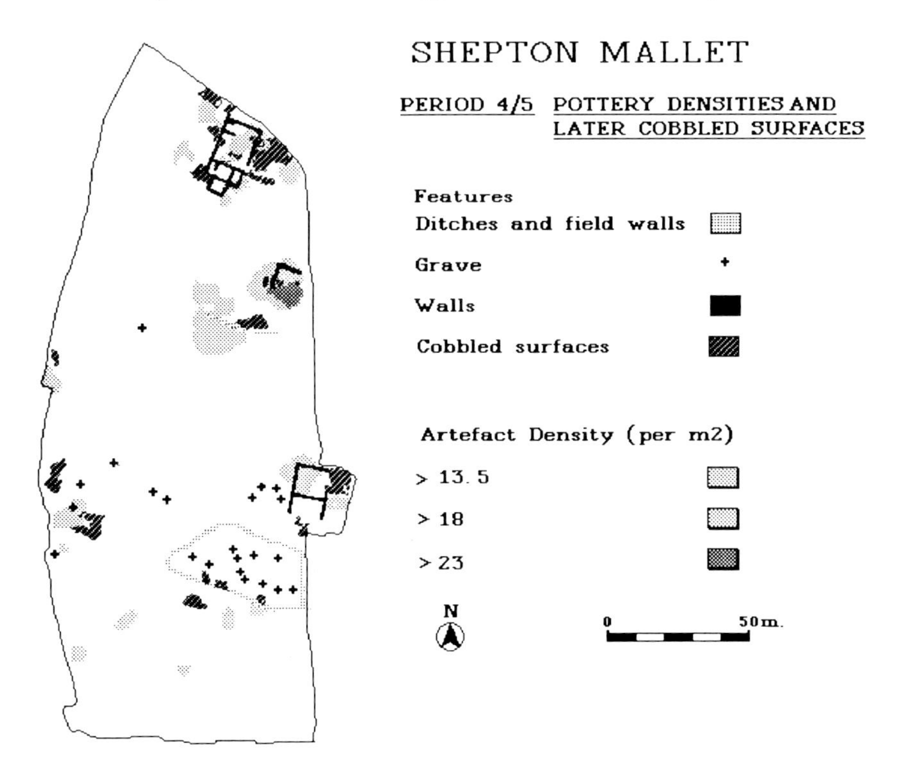

Figure 20.9 Period 4/5: Pottery densities and later cobbled surfaces.

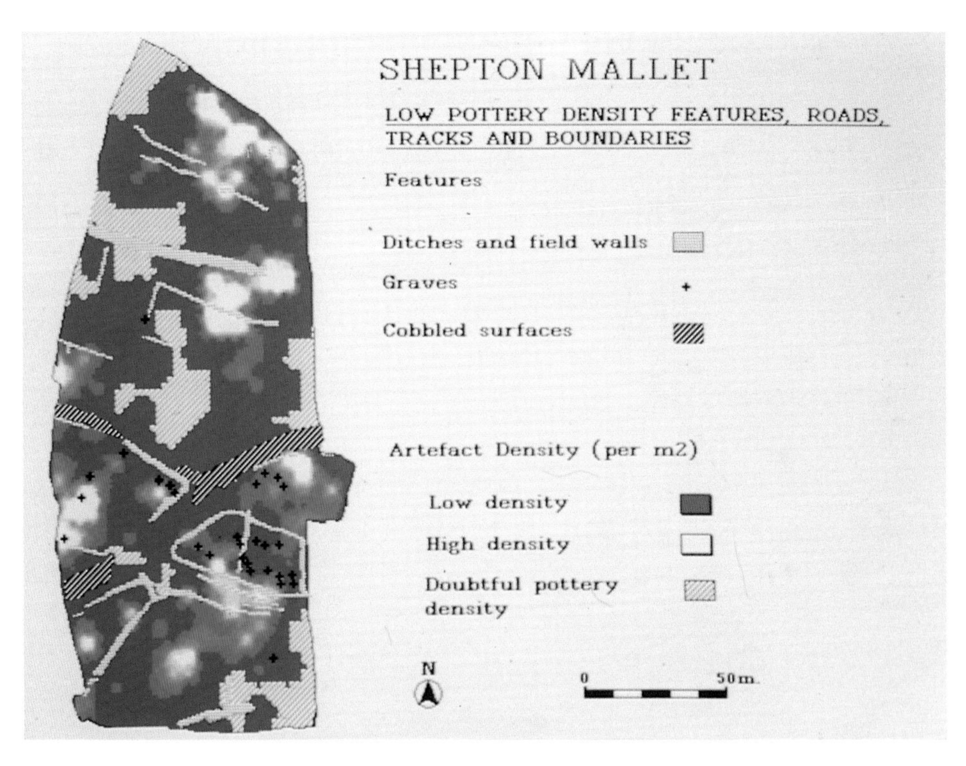

Plate 1. Shepton Mallet, Somerset, England. Coins and fineware in Period 4/5.
Plate 2. Shepton Mallet, Somerset, England. Low density artefact distributions, graves and boundaries.

during the period. The distribution of coins and finewares, however, are more closely associated with structures and principal pottery foci than in early periods and it must be considered that this may indicate some form of abandonment behaviour (Plate 1).

While distributional information of this sort is extremely useful, and goes a considerable way to fulfilling the primary aims of the analysis listed above, such analyses do not utilize the full potential of GIS for site-based analysis. Some further insight into site structure and development over time can be provided by further use of the simple overlay and additional modules which are part of the IDRISI GIS ring.

The data held within the IDRISI graphic databases can be used to illustrate that discard behaviour and presumably activity, becomes more intense between early and later periods. Figure 20.10, which illustrates the positions of major pottery foci during this period, also suggests that there is increasing evidence for small-scale activity areas in the later period.

Although it might be suggested that some of these small, late, artefact foci are simply dumps of rubbish within field plots, analysis of these foci with the distribution of industrial artefacts and debris suggests that this is not the case (Figure 20.11). There is a close association between many of the later pottery concentrations and concentrations of industrial material primarily relating to metalworking.

This data can also be used to contrast the distributions of data relating to domestic activities and those relating to industrial activity (Figure 20.12). This suggests that there is a bipartite—domestic/industrial—division on the site. The northern half of the sites tends to be associated with structures and aretefacts which are

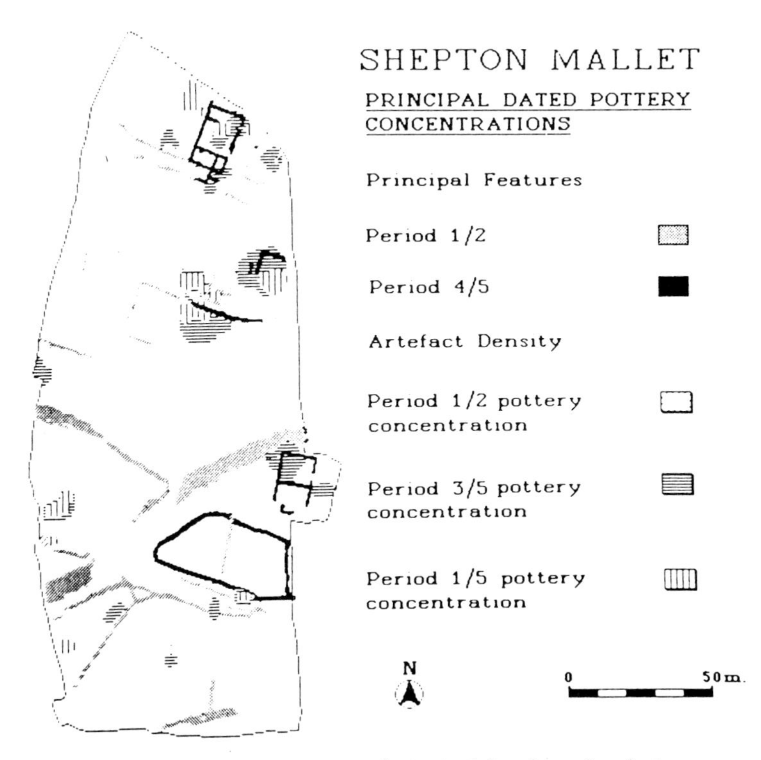

Figure 20.10 Comparison of principal dated artefact foci.

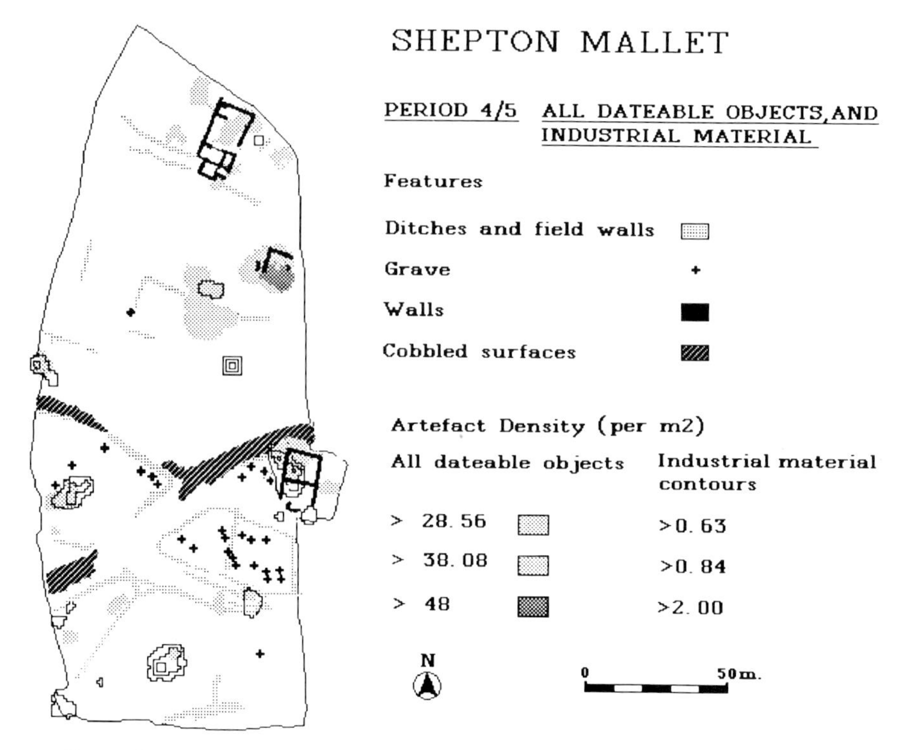

Figure 20.11 *Period 4/5 artefact foci and the distribution of industrial materials.*

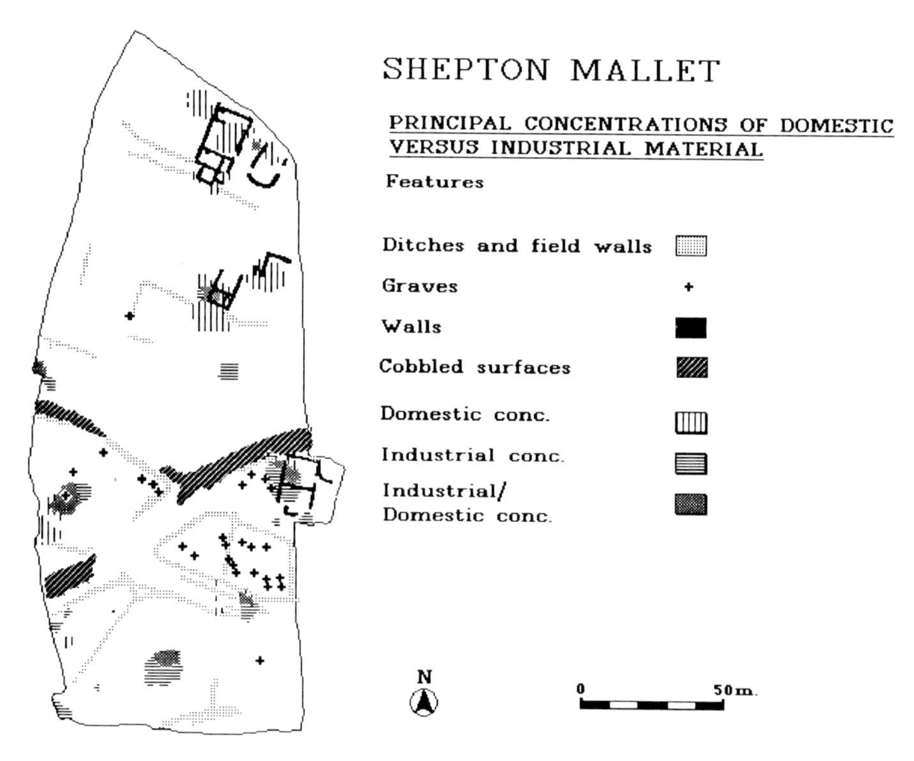

Figure 20.12 *Distribution of domestic and non-domestic artefacts at Shepton Mallet.*

primarily related to domestic activity. This contrasts with the south where non-domestic activities including smithing and burials were taking place. It may, perhaps, be an aside, but the association of burials and smithing may indicate more than simply a functional desire to place potentially dangerous activities in safe areas. Smithing is, in many societies, a liminal occupation associated with impurity, and it is to be wondered whether this is being reflected in the association of industrial activities and burials at Shepton Mallet.

This information is, of course, useful and, as has already been emphasized, essentially fulfils the basic goals set for the project and outlined above. However, the use of GIS within site-based analyses should result in more than a basic distributional analysis. Would the Shepton Mallet data be robust enough to sustain more than this? The answer is probably that it would not. The potential limitations of the Shepton Mallet data become apparent when we attempt to look at the data in real detail (Plate 2). Rather than looking at high artefact concentrations, low artefact densities were studied in order to investigate the relationship of artefact discard to tracks. The major road through the site is relatively artefact free. Several minor boundaries also appear to follow this patterning. Tantalizingly, the relationship of a number of graves with property boundaries and, perhaps, low-level artefact densities which might accord with paths, suggests that the data might be used to indicate the presence of minor paths associated with low artefact densities. Unfortunately, interpretation at this level of detail is difficult to justify. Scrutiny of the data reveals significant areas where there is little or no recorded pottery, suggesting that there is cause for real concern in attempting such analyses.

Of particular importance in such a discussion is the ability of GIS to incorporate negative data within analyses. The real power in many site analyses may be the ability to contrast areas which have things or properties with areas that do not (Kvamme, 1989). Unfortunately, the truth is that excavations on any significant scale rarely provide for collection of such data. We tend to dig where we find things—even within sites. Sampling strategies which allow some control over negative data are rarely applied, even on large sites (including that at Shepton Mallet), where such data would be useful and could be used to some effect.

At this juncture, we reach the final and perhaps most telling point about the use of GIS within site-based archaeology at least in Britain, and probably elsewhere. The potential for the use of GIS is clearly enormous. The spatial element of archaeology has never been so accessible as at the present time. Hardware and software costs are so low that virtually every individual archaeologist—never mind institution—can have access, if they wish, to some sort of GIS. But will the use of GIS be effective? The Shepton Mallet assemblage is in many ways excellent, yet in GIS terms, it is severely limited by data collection strategies which were primarily the result of non-archaeological constraints, rather than archaeological necessity. The result of this situation is, in the case of the Shepton Mallet study, an analysis whose main thrust has been distributional and does not utilize the full power of GIS technologies. It is possible, as work on the Shepton Mallet assemblage proceeds, that more useful information will be extracted and utilized through the unit GIS. The incorporation of environmental data might have opened significant analytical opportunities. Unfortunately, the environmental sampling procedures on the site and the survival of such data do not appear to be adequate for such analysis.

We can conclude that Shepton Mallet may not have been the perfect site for such work, but it does indicate that some quite elaborate functional analyses could be carried out using GIS on site-based data. The ability to combine disparate artefact groups within spatial analyses is a useful function, and there is no reason why the thoughtful combination of artefactual, environmental and structural databases cannot be used to contribute significantly to our understanding of conceptual arrangements within sites, and GIS still seems the best vehicle to carry out this work on a day-to-day basis. We may go further and suggest that some of the most thought-provoking suggestions for the analysis of site-based materials may not even be possible without GIS. Within the sphere of Roman archaeology, the need for volumetric analysis of pottery data as a guide to pottery supply and access, and Tomber's (1993) appeal for unified excavation and post-excavation strategies for the purposes of producing quantifiable ceramic data are examples of goals which will ultimately require the application of advanced database and GIS analytical techniques if they are to be achieved.

However, it is difficult to see how such a development will take place if we are unable to provide decent two-dimensional data. Certainly, collecting adequate data is not going to be cheap or easy. It may well demand a stricter approach to site recording than we have been wont to do. It will certainly demand a re-evaluation of how we collect spatial data. A simple sampling strategy which demands, for example, 50 per cent sampling of located features, but which pays no attention to excavated areas which do not have sub-surface remains, is not likely to be an adequate response if we intend to use GIS within our excavation strategies.

20.3 Conclusions

While there is no doubt that GIS has a place in archaeology in general, in Britain at least, its role within salvage archaeology is not assured. This situation could be changed if we re-assess our current approaches to the collection of spatial data from excavation. The potential return in archaeological information that might flow from an increased emphasis on spatial data, as a complementary source to more traditional chronological or functional information is significant. The alternative is the application of increasingly sophisticated analyses to data which, ultimately, are not adequate for the task.

Acknowledgements

The authors would like to acknowledge the work of all the individual specialists who have worked on the Shepton Mallet data as well as our colleagues at the Birmingham University Field Archaeology Unit who read and commented on the text. Funding for the project was provided by Showerings Ltd., English Heritage, Somerset County Council and Mendip District Council.

References

Allen, K. M., Green, S. W. and Zubrow, E. B. W. (Eds), 1990, *Interpreting Space: GIS and Archaeology*, London: Taylor & Francis.

Association of County Archaeologists 1993, *Model Briefs for Archaeological Assessments and Field Evaluations*, London: Association of County Archaeological Officers.

Brereton, S., nd., 'A survey of the results of field assessments undertaken by the Oxford Archaeological Unit and Wessex Archaeology in Southern England', unpublished manuscript, Oxford Archaeological Unit.

Gaffney, V. and Stančič, Z., 1991, *GIS Approaches to Regional Analysis: A case study of the Island of Hvar*, Ljubljana: Filozofska fakultera.

Institute of Field Archaeologists, nd., Standard and Guidance for Archaeological Field Evaluation, Institute of Field Archaeology.

Kvamme, K. L. and Kohler, T. W., 1988, Geographic Information Systems: Technical aids for data collection, analysis and display, in Judge, J. W. and Sebastian, L., 1988, *Quantifying the Present and Preserving the Past: Theory, Method and Application of Archaeological Predictive Modelling*, Denver: US Department of the Interior, Bureau of Land Management, pp. 493–548.

Kvamme, K. L., 1989, Geographic information systems in regional archaeological research and data management, in Schiffer, M. B. (Ed.), *Archaeological Method and Theory 1*, Tucson, pp. 139–203.

Leach, P., 1991, *Shepton Mallet: Romano-Britons and Early Christians in Somerset*, Birmingham.

Lobb, S., 1988, 'The Kennet Valley Survey 1982–7: A review of evaluation and fieldwalking methodology', unpublished manuscript, Wessex Archaeology.

Lock, G. R. and Harris, T. M., 1991, Visualising spatial data: the importance of geographic information systems, in Rahtz, S. P. Q. and Reilly, P. (Eds), *Archaeology in the Information Age*, London: Routledge.

Shott, M. J., 1987, Feature discovery and the sampling requirments of archaeological evaluations, *Journal of Field Archaeology*, **14**, 359–71.

Spoerry, P., 1992, *The Structure and Funding of British Archaeology: the RESCUE Questionnaire 1990–91*, Warwick: Council for British Archaeology.

Tomber, R., 1993, Quantitative approaches to the investigation of long-distance exchange, *Journal of Roman Archaeology*, **6**, 142–66.

21

Spatial relations in Roman Iron Age settlements in the Assendelver Polders, The Netherlands

M. Meffert

21.1 Introduction

Over the past decade, regional studies of prehistoric site locations have become a common phenomenon in archaeology. In the last five years, geographic information systems (GIS) have become popular among archaeologists dealing with inter-site research. This chapter attempts to introduce GIS to the level of intra-site analysis and discusses the role of GIS within the context of settlement excavations. It presents the results of an intra-site analysis based on the data from two excavations in the Assendelver Polders in the west of The Netherlands (Figure 21.1).

The main goal of the research is to trace human activity within the settlements based on the locations of finds and features but in particular on indices of finds (e.g. proportions of pottery wares), densities and indices of find dimensions (e.g. the level of pottery fragmentation). Finds, find indices and features can provide information on specific human activities. Indices of finds' dimensions and also proportions of finds compared to each other can tell us more about depositional and post-depositional processes. The identification into primary, secondary and *de facto* refuse of several find categories gives more information about the activities in settlements.

One of the aims of this research is to investigate the utility of GIS for an intra-site analysis in a non-urban context. Large-scale excavations of the last decade have produced so much data that it is almost impossible to master them all in traditional ways of presentation. It is very difficult to manage the large body of information in tables, graphs, figures and text and obtain a coherent picture of a specific situation. Therefore, several kinds of data should be transformed into a single image, as is possible with a GIS.

Combinations of a large number of finds, features, indices of finds and indices of find dimensions can only be achieved by a GIS in which the raster component is fully integrated with the vector part of the program, and which easily combines and dissects features, attributes and phases. Outcomes of bivariate analyses, vector maps of features, discrete and continuous information about depositional and post-depositional processes are all integrated and give more archaeological information than could be generated with traditional methods. Extrapolations from these variables are used for the reconstruction of find patterns on the former occupation surface, and also give an idea about the relative meaning of the finds data.

Figure 21.1 The research area of the Assendelver Polders.

21.2 The Assendelver research project

Several sites have been excavated in the Assendelver Polders between 1978 and 1982. The large majority of these sites date from the Roman Iron Age. The archaeological project wherein the pilot study on GIS is embedded, is in the tradition of settlement research in The Netherlands (Brandt *et al.*, 1987).

In Roman times the Assendelver Polders were part of the Rhine estuary in the West of The Netherlands and inhabited by the Frisii Minores. Around AD 1, the mouth of the river Oer-IJ had silted up almost completely, although the gulleys which previously served as drainage channels for the bog remained as relicts in the landscape. The southern part of the Oer-IJ became a freshwater lake. Five main zones which were used differently can be distinguished in the region (Figure 21.2).

1. **The dunes in the west** were permanently inhabitable. Agriculture and horticulture were possible (Brandt and van Gijn, 1986).
2. **Tidal flats** could be occupied from the first century AD onwards. Not many settlements have been found there. Perhaps the tidal flats were used as pasture for cattle.
3. **The sandy banks of the creeks** (levees) were the location of most of the settlements. Agriculture and horticulture were possible on the levees.
4. **The western edge of the peat, the reed-peat zone** was only briefly occupied. The settlement pattern was one of dispersed single farmsteads.
5. On **the Oligotrophic peat in the east** permanent settlement was impossible. It may have been used as pasture for animals.

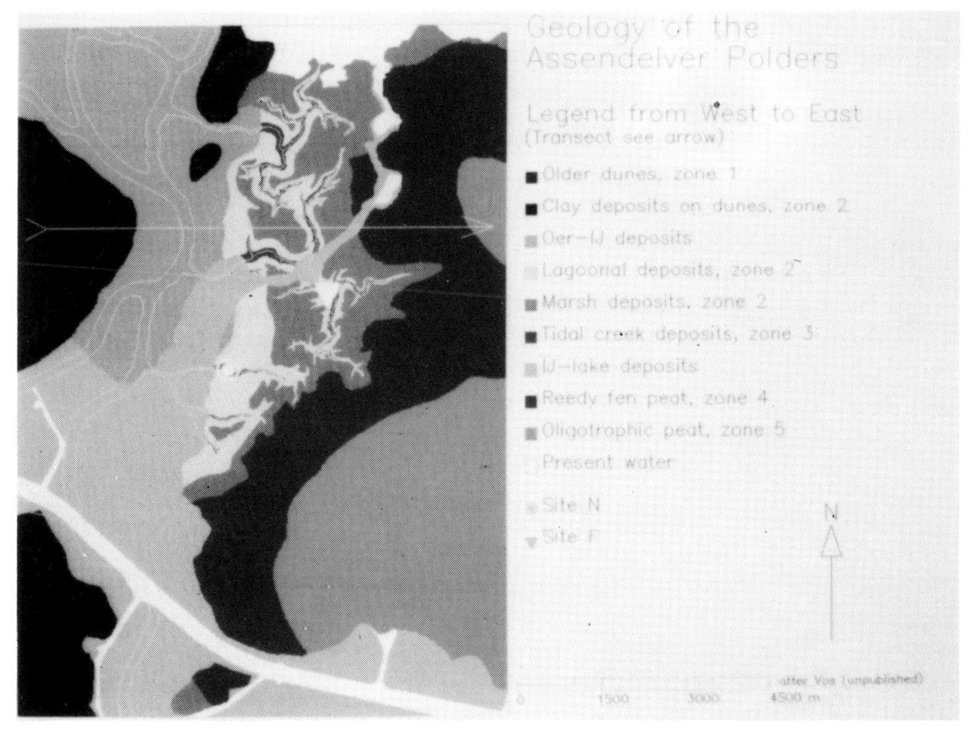

Figure 21.2 The Assendelver Polders with environmental zones.

The sites used in this chapter are situated on levee deposits. The sites comprise one or two farmsteads and contain two kinds of houses: three-aisled houses and wall-ditch houses constructed with sods. In the three-aisled houses, the living area with the hearth was probably separated from the stable which possibly contained cattle boxes. The wall-ditch houses have no internal division of space, and their total inner surface equals that of the living quarters of the three-aisled houses (Therkorn, 1987).

21.3 Sites and sampling strategy

The research strategies of the Assendelver Polders project were aimed to collect data on three levels:

1. regional,
2. the settlement, including their houses, granaries, pits, drainage systems, etc., and
3. the individual buildings and other components of the settlement and their immediate environments.

The first of these levels included a systematic network of borings over the whole polder. The second included narrow excavation trenches across a settlement, fields and other features belonging to it, while the third was researched by means of full-scale excavations (van der Leeuw, 1987).

This chapter focuses on two sites: sites N and F, both of them stratified. The sites were dated to the first two centuries AD on the basis of finds which include large quantities of indigenous pottery, fibulae, Roman ware including Samian, and also a lot of bones and wood. The sites were excavated in five pits at most. The collection of finds on the excavations was systematic with the trenches being overlaid with a grid of squares, the size depending on the situation (e.g. 1 × 1 m or 2 × 2 m).

Finds in features were collected from box-sections, using samples with a known volume. The finds in larger features, mainly trenches and ditches, were generally sampled rather than being totally collected. For this purpose, box-sections were made at regular points along them. These samples were considered to be representative. In the larger features, at least three box-sections were excavated totalling at least 20 per cent of the feature (van der Leeuw, 1987). This method of excavation can be used for relational distribution plans, as will be explained below.

21.4 The bricks to construct a GIS

The visualization of several components at a location can give a better insight into what is going on there and large quantities of information can be shown in one image. A GIS can combine the information from tables, graphs, together with the descriptions of finds and features within a spatial framework. Problems of depositional processes can be solved using map overlays. A raster GIS is used for continuous find information and a vector GIS for connecting the artefactual database with features, squares and point locations. This combination of GIS tools gives the opportunity to relate depositional patterns to find distributions.

The program used here is mainly a vector GIS but has fully integrated raster routines (Genasys, 1991) which are used to reconstruct occupation surfaces for interpreting the distribution patterns of finds using find densities, indices of dimensions of finds, and dumping patterns. Vector applications are mainly used for distributions, indices, densities, and fragmentation of finds at the level of features.

Excavation plans and plans created from databases which contain information on features, locations and sampling methods form the basis of the analysis. The data analysis combines these plans with the attribute database, in which not only artefactual information but also information about period and features is stored.

For the successful use of a GIS at the site level, at least two conditions must be met by the finds administration. All finds must be administered by location (point or square), as well as feature. The graphical data are partly digitized from the excavation plans scale 1:50 and partly extracted from the database to regenerate points, squares and box-sections. After linking these graphical data with attribute information, both find and feature information can be retrieved. The attribute information is stored in several databases that are linked.

21.5 The pilot study

21.5.1 Depositional processes

Previous to this GIS analysis, using simple scatter-plots and cross-tables gave some indication that the features are not completely disturbed by showing how find densities and fragmentation of materials relate to the features, making discarding and dumping patterns much clearer.

Some patterns are evident:

1. on average, there seem to be systematic differences in pottery densities in various features;
2. bone densities have a more or less positive correlation with pottery densities; and
3. two feature types—house floor and ditches—are not typical of this pattern.

House floors have very few bones and many pottery sherds suggesting that the house was kept clean of bone but that sherds were used for consolidation of the floor. Ditches, which contain relatively many bones may be interpreted as dumping areas for bone away from the houses. These preliminary conclusions based on features and find dimensions indicate that it is possible to trace spatial patterns in find distributions arising from the systemic context.

A combination of this information shows that layers and parcelling trenches have low densities of pottery and bone which are highly fragmented, indicating that there was no dumping in these features but that the degree of trampling was very high. Pits have the highest densities of pottery and bone in association with low fragmentation. The house floor is full of pottery, as can be expected, with a relatively high degree of fragmentation through trampling. The problem with results such as these is that the location and the position of the features are not considered making it unclear how features relate to their neighbouring features. All this discrete information is without any spatial relations, whereas the people who lived there dealt with an occupation surface where features were related to each other. There was a smooth transition between occupation layers and most of the features, and to be able to reconstruct the activities of the inhabitants we have to reconstruct this occupation sitescape as an integrated whole of features and layers.

21.5.2 The archaeological challenge

One of the archaeological aims of this research is to investigate the distribution of finds in relation to their context. Find context means information about find densities, find indices and about features and other find categories. Are there relations between find contexts and the distribution patterns of finds? Answering the question is necessary before making interpretations about activity areas based on finds spectra visible in distribution plans.

Raster maps which contain information about depositional processes are created from the continuous information of densities and dimensions of finds. These maps give an insight into which parts of the occupation surface were used for walking and which for dumping and thus can be employed to interpret the distribution of finds and find indices. This information shows which distributions are a result of dumping or chance, and which distributions of finds give an opportunity to interpret finds as primary or as *de facto* refuse.

21.6 Intra-site analysis

Analysis can be made between finds and domestic structures and finds and square and/or point locations. For both, the finds attribute information and feature data

(a)

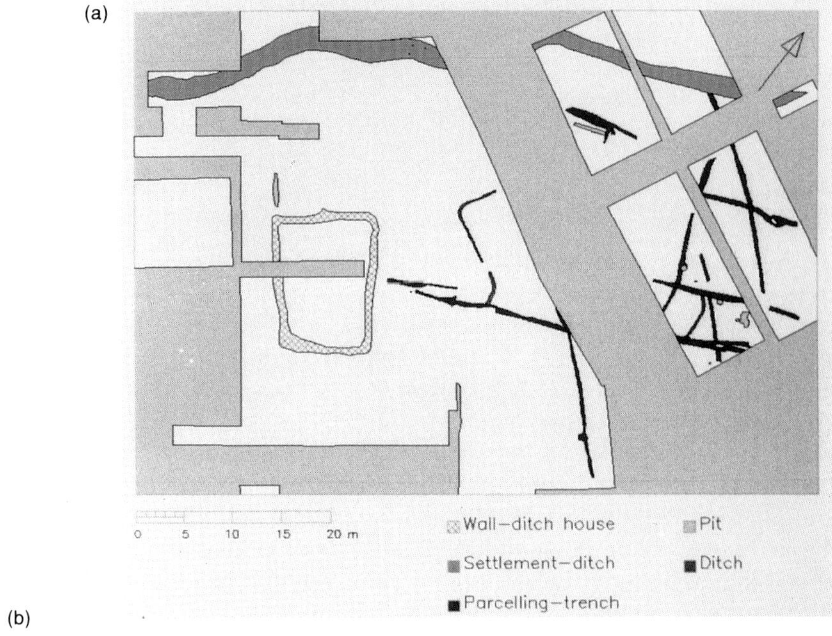

(b)

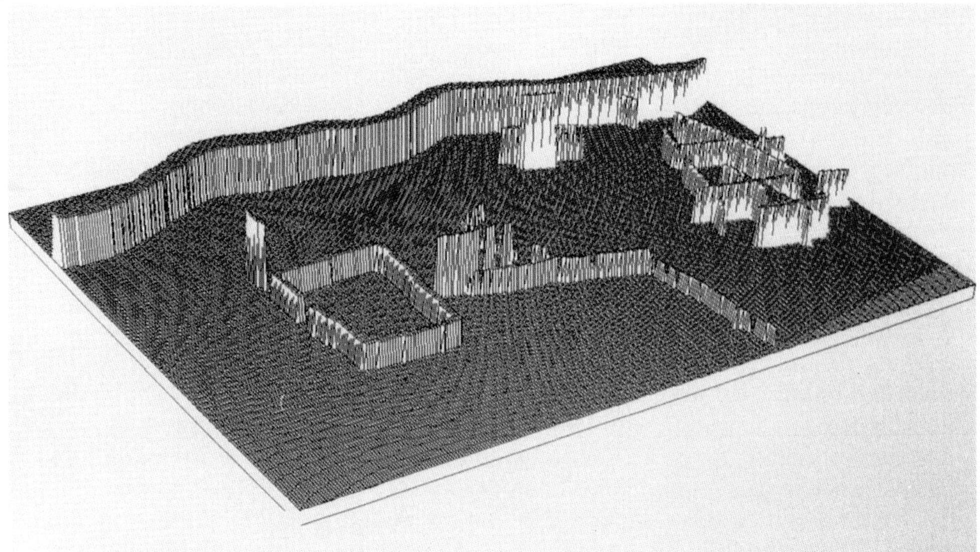

Figure 21.3 Site N, phase 5. Fragmentation of pottery in features in three-dimensional projection: (a) key to the features; and (b) features displayed as amount of fragmentation.

must be linked to cartographic information. This information, combined with graphical information in GIS format, forms the basis of the analysis whereby several maps are combined and data layers are projected on top of each other. From these projections, human activity and depositional processes can be extracted. Questions of primary and secondary deposition and *de facto* refuse can be answered in favourable situations with artefact traps such as postholes being used as checkpoints. There are, of course, many natural and cultural processes that disturb these patterns including post-depositional processes such as trampling and pedoturbation which smear deposition. Primary deposition in permanent settlements is difficult to reconstruct as sedentary communities usually clear away their refuse. It will only be found in its primary position when small in quantity, so that it will only be traced when archaeologists sieve their samples. Secondary deposition, on the other hand, is much easier to trace although the problem is that association between artefacts is not necessarily related to systemic context. Looting, post-abandonment processes and recycling of material make *de facto* refuse disappear from its original position or transform it. The greatest destroyer, however, is the site location itself; if a location is favourable for occupation, it will be used over and over again with several generations building, demolishing, levelling and rebuilding their settlement. A good site is a palimpsest, so that the last occupation surface is usually the only one that can be reconstructed with its features and occupation layers being minimally damaged by later phases, and for which the chance to find *de facto* refuse is at its greatest. For earlier phases, find analysis is only possible within features. In the Assendelver Polders, the latest occupation surface is phase 5; afterwards the sites were covered with a blanket of peat and abandoned for about 1000 years.

21.6.1 GIS approaches to depositional processes in features

Visualization of discrete information in features is one of the possibilities with GIS. In Figure 21.3(b) (a three-dimensional projection) and Figure 21.4 (two-dimensional projection) two examples of visualizing the fragmentation of pottery are presented.

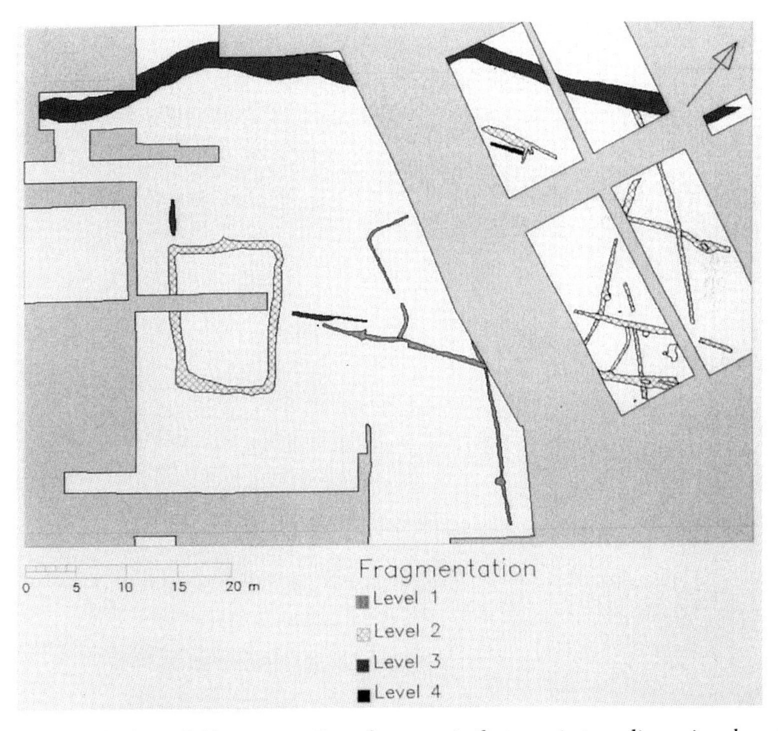

Figure 21.4 Site N, phase 5. Fragmentation of pottery in features in two-dimensional projection.

In Figure 21.3(b), features are displayed showing the amount of fragmentation—the higher the feature, the lower the fragmentation. It is clear that the pits and the settlement ditch have low fragmentation and that the parcelling trenches a more or less medium degree of fragmentation; Figure 21.3(a) shows the various types of features. Figure 21.4 seems easier to interpret, but the interpretation is restricted by the selected class boundaries as shown in the key. Throughout the rest of this chapter the levels of pottery fragmentation are defined as follows.

- Level 1: sherds 1–10 g.
- Level 2: 10–20 g.
- Level 3: 20–30 g.
- Level 4: sherds heavier than 30 g.

In the three-dimensional projection, on the other hand, fragmentation has been visualized as a continuous variable, which is very useful when seeking an indication of how amounts of fragmentation are distributed throughout the settlement.

To visualize more than one variable at once, other methods are necessary. Figure 21.5 shows the fragmentation of bone and of pottery. To be able to compare the fragmentation of different kinds of find categories, the values have to be scaled with the maximum value of a variable displayed as $\frac{1}{6}$ of the image height, and the minimum as a square. When the levels of fragmentation within the range of each variable are roughly equal, the bars have more or less the same height. Figure 21.5 shows that the fragmentation is approximately equal for both categories throughout the settlement with the exception of the parcelling ditches.

The fragmentation level of pottery in combination with pottery and bone densities provides an insight into the depositional processes of phase 2 (Figure 21.6).

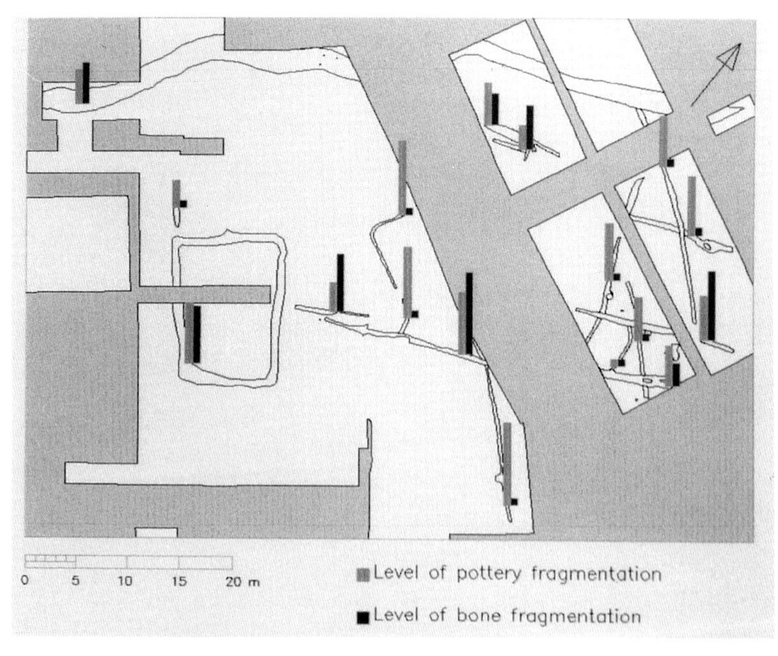

Figure 21.5 Fragmentation of pottery and bone in features shown as scaled proportions (Site N, phase 5).

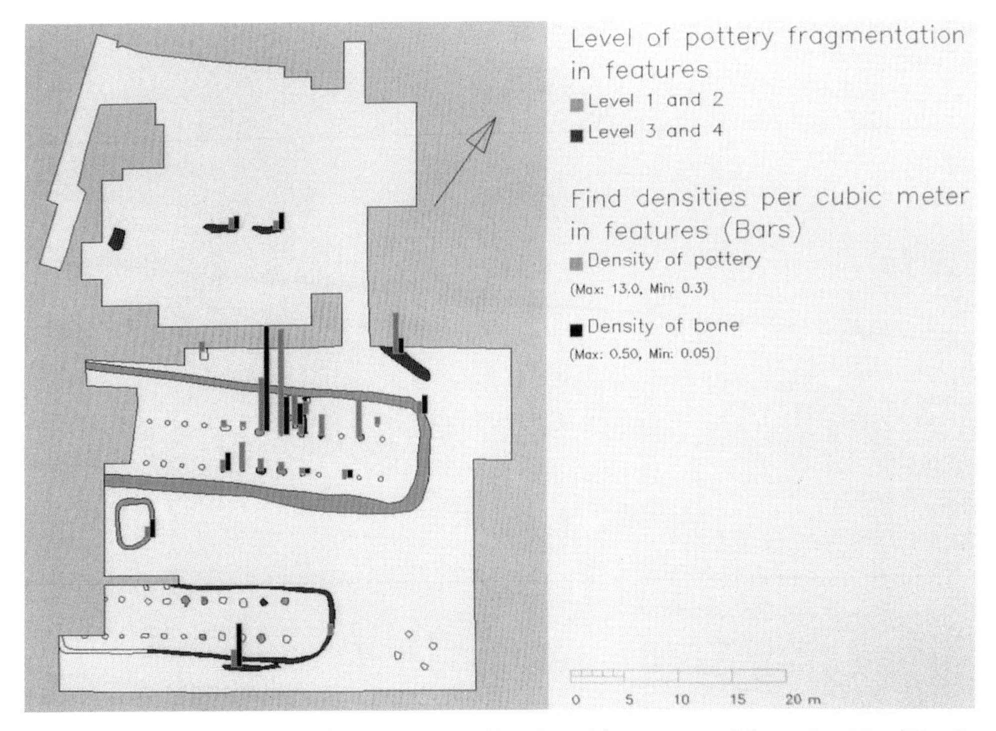

Figure 21.6 Fragmentation of pottery in combination with pottery and bone densities (Site F, phase 2).

The high fragmentation in the central part of the excavation and most of the post-holes is striking, whereas pits and the southern house show relatively low fragmentation. The densities of bone and pottery vary considerably. Another example of the fragmentation of pottery in combination with the pottery and bone densities is given in Plate 1 which shows the situation on phase 4. In this phase a terp ditch with very low density scores for pottery and bones encloses a dwelling mound (terp). Fragmentation is very low in the terp ditch and the pits, while other features contain highly fragmented pottery.

21.6.2 GIS approaches to absolute and relative occurrences of finds

Figure 21.7 shows the combination of absolute numbers, densities and indices of artefact types in features. The first and the third bars from the left represent absolute numbers of two kinds of pottery: coarse and smooth wares. The second and the fourth represent densities of the wares and the fifth bar gives the index of these two. It can be seen that the settlement ditch has a relatively low density of smoothed wares but, in relation to the other features, it has the highest absolute number. Also, the pits with small absolute numbers of smoothed wares have relatively high densities of these wares but the indices are very different. This way of visualizing data shows the relationships of find categories within and between features, and to each other.

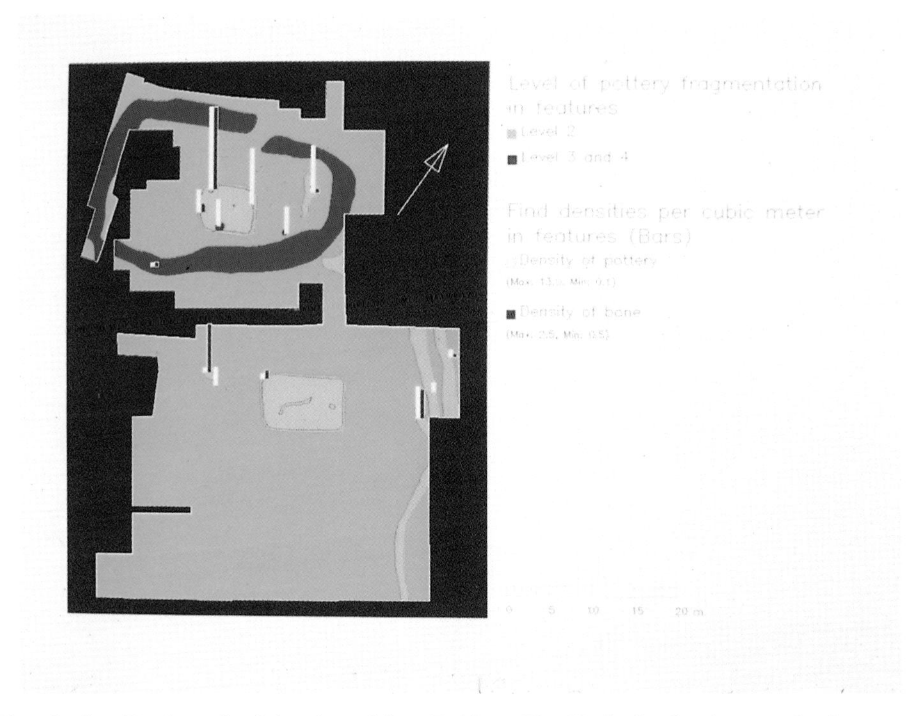

Plate 1. Site F, phase 4, of the Assendelver Polders, The Netherlands, showing the fragmentation of pottery in combination with pottery and bone densities.

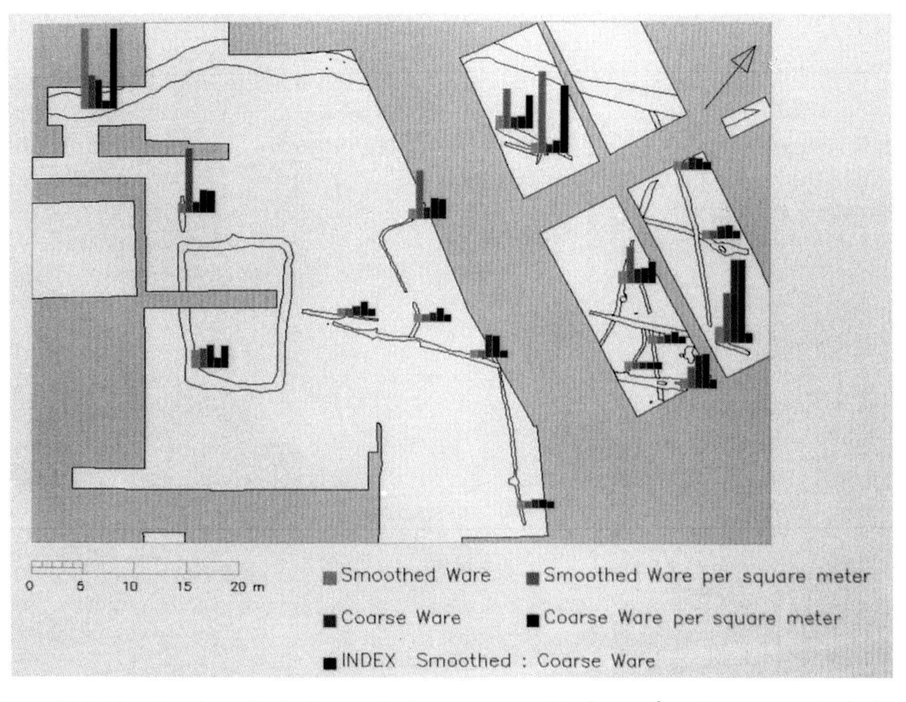

Figure 21.7 Combination of absolute, relative counts and indices of pottery wares in features (Site N, phase 5).

21.6.3 GIS approaches to depositional processes and the distribution of finds

Continuous maps with information from squares and/or boxes and/or point information can be created by a GIS and, for the last phases of these sites, it is possible to combine such continuous cartographic information. For instance, the map with pottery densities (Figure 21.8(a)) can be combined with the pottery fragmentation map (Figure 21.8(b)), resulting in the occupation surface with dumping patterns (Figure 21.8(c)). Locations with high fragmentation and low densities of pottery will have been walking surfaces, the combination of high fragmentation and high densities indicates trampled dumping zones, and low fragmentation and high pottery densities represent dump areas.

21.6.4 Relational distribution plans

Figure 21.9 is an example of the mapping of a find category in its broader context and presents the combination of the level of pottery fragmentation and complete pots on site F. It can be seen that complete pots were all found in areas with low fragmentation, and this association counteracts the interpretation that complete pots are *de facto* refuse.

Figure 21.10 shows dumping areas in combination with slag and quern-stones and is divided into three kinds of surfaces: dumping areas; dumping zones where pottery is trampled; and areas where pottery is almost absent, the walking surfaces.

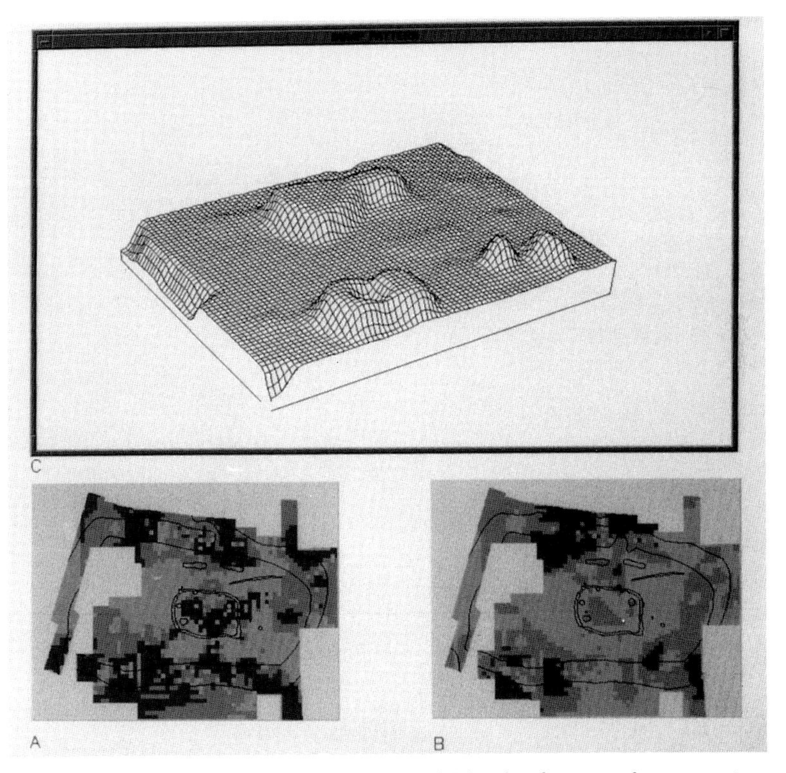

Figure 21.8 Site F, phase 5: (a) pottery densities; (b) levels of pottery fragmentation; and (c) dumping patterns.

Figure 21.9 Complete vessels in relation to levels of pottery fragmentation (Site F, phase 5).

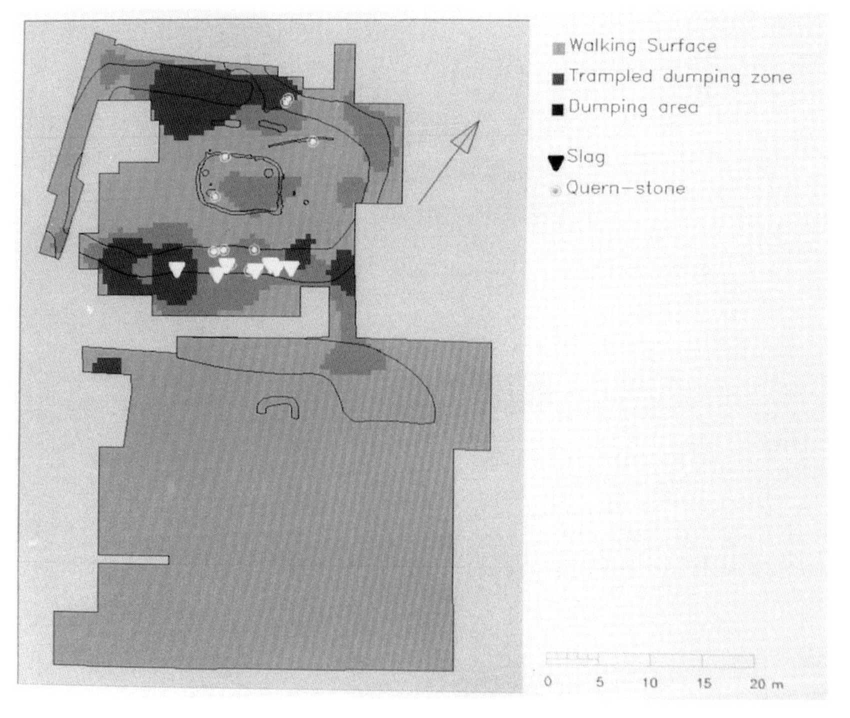

Figure 21.10 Distribution of slag and quern-stones in relation to dumping patterns (Site F, phase 5).

Slag is associated with dumping zones and quern-stones are not; the latter are probably, certainly partly, *de facto* refuse, whereas slag is found mainly in trampled dumping zones and does not provide conclusive evidence about the location of metalworking.

21.7 Conclusions

In this chapter, an attempt has been made to put finds and their dimensions into a locational context. Relational distribution plans show, better than ordinary distribution plans, which processes may have been responsible for the deposition of find categories and the reason of their deposition in certain locations. Based on such plans, decisions as to whether finds are primary, secondary or *de facto* refuse can be made. The examples given show how find categories and dimensions relate to other categories and dimensions of finds within and between features. An attempt has been made to reconstruct depositional processes for a more accurate interpretation of finds in relation to their location.

The conclusions of this research, which took place within the context of investigations of rural settlements are still preliminary and an all-embracing interpretation of find dimensions and their archaeological context cannot yet be presented. One thing is clear however: a GIS is very useful for intra-site analysis. One of the major benefits is a more accurate interpretation of find deposition as a reflection of cultural processes in rural settlements. The use of a GIS provides an insight into protohistorical activities. The final advantage is an improved visualization of the results, not only for our colleagues but also for a wider public.

Further research and improved excavation methods are recommended and necessary to increase efficiency, with the scale of excavation on archaeological sites being very important. Not only features must be excavated but also wide areas surrounding them. At the same time, it is necessary to increase the number of controlled samples; more finds and more exact information on location, sample volume and sampling strategy will improve results. The qualitative aspects of these recommendations are probably feasible, whereas the quantitative aspects are a matter of finance.

References

Brandt, R. W. and van Gijn, A. L., 1986, Bewoning en economie in het Oer-IJ estuarium, *Gedacht over Assendelft*, Working paper **6**, 61–75.

Brandt, R. W., Groenman-van Waateringe, W. and van der Leeuw, S. E. (Eds), 1987, *Assendelver Polder Papers 1, Cingula*, **10** Amsterdam.

Genasys, 1991, *Genamap Reference Manual*, version 5.1, Genasys Integrated Solutions–GIS with a difference, Manchester.

Therkorn, L. L., 1987, The structures, mechanics and some aspects of inhabitant behaviour, Assendelver Polders, *Assendelver Polder Papers 1*, 177–224.

Van der Leeuw, S. E., 1987, Outline of the project and first results, Assendelver Polders, *Assendelver Polder Papers 1*, 1–21.

Vos, P. C., no date., 'Geologische kaart van de Assendelver Polders en de Uitgeesterbroekpolder', unpublished, Universiteit van Amsterdam.

22

Another way to deal with maps in archaeological GIS

R. Wiemer

22.1 Introduction

At the State Service for Archaeological Investigations in The Netherlands (ROB), the Centre of Expertise ARCHIS is responsible for building and maintaining an archaeological database linked to a GIS. Apart from cultural resource management (CRM), the GIS (GRASS) is used for all kinds of spatial analysis (Roorda and Wiemer, 1992a; 1992b).

Analysis is done on excavation data, field survey data and site distributions. The main purpose of these analyses is to find or predict and understand patterns in the distribution of artefacts, features and sites, possibly related to environmental factors and thus providing a new tool in CRM. It might improve the planning of survey and excavation schemes or coring programmes, which eventually can either save money or prevent archaeological sites from being damaged. The basis for most of the GIS work done is, therefore, not merely scientific but rather practical.

22.2 Background

Until recently, digital maps were being used in the same way as their paper equivalents. The map units were digitized and the resulting polygons, lines and points could be used to find relations between the distribution of the points and the digitized physical environment. A major drawback of this way of looking at maps is that it is impossible to deal with inaccuracy in the maps, with uncertainty in the data or with edge effects. A distribution map of archaeological sites can be used as the basis for predictive modelling, but the degree of uncertainty is usually very high since lots of sites have not been discovered yet or have disappeared due to erosion.

When an archaeological site is located close to the boundary of a map unit (such as a geological formation, circumscribed by a polygon), a GIS can tell you whether it is inside or outside the polygon; it will provide no information about the distance to the boundary of the unit, nor about map units in the surrounding areas. Finally, nature itself is merely a complex of gradual transitions between different soils, geomorphological units, etc., which are translated into the areas, points and lines which make up the paper map in order to give a more or less general idea (depending on the scale of the map) of the real situation.

To overcome these limitations of 'classical' maps a transformation to continuous models (or 'trend surfaces') might be a solution. In a continuous model, each point in a map has a meaningful value, comparable to a digital elevation model (DEM), in which the value at each point represents the real elevation at that point.

The question is how to transform polygons, lines and points to continuous surfaces or how to deal with 'fuzziness' as it is sometimes called (Burrough, 1989). For each of the three types of data-points, polygons and lines, an easy to use procedure is presented, although this does not mean that these methods are the only possible solutions to the problem.

22.3 Points

Numerous methods exist for interpolating XYZ data to continuous surfaces, e.g. weighted distance, kriging, etc. These algorithms can give good results when used with care (Haigh and Kelly, 1987; Kvamme, 1990; Warren, 1990:211; Watson, 1992) but they are not applicable in the case of point distributions—or events as they might be called—which always have a Z-value of 1. Usually events are transformed to a continuous surface by recording cell frequencies, followed by applying an $n \times n$ moving average as a smoothing filter (Johnson, 1984). On the other hand, if there is a Z-value but the number of samples is low and values fluctuate widely, the usual interpolation techniques can give unsatisfactory results. Another disadvantage of many interpolation techniques is that they are often difficult to understand for non-statisticians. For this reason, a simple interpolation technique is introduced here, based on probability, reflecting an uncertainty related to each measurement or event. The proposed interpolation method is based on three assumptions:

1. The probability that an event occurs at a specific location is related to the occurrence of events at nearby locations.
2. The fact that a certain measurement at a certain location results in a specific value is in many cases to a certain extent the result of coincidence. For example, the phosphate content in the soil fluctuates highly and therefore the measured value in a sample is to a large extent dependent on the location of the sample point. If one were to take a sample just tens of centimetres away from the chosen location, one would probably measure a completely different phosphate level.
3. The probability of measuring the same value decreases with increasing distance from the sample location.

From these three assumptions, it must be clear that there are two different measures of probability. The first one deals with the uncertainty of the location of a point, the second one deals with the uncertainty of the value at a point.

22.3.1 Uncertainty relating to location

Events are points of special interest which have no Z-value. The occurrence of an event in an archaeological sense can be the presence of an archaeological site or an artefact. A site is there, or it is not there. The locations of these events have a certain degree of uncertainty. One can compare this with the occurrence of lightning strikes. Probably the distribution of lightning strikes is related to physical properties of the environment and the atmosphere, but the fact that a strike occurs at a specific location at a specific moment is merely a matter of chance. However, when looking at the distribution pattern of lightning strikes monitored over several years, it is possible to recognize areas where lightning strikes occur more often, due to environmental and atmospheric factors.

The finding of an archaeological site or artefact can easily be compared with the occurrence of lightning strikes. Usually, sites are found in areas which were most suitable for habitation. The reason that a site is exactly at the location where it is found, is not always clear. It is very likely that the site could just as well have been situated at a certain distance (100 m, 1 km, maybe even 10 km, depending on the landscape and the type of site) from the found location. Since the distribution pattern we see now is the result of many generations of occupation, the areas in which many sites are found are probably favoured areas. The chance of finding more sites in those areas is therefore greater than in areas where no or very few sites are found. The chance of finding additional sites is, among other factors, related to the distance from known sites; the probability is highest at the location of the sites and diminishes the further you move away from them and eventually it will become (almost) 0. Of course one always has to take into account that the distribution pattern we see now is influenced by transformations which have occurred in the environment.

22.3.2 Uncertainty relating to the value of points

Data resulting from sampling have a similar uncertainty. The probability of measuring a certain value is largest at the point at which that specific value has been found and is related to the distance to the sample point. This is especially true when the data are highly fluctuating. The phosphate level in a soil sample for instance is to a large extent dependent on the chosen location of the sample. Just a few centimetres away, the level can be very different. The same is true for artefact counts in cores, resulting from systematic augering, a commonly used practice to determine the boundaries of a site. The probability of measuring a high value is also related to the number of samples in the neighbourhood that resulted in a high value.

The algorithm for the interpolation is quite simple but by no means new (Cressie, 1993). Each data point, which can be either a sample point or an event, is transformed into a cone and the size of the base of the cone (the 'search-radius') has a user-defined value. It can be based on the average distance between neighbouring points, although this might not always be the best value. The larger the value, the more smoothed the eventual surface will be. No matter how one chooses the search radius, it has to be large enough for neighbouring cones to overlap. The value of the centre of the cone is calculated by multiplying the search radius and the Z-value at the sample point, or by multiplying by one in the case of events. Intermediate points (points between the base and the top of the cone) are the product of the distance to the base of the cone and the Z-value. The final step is to sum all cones. In the case of events (archaeological sites, $Z = 1$), the result is the final model. Mathematically, the algorithm can be represented as follows:

$$V = \sum_{i=1}^{i=n} (R - d_i)$$

where V is the calculated value, R is the search radius, n is the number of events within search radius, and d_i is the distance from event i.

A simple example can illustrate the modelling process. Suppose there are nine ($n = 9$) events (Figure 22.1) and the search radius has been defined as five ($R = 5$). Each individual event is transformed to a cone, the top of the cone having a value of 5, the base has a value of 1. Corresponding cells in the 9 individual cones are summed, resulting in the eventual probability model (Figure 22.2).

If the points used in this example are samples (Figures 22.3), which means they have a value, and their location is chosen, the cells of the individual cones are multiplied with the value of the sample itself, and then summed. In the case of samples, the resulting model is to a large degree dependent on the distribution of the sample points. Where two sample points are close, the probability values near those two sample points will reflect this. The solution to this is to calculate the same model with the same search radius, as if the samples were events, so with a Z-value of 1. The model that results from this procedure in fact reflects the probability that a sample has been taken. This latter model can be used to filter out the (irregular) distribution pattern of the samples from the probability model. The division of the sample model and the corresponding event model results in the final sample model:

$$V = \frac{\sum_{i=1}^{i=n} V_i(R - d_i)}{\sum_{i=1}^{i=n} (R - d_i)}$$

where V, R, n and d_i are defined as before, and v_i is the measured value of sample i.

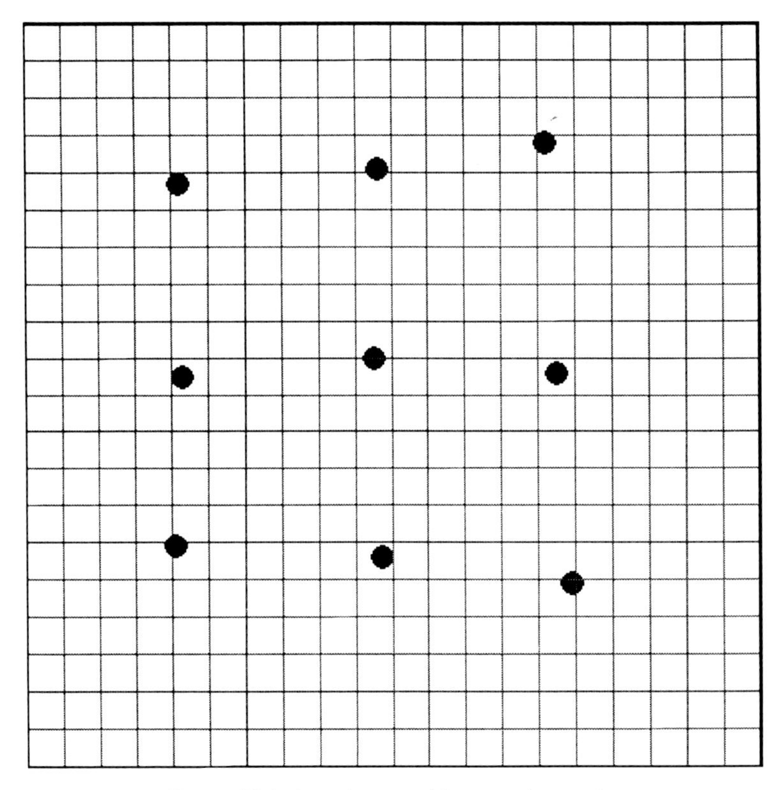

Figure 22.1 Distribution of 9 points (events).

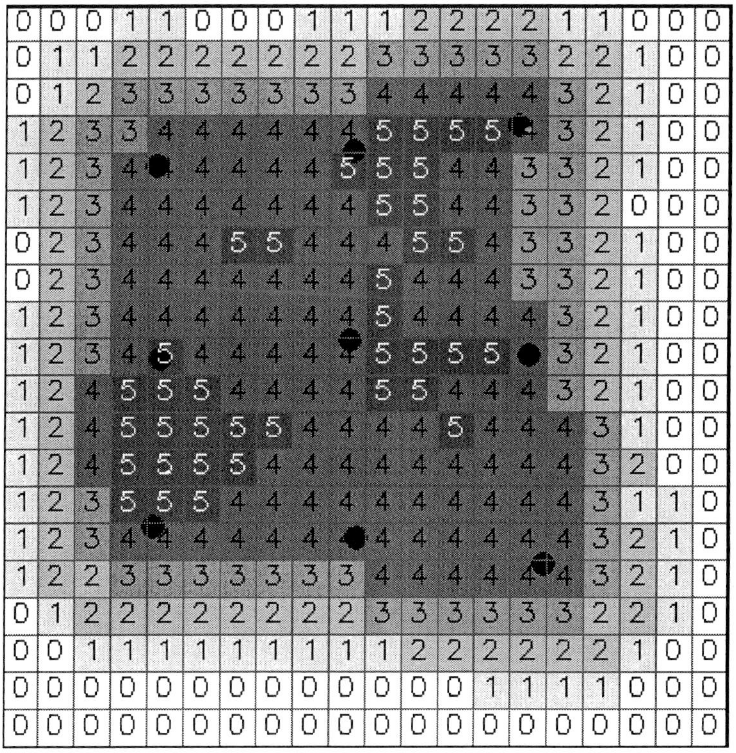

0	0	0	1	1	0	0	0	1	1	1	2	2	2	2	1	1	0	0	0
0	1	1	2	2	2	2	2	2	2	3	3	3	3	3	2	2	1	0	0
0	1	2	3	3	3	3	3	3	3	4	4	4	4	4	3	2	1	0	0
1	2	3	3	4	4	4	4	4	4	5	5	5	5	4	3	2	1	0	0
1	2	3	4	4	4	4	4	4	5	5	5	4	4	3	3	2	1	0	0
1	2	3	4	4	4	4	4	4	4	5	5	4	4	3	3	2	0	0	0
0	2	3	4	4	4	5	5	4	4	4	5	5	4	3	3	2	1	0	0
0	2	3	4	4	4	4	4	4	4	5	4	4	4	3	3	2	1	0	0
1	2	3	4	4	4	4	4	4	4	5	4	4	4	4	3	2	1	0	0
1	2	3	4	5	4	4	4	4	4	5	5	5	5	[illegible]	3	2	1	0	0
1	2	4	5	5	5	4	4	4	4	5	5	4	4	4	3	2	1	0	0
1	2	4	5	5	5	5	5	4	4	4	4	5	4	4	4	3	1	0	0
1	2	4	5	5	5	5	4	4	4	4	4	4	4	4	4	3	2	0	0
1	2	3	5	5	5	4	4	4	4	4	4	4	4	4	4	3	1	1	0
1	2	3	4	4	4	4	4	4	4	4	4	4	4	4	4	3	2	1	0
1	2	2	3	3	3	3	3	3	3	4	4	4	4	4	4	3	2	1	0
0	1	2	2	2	2	2	2	2	2	3	3	3	3	3	3	2	2	1	0
0	0	1	1	1	1	1	1	1	1	1	2	2	2	2	2	2	1	0	0
0	0	0	0	0	0	0	0	0	0	0	0	0	1	1	1	1	0	0	0
0	0	0	0	0	0	0	0	0	0	0	0	0	0	0	0	0	0	0	0

Figure 22.2 Probability model applied to the points of Figure 22.1 (search radius = 5).

Because each calculated value has to be based on more than one sample point, the search radius has to be large enough to include more than one. Usually the areas outside the sampled region cannot be considered accurate. Figure 22.4 shows the result of this algorithm after being applied to the nine sample points from Figure 22.3. The zero-value cells are outside the sampling region and in every other cell the interpolated value is multiplied by ten. To illustrate the effect of increasing the search radius, the same area has been interpolated with a search radius of eight. The image in Figure 22.5 is clearly more smoothed than the previous one.

The interpolation resembles very much the usual weighted average interpolation techniques. The main difference is that weighted average interpolates on a fixed number (usually eight) of nearest points, whereas this method uses all points within a fixed distance from the point to be calculated. Especially when there is no need for the prediction surface to reproduce the original data, and the sample points are evenly distributed over the area, this method gives very satisfactory results. For this reason, this method is very useful in a number of archaeological applications, where the degree of uncertainty is very high, as in artefact and site distributions.

22.3.3 An example

In The Netherlands, many archaeological remains are covered either by natural deposits or through human activities. In these cases, it can be very hard to distinguish activity areas within a site, since surface finds are very rare or even absent.

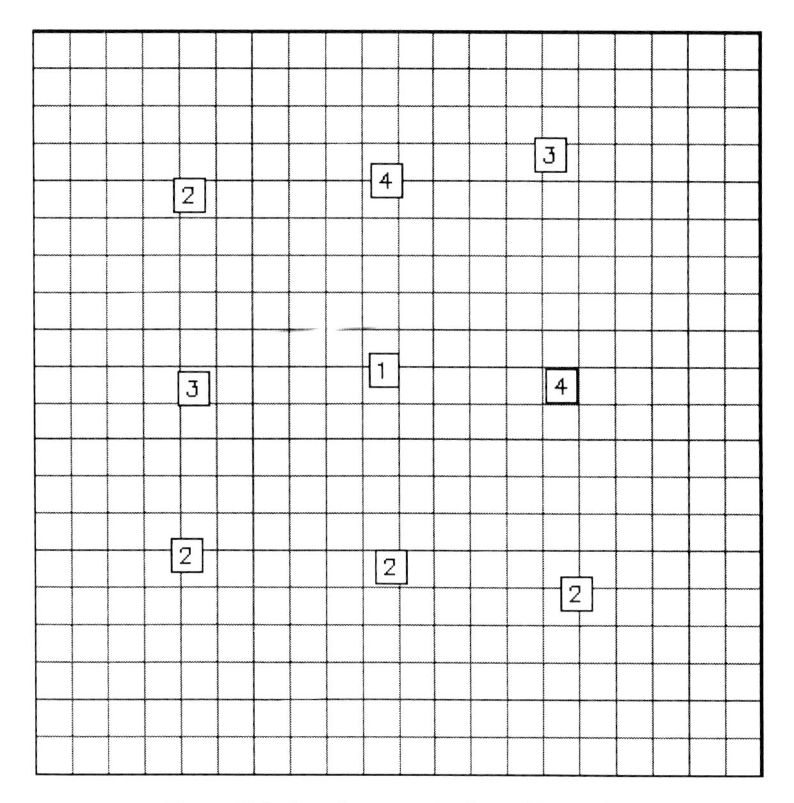

Figure 22.3 Distribution and values of 9 samples.

The solution to this is a 20-cm-wide augering device. Applying the probability model to artefact counts from the auger holes can give an indication of the internal structure of the site. An example of this is a Bronze Age site, covered with 120-cm-thick deposits of sand and clay. The resulting model (Figure 22.6) clearly shows a zone of higher probability values. The model can be checked because the site has been excavated (Figure 22.7) and it is clear that the predicted limits of the site fit rather well with the excavation plan (Groenewoudt, 1994).

22.4 Polygons/areas

A polygonal map can be either a classification of a continuous surface (where each class represents a certain range of the original data) or a map with nominal data (where the assigned categories have no mathematical relation to each other). In the first case, the area edges can be seen as contour lines, which can be interpolated using a regular interpolation technique.

In order to convert a polygonal map with nominal data into a continuous surface, the map has to be split into its distinct units. Thus a map with three units results in three maps, each one representing one specific unit. Each map shows all areas that belong to this specific unit. Since elements in a digital map must have a certain value, the areas within the polygons can have a value of one, while the areas outside can have a value of zero. GIS can be used to calculate for each point on the

0	0	0	0	0	0	0	0	0	0	0	0	0	0	0	0	0	0	0	0
0	0	0	0	0	0	0	0	0	0	0	0	0	0	0	0	0	0	0	0
0	0	0	0	0	0	0	0	0	0	0	0	0	0	0	0	0	0	0	0
0	0	0	0	0	0	0	0	0	0	0	0	0	0	0	0	0	0	0	0
0	0	0	0	[illegible]	25	30	35	40	36	38	36	42	37	38	0	0	0	0	0
0	0	0	0	25	27	30	32	35	37	32	30	40	35	40	0	0	0	0	0
0	0	0	0	27	25	24	24	27	30	27	28	30	35	43	0	0	0	0	0
0	0	0	0	30	27	25	25	22	22	20	30	32	35	46	0	0	0	0	0
0	0	0	0	32	30	25	17	17	15	16	25	32	37	40	0	0	0	0	0
0	0	0	0	[illegible]	30	25	20	15	10	16	22	28	34	[illegible]	0	0	0	0	0
0	0	0	0	28	26	25	20	17	12	16	22	32	37	40	0	0	0	0	0
0	0	0	0	26	24	22	18	17	15	20	25	26	35	35	0	0	0	0	0
0	0	0	0	24	22	22	22	17	17	17	25	30	30	30	0	0	0	0	0
0	0	0	0	22	20	22	20	20	20	20	20	25	27	25	0	0	0	0	0
0	0	0	0	20	20	20	20	20	[illegible]	22	22	22	22	22	0	0	0	0	0
0	0	0	0	0	0	0	0	0	0	0	0	0	0	0	0	0	0	0	0
0	0	0	0	0	0	0	0	0	0	0	0	0	0	0	0	0	0	0	0
0	0	0	0	0	0	0	0	0	0	0	0	0	0	0	0	0	0	0	0
0	0	0	0	0	0	0	0	0	0	0	0	0	0	0	0	0	0	0	0
0	0	0	0	0	0	0	0	0	0	0	0	0	0	0	0	0	0	0	0

Figure 22.4 Interpolated surface (search radius = 5).

map for which the value of the area equals one the nearest distance to the boundary of the area. For each point of the value equals zero areas the inverse distance to the areas can be calculated, so that points at a larger distance have a more negative value than closer points, and a distance of zero is represented by a value of zero. In this case, there is a linear relationship between point value and distance to unit boundary. The relation can also be exponential or logarithmic. The resulting image shows mountains on the value equals one areas and depressions on the value equals zero areas. The map is transformed into a continuous surface and can be treated like a DEM.

In Figure 22.8, the grey areas represent Coversand areas (Pleistocene sandy deposits). Applying the above-mentioned linear buffering technique, the map is transformed to a continuous model. A rescaled version of the resulting image is presented in Figure 22.9. Figure 22.10 shows a three-dimensional model, in which the valley bottoms (no Coversand) and hilltops (Coversands) are flattened due to logarithmic buffering, whereas the steeper zones represent transition zones (probably Coversand).

22.5 Lines

The transformation of lines, like streams and roads, to continuous areas is much easier. In fact, this is a common practice in predictive modelling. It can be done by calculating for each point on a map the distance to the nearest line. As with polygon

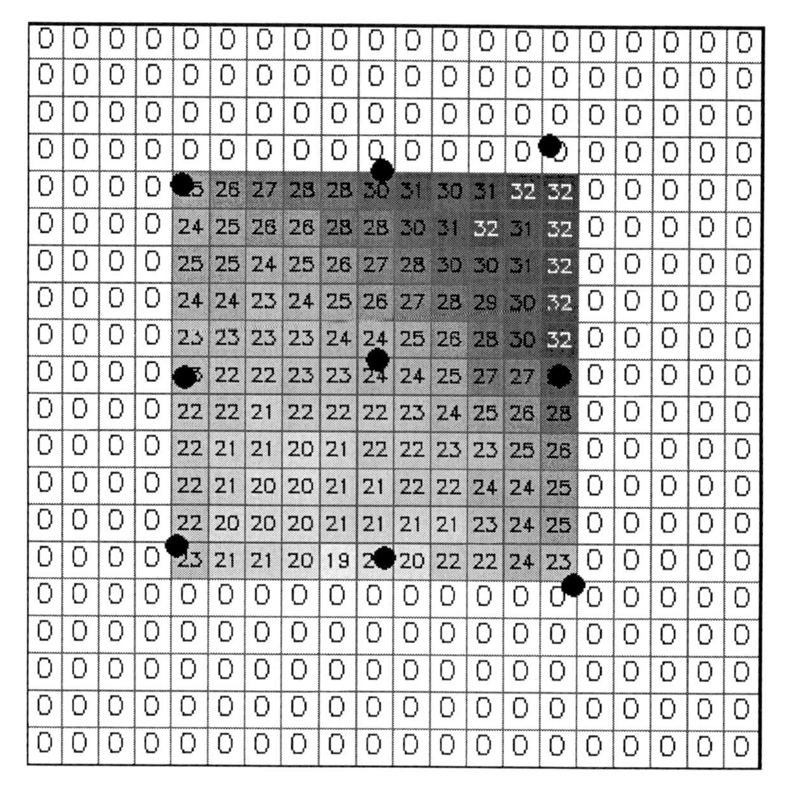

Figure 22.5 Interpolated surface (search radius = 8).

data, the result is a model where each value has a linear relation to the distance to the nearest line. Again, a logarithmic or exponential relation is also possible and, when the distance between lines is large, inevitable. For the transformation of contour lines to a DEM, many GIS packages offer standard procedures, which

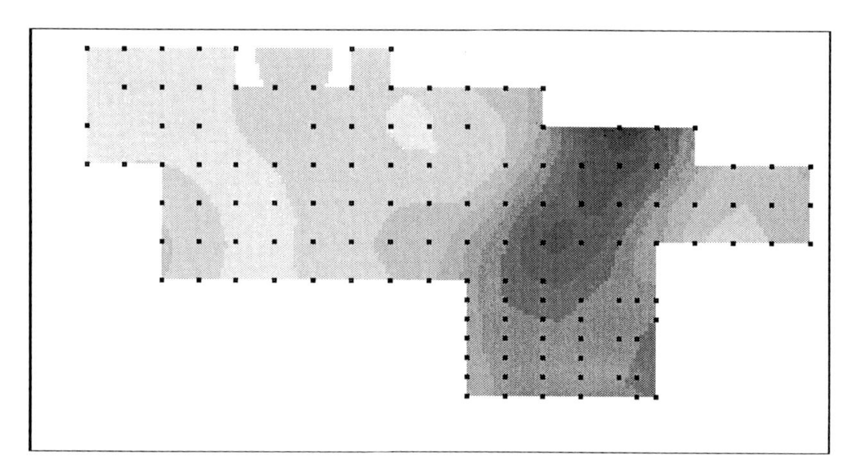

Figure 22.6 Probability model applied to artefact counts from auger holes on a Bronze Age site.

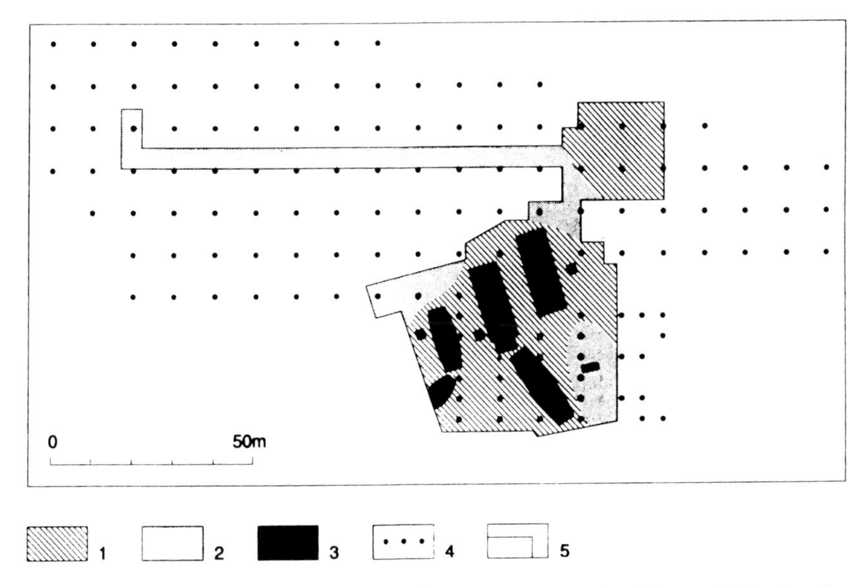

Figure 22.7 Simplified excavation plan of the same site as in Figure 22.6: 1 = features; 2 = nothing found; 3 = building remains; 4 = auger holes; and 5 = limits of excavated area.

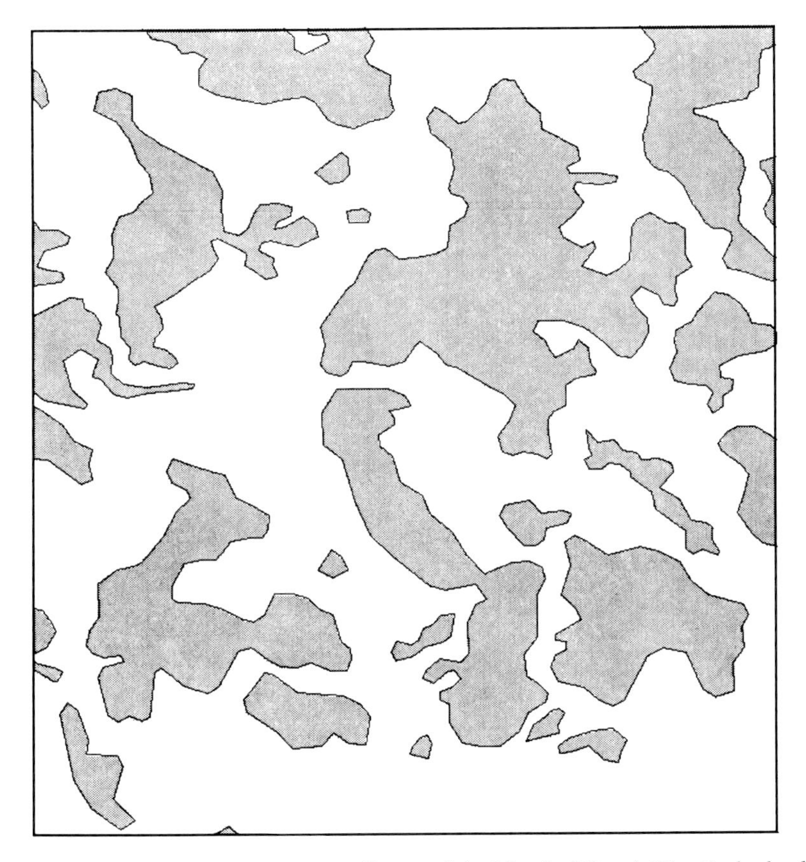

Figure 22.8 Coversand areas in a small area of the island of Texel, The Netherlands.

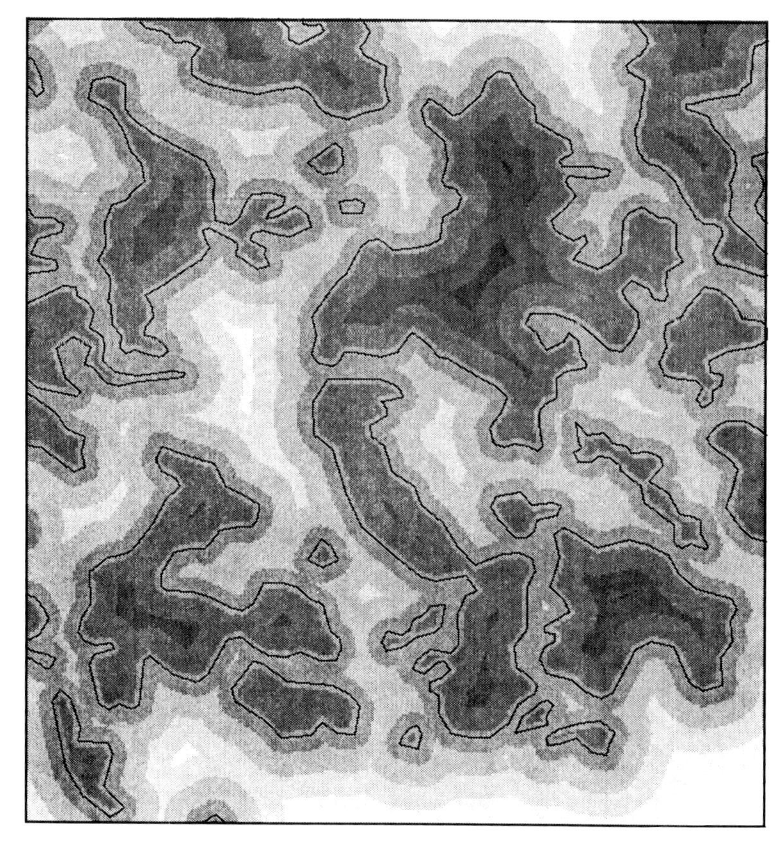

Figure 22.9 *The Coversand areas transformed to a continuous area, using linear buffering.*

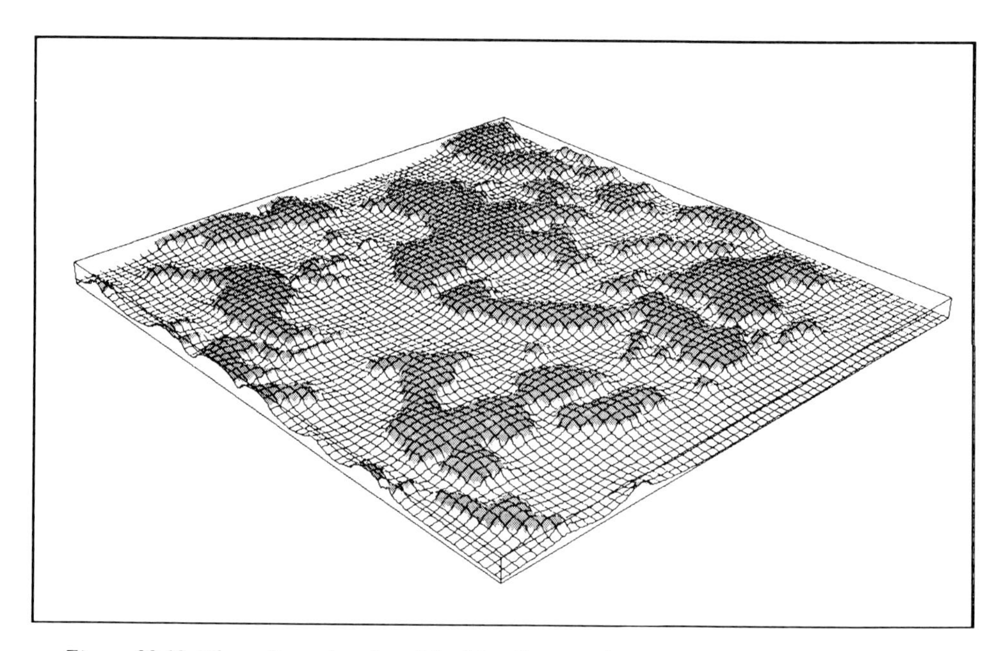

Figure 22.10 *Three-dimensional model of the Coversand areas using logarithmic buffering.*

interpolate the contour lines pretty much the same as one would do by hand: for each point on the map the nearest two contour lines are searched for, and the value of the point is calculated from the elevation values of the contour lines and the distances to each of those two contour lines (Ehlschlaeger, 1991).

22.6 Concluding remarks

Continuous surfaces contain more information than traditional maps. They reduce the effect of inaccuracy in maps and are a better representation of reality. A distinction has to be made between line and polygon data on the one hand, and point data on the other. The probability method for interpolation of events and samples can be used as an analytical tool itself, whereas the models resulting from polygon and line data can be the basis for further analysis like predictive modelling.

References

Burrough, P. A., 1989, Fuzzy mathematical methods for soil survey and land evaluation, *Journal of Soil Science*, **40**, 477–92.

Cressie, N. A. C., 1993, *Statistics for spatial data*, New York: John Wiley & Sons.

Ehlschlaeger, C., 1991, r.surf.contour, in *GRASS version 4.0, User's reference manual*, Champaign, Illinois: US Army Corps of Engineers Research Laboratory, pp. 307–8.

Groenewoudt, B. J., 1994, *Prospectie, waardering en selectie van archeologische vindplaatsen*, Amersfoort: Nederlandse Archeologische Rapporten 17.

Haigh, J. G. B. and Kelly, M. A., 1987, Contouring techniques for archaeological distributions, *Journal of Archaeological Science*, **14**, 75–80.

Johnson, I., 1984, Cell frequency recording and analysis of artefact distributions, in Hietala, H. (Ed.), *Intrasite Spatial Analysis in Archaeology*, Cambridge: Cambridge University Press, pp. 75–96.

Kvamme, K. L., 1990, GIS algorithms and their effects on regional archaeological analysis, in Allen, K. M., Green, S. W. and Zubrow E. B. W. (Eds), *Interpreting space: GIS and archaeology*, London: Taylor & Francis, pp. 112–25.

Roorda, I. M. and Wiemer, R., 1992a. The ARCHIS project: Towards a new national archaeological record in The Netherlands, in Larsen, C. U. (Ed.), *Sites & Monuments. National Archaeological Records*, Copenhagen: The National Museum of Denmark, pp. 117–22.

Roorda, I. M. and Wiemer, R., 1992b, Towards a new archaeological information system in The Netherlands, in Lock, G. and Moffet, J. (Eds), *Computer applications and quantitative methods in archaeology 1991*, BAR International Series S577, pp. 85–8, Oxford.

Warren, R. E., 1990, Predictive modelling of archaeological site location, in Allen, K. M., Green, S. W., and Zubrow, E. B. W. (Eds), *Interpreting space: GIS and archaeology*, London: Taylor & Francis, pp. 201–15.

Watson, D. F., 1992, *Contouring. A guide to the analysis and display of spatial data*, Oxford: Pergamon Press.

23

Humans and PETS in space

S. Stead

23.1 Introduction

This chapter considers the archaeological implications of three classes of models or surfaces generated by GIS: perception surfaces, effort or cost surfaces and time surfaces. Definitions of each are offered and assessed in the light of their usefulness to archaeology. Possible uses of each type of surface are explored together with possible methodologies for their derivation from standard GIS layers. The chapter concludes with a review of the author's research on effort modified Thiessen polygons.

23.2 Perception surfaces

Perception surfaces are images of how people view their world. The world that people live in, is not always delineated by the realities of physics. Distance is not always perceived as linear or indeed, 'as a cost' in calories or effort, but often in 'roughness' and difficulty. Consider the journey from your home to the shops. The tendency is to break the journey into 'chunks' or segments: a process known as segmentation. These segments are not of equal length or effort. They tend to be divided up by line of sight or complexity as well as distance. So for example the journey from the author's home to the railway station may be divided into 13 segments. The segments vary from long straight segments of road to short, complex underpass and stair segments. In linear distance terms, they vary by a factor of 6:1, and in effort terms by about 4:1.

Perception surfaces must be of use to archaeologists, as they are a model of a population's interpretation of the landscape within which they act. Archaeologists trying to understand a people's use of a landscape must have such perception models, as the use of the landscape will have been conditioned by the population's perception of it.

However, the above segmentation model may be inappropriate for other times and cultures. Wheatley (1993) has reported on Firth's work, which illustrates this principle with the example of the Isle of Wight. In this example, early map forms provided the evidence for the interpretation that a maritime culture's perception differed from Cartesian 'reality'. This insight into a society's perception will be more problematic were no map evidence exists.

Once archaeologists have been successful in gaining models of people's past perception of their world, they should use these models to provide the framework for current research. Wheatley (1993: 137) has argued that 'a normativist approach' to framing study areas must use social or political entities that existed in the past. This can be extended to the landscape within which studies are conducted. The perceptual frameworks would be used to distort Cartesian maps into an ethnocentric map base for the study period and culture.

The development of techniques to derive these surfaces is a major task. Contemporary map data may be used in one of two ways. Correction factors for our modern map base may be derived directly by transforming known points using rubber-sheet geometry. Alternatively, the map data can be used as the basis of a model, from which 'correction' factors can be derived. At points in time and space where such map data are not available, archaeologists will have to generate similar models. The models can then be tested against archaeological data, although there is a grave danger of producing self-fulfilling or cyclically reasoned models by this method.

23.3 Effort

Many current effort surfaces are based on summing slope values across the study area. This approach does not seem to provide sufficient consideration for actual or perceived effort. More sophisticated approaches reclass raw slope values into bands, with low values changing little and higher values being distorted more (1, 4, 9, 25, etc.). This is similar to Wheatley's (personal communication) suggestion that the relationship between slope and time taken to traverse the slope, is an exponential function. However, even these more sophisticated approaches do not take sufficient cognisance of the linear distance component of the 'effort' or of barrier effects.

A further interesting possibility is the idea that as the human body, in non-athletic condition, can be regarded as a constant effort producer, the 'effort' could be mapped to calories. Calories may be a useful unit of measure as there is the possibility of calibrating against diet to give ideas about 'break-even' exploitation zones.

Perception must play a role in the calculation of realistic effort surfaces. Steep slopes are perceptually harder work. Detouring to avoid the 'big hill' is common practice even if this actually adds to the journey time or effort. This may be a function of the effort release mentioned above and further work is needed in this area.

The move to a harder basis for effort must be of great benefit to the consideration of landscape. Thiessen polygons, for instance, can no longer be based on linear distance. The move towards effort, perceived or actual, being our measure for 'distance' in landscapes, must be made.

The derivation of effort surfaces presents some new and interesting problems. Both positive and negative barrier effects, for instance streams and roads, and directional and non-directional elements, for instance slope or waterflow, must be considered. There is little work being done in this area, but the work of Zahn *et al.* (1993) and Stead and Kvamme (forthcoming) show approaches to the problem.

23.4 Time

There are four readily identifiable types of time surface:

1. calibration,

2. process or model surfaces,
3. perception surfaces, and
4. artificial or postulated landscapes.

Calibration surfaces are reclassified effort surfaces. The reclassification is based on the time taken to traverse a certain amount of 'effort'. The calibration is often done by experimentation as in the case study of the island of Hvar (Gaffney and Stančič, 1991).

Process or model surfaces show time as a variable across the surface. This is often used to illustrate diffusion models, for example Zubrow (1990) and Allen (1990).

Perception surfaces are always set in a particular era. The society that perceived them existed during a band of time and therefore the perception surface is only valid during that period of time. To consider the surface outside that time span is fraught with danger, although such a consideration may be of particular interest when considering transitions between one cultural perception and another.

The final group is **artificial or postulated landscapes**. These can be divided into two classes. The first class consists of 'time slices' in which a landscape of a particular period is generated at a particular time. In this type of representation snapshots of the landscape are stored and may be used as interpretive backdrops. This presents problems in that the snapshot is only one possible interpretation of the data and the interpretation will tend to be fossilized due to the high 'cost' of compilation.

This approach also finds processual changes difficult to represent. Change is represented not by process but by images of the results of process. 'Time-slicing' temporal GIS are like movie cameras. They present a series of images which if 'sliced' fine enough give the illusion of temporal movement or change. In effect, one is dealing with state changes. To investigate a time between two known states, one has to start the interpretive process again with the aim of defining a new state.

The second class includes dynamic landscape models, which do not, in theory, require the same level of resources to complete and maintain. These models represent known data elements in four dimensions (x, y, z, t), instead of the conventional three. These data can then be mapped, using analogues of real world processes, from their known positions into predicted, or modelled, positions. A set of these modelled positions, with the same value for 't', would be the equivalent of a 'still' from a 'time-slice movie'. However, these models are currently only on the drawing board.

Effort and perception surfaces will probably have close links with artificial landscapes. This is due to the need for the effort and perception surfaces to be based upon landscapes that are contemporary with the societies for which they are generated. It would be very unsatisfactory to generate effort and perception surfaces from a modern landscape that did not reflect the true setting within which the culture existed.

23.5 Effort modified Thiessen polygons

The Ruatha teaching data set was chosen for this study. Although this data set is false and has many faults, it was freely available and well understood at the time. Thiessen polygons were generated for the three villages in the data set. The villages

Figure 23.1 Thiessen polygons for the three villages in the Ruatha teaching data set, showing contours and the boundaries of the Thiessen polygons modified by the friction surface.

are marked by crosses in Figure 23.1 with the polygons shown in contrasting shading. A friction surface was generated using a combination of linear distance and twice the local elevation change. This was further modified by adding positive barrier effects for water bodies and negative effects for roads and tracks. The cost surfaces from each of the three villages were then generated. These were then overlaid using a 'minimum' function to generate a final cost surface. Each cell was then allocated to the nearest village across the cost surface and the edges of the resulting polygons were then vectorized. These vectors have been superimposed on Figure 23.1 to show how the Thiessen polygons have been altered by the friction surface. It should be noted that the cost surface was non-directional but the general principle has been illustrated.

22.6 Conclusion

The inclusion of sophisticated models of human perception of landscape, movement within landscape and their inter-relationship within GIS, is both challenging and essential. Without the elements of human interaction with each other and with their environment, archaeologists run the risk of divorcing themselves from the very people they are trying to study. It is also important that studies of the past are conducted within landscapes that are contemporary with the cultures under scrutiny. If GIS are not to become another abused tool, we must tackle these issues.

References

Allen, K. M. S., 1990, Modelling early historic trade in the eastern Great Lakes using geographic information systems, in Allen, K. M. S., Green, S. W. and Zubrow, E. B. W. (Eds), *Interpreting Space: GIS and Archaeology*, London: Taylor & Francis, pp. 319–29.

Gaffney, V. and Stančič, Z., 1991, *GIS approaches to regional analysis: A case study of the island of Hvar*, Ljubljana: Znanstveni Institut Filozofske fakultete.

Stead, S. D. and Kvamme K. L., forthcoming, A new approach to cost surface analysis, paper presented at CAA94.

Wheatley, D., 1993, Going over old ground: GIS, archaeological theory and the act of perception, in Andresen, J., Madsen, T. and Scollar, I (Eds), *CAA92 Computer Applications and Quantitative Methods in Archaeology*, Aarhus: Aarhus University Press, pp. 133–8.

Zahn, C., Menon, S. and Gao, P., 1993, A directional path distance model for raster distance mapping, in Frank, A. U. and Campari, I. (Eds), *Spatial Information Theory—A Theoretical Basis for GIS*, Lecture Notes in Computer Science 716, Berlin: Springer-Verlag, pp. 434–43.

Zubrow, E. B. W., 1990, Modelling and prediction with geographic information systems: a demographic example from prehistoric and historic New York, in Allen, K. M. S., Green, S. W. and Zubrow, E. B. W. (Eds), *Interpreting Space: GIS and Archaeology*, London: Taylor & Francis, pp. 307–18.

24

How to look good and influence people: thoughts on the design and interpretation of an archaeological GIS

P. Miller

24.1 Archaeological GIS grows up?

From humble beginnings only a few years ago (e.g. Brown and Rubin, 1982), geographic information systems (GIS) are now becoming more commonplace in archaeology, both in Europe (Lock and Moffett, 1992; Andresen, *et al.*, 1993) and in the USA (Allen *et al.*, 1990). The rapid adoption of GIS has meant that advances in methodology and theory relating to the manipulation and understanding of output from these systems often lag behind. In some cases, this has led to the creation of pretty, but sadly unusable applications.

With archaeological use of GIS now approaching maturity, it is high time for us to consider complex problems relating to how our GIS displays actually look, what information they impart to the viewer, and how this process of information transfer may be improved. This chapter explores some of the issues of display design as raised in disciplines other than archaeology and shows how these concepts are of relevance to our very peculiar subject where so little data is truly of a type that may be categorized and displayed with the precision offered by computer-based systems.

This chapter is written as a result of experience gained while working on the pilot phase of the York Archaeological Assessment (YAA) (Miller, forthcoming). During this work, the importance of well-designed formats for the presentation of data became readily apparent, and it was also clear from the literature that fairly basic design considerations were often overlooked in other projects. The suggestions and guidelines outlined below are based upon reading from the human–computer interaction (HCI) field of psychology, and from experiences within the YAA. They outline solutions that appear to work for the problems being tackled by the project and are hopefully of more general use to other GIS projects.

Despite the commendable move among GIS companies worldwide towards making their packages more alike, and the blurring of the distinction between raster and vector systems (Burrough, 1986), some of the presentation-oriented aspects of GIS software remain very different. In certain packages, for example, a particular display effect may be easily achieved with a click of the mouse, while the same effect may require extensive programming in another package. For this reason, it is worth stating that the basis for many of the comments below is work with the ARC/INFO

GIS through all incarnations from 5.0 to 6.1. Data and output from the early stages of the YAA are used as illustrations throughout.

24.2 The York Archaeological Assessment (YAA)

The YAA is a research project based in the University of York (UK), which is investigating the use of GIS techniques to model archaeological deposits beneath the ancient city of York. Data for the project comes from a variety of sources, including the archives of the York Archaeological Trust (who provide part of the funding for the project) and the Ove Arup (1991) database. An important aspect of the work is examining ways in which complex multi-dimensional data may be clearly displayed and understood both on and off the computer screen.

Maps produced by the YAA include traditional plan views of the modern city centre (derived from 1:1250 Ordnance Survey digital maps) showing a variety of archaeological and management related data including archaeological interventions, modern land use, statutorily protected areas, etc. and more complex representations of three-dimensional data such as the modern topography and archaeological deposits.

The completed system will allow detailed querying of archaeological interventions within the city in relation to modern land use and cityscape, as well as the mapping and prediction of subsurface archaeology. Although an understanding of deposits and depositional processes may be considered essential in modern archaeological excavation, little work of this nature has been undertaken. The YAA will answer questions of deposit creation and survival within the city, and will look at how these deposits reflect past human activity.

24.3 Problems with archaeological data and with GIS

As anyone who has ever looked at a site archive will be aware, archaeological data are intrinsically difficult to map and quantify in any meaningful fashion. While on first consideration GIS and other location-based applications may appear to be the answer to many archaeological problems, they often create many more.

GIS technology makes use of the inherent precision of computers, and allows the accurate positioning of sites and artefacts within space. This is useful with accurately located components, but many archaeological archives are sadly incomplete, and, where data are present, they may well be inaccurate. With antiquarian work, excavations may only be located within the parish, and even modern excavation archives of the 1970s and 1980s are riddled with inconsistencies and errors. A major problem facing archaeologists is the search for a means by which error and uncertainty may be displayed clearly, but without cluttering already complex displays. This problem becomes still more pressing with the increased use of computer-generated maps as it appears that images generated by different means are accepted to varying degrees by users, with scribbles on paper being accepted with a degree of scepticism, while computer-generated images are inexplicably almost universally accepted without question. The map that was understood to be of dubious quality on paper becomes inviolate once reproduced on screen. These problems are further compounded by the cavalier approach certain users take to scale on digital maps; a map digitized at 1:50 000 is not suitable for detailed measurement of the distance between ramparts in a hillfort, nor even for showing the shape of the fort. Poorly digitized maps are of limited use even at the scale used to digitize them, and no map should be used for analysis at scales greater that at which it was digitized, as errors rapidly begin to creep in, often without the user being aware, until it is too late.

In York, for example, the recent York Archaeological Trust (YAT) programme to digitize the outlines of their excavations had to face the problem of defining the outlines of some early excavations where consistent measurements were not available. What should be done in these circumstances? Should the digitizer make a best guess? Should the site not be digitized at all, as it is of poor quality anyway? Should a dashed line be used, to denote uncertainty? Should different possible outlines be given? All of these techniques have been used in different situations by different people, and each solution has its advantages and drawbacks. Like so many solutions, one does not necessarily appear to be better than the others for all situations, and it would appear that any of the above may be used, so long as the user is consistent, and makes it clear to any potential audience what has been done.

One of the greatest problems with the use of archaeological data in today's GIS is the conflict between multi-dimensional data and essentially two-dimensional display technology. Complex multi-variate archaeological data reside in threespace[1] and it is difficult to clearly map these to representational flatland[1] (Tufte, 1990: 12). Even a simple three-dimensional scatter plot (Tufte, 1990: 115) is difficult to realize in flatland, and fairly high level spatial skills are required before a user may comfortably navigate in threespace in the absence of familiar perceptual cues such as the ever present ground. Without these cues, even language begins to fail the viewer, as the familiar concepts of up, down, forward, backward become merely relative, and dependent entirely upon viewpoint.

A discipline facing problems similar to those of archaeology is geology, where geological strata need to be visualized in much the same manner as archaeological contexts. The excellent *Three Dimensional Applications in Geographic Information Systems* (Raper, 1989) outlines some of the solutions adopted in this discipline, but even here where the problems have been under consideration for longer than in archaeology, the solutions are often far from satisfactory. The main problem is the sheer volume of data that must be represented in an archaeological display—a volume that flatland may well be incapable of supporting. A typical map that the YAA would hope to produce, for example, would include the modern terrain model and townscape, a representation of the various periods hidden beneath the surface (Richards, 1990), a measure of the nebulous concept of deposit quality, a measure of deposit waterlogging (including, possibly, the water table itself), deposit preservation (not the same as quality), archaeological interventions in the area (trench outlines, etc.), a representation of error (potentially in all elements of the display, but realistically in only one of two at any time) plus, possibly, textual data and more detailed site-specific images. It is obvious that pseudo-threespace representation must be used in displaying such an image, but conventions must still be developed to allow this mass of information to be understood and manipulated in any meaningful fashion.

As yet, no truly three-dimensional commercial GIS exists and while it may be possible to record a third dimension in current GIS packages, manipulation of that third dimension (and those dimensions beyond) is not so easily undertaken as with the first two. A three-dimensional GIS would allow the well-known GIS buffer zone to be extended into threespace, creating cylindrical buffers. Voronoi tessellations in a three-dimensional GIS would take the value of z into account when constructing polygons, and visualization of threespace data structures would become intuitive and accessible. Manipulation of the time component of any dataset would also become easier. Until these functions, and others, become available, our ability to comprehend and manipulate the complexities of threespace will be severely handicapped. However, this lack of software functionality may not be as much a disadvantage as it appears. As suggested in this chapter, current methodological and cognitive constructs are strained to their limits by the abilities of current software. The addition of flexible access to threespace may well be more than can currently be accommodated, leading to an even wider gap between perception and true understanding than that faced at present.

24.4 Designing the display

Graphical output from GIS must be adaptable for use in a variety of circumstances ranging from on-screen query to both colour and monochrome hardcopy. These different forms of display must be borne in mind at all times during the process of designing a style, as the most beautiful of colour displays may be rendered unintelligible when reproduced in the black and white employed by most publications. Unless the designer is prepared to use different formats for colour and monochromatic displays—surely an unnecessary waste of effort except in exceptional circumstances—it is necessary to consider all forms in which a map may be displayed right from the design stage.

In creating an archaeological GIS, it is important to consider from the beginning the fundamental factors of database design (Chartrand and Miller, 1994); coverages (how should the map data themselves be subdivided and archived?); and analytical destiny (to what sort of use will the data be put?).

I would argue that consideration of a design for the 'visualization engine' (VE)[1] should also take place at this time. The VE is the name given to the whole process of displaying maps and results, whether on a VDU, on paper, or through another medium. As such, the VE includes the mechanics by which a display is created and the design considerations which make the display distinctive and interpretable.

The importance of good design for databases and the GIS engine itself are clearly recognized but display is often considered only as an afterthought, if truly considered at all, leading to a proliferation of poorly considered presentations of the type so denigrated by the likes of Edward Tufte (1983; 1990). Without a well-designed, ergonomically structured VE, results from even the most excellent GIS fail to realize their full potential to impart data to the viewer. In extreme cases, these results become a meaningless jumble or even, most worrying, positively misleading. The most greatly abused and, to some, the most powerful aspect of any VE is colour. When used correctly, colour can greatly add to an image, but overuse of colour may result in information overload (Travis, 1991), or may detract from the true message of the graphic (Travis, 1991: 116).

> The purpose of using color is to emphasise certain relationships, or distinguish certain areas, and not to give a global structure

Displays for GIS and other applications are often designed using colour from the outset, with the colours thus becoming an essential part of the image. Reproduced in monochrome, or seen by a visually impaired viewer, such an image is largely meaningless. Most work on the subject would appear to agree in insisting that, wherever possible, colour should be used independently of the formatting—rather than being an essential part of the design, colour should add to a message that would still be present without the use of colour. In other words, a GIS map should be as informative, if slightly less pretty, in monochrome as it would be in colour. Travis' golden rule should be remembered at all times: 'Design for monochrome first' (Travis 1991: 117).

There has been a great deal of work regarding the use and perception of colour within psychology (Itten, 1961; Rowell, 1988; Boynton *et al.*, 1989; Travis, 1991), and much of this is applicable to the design of the VE.

As Travis (1991) states, colour may be used for four main reasons:

1. aesthetics—precise colours are unimportant so long as it looks good;
2. formatting—colour is used to segregate or group types of visual information;
3. coding—colour is used symbolically, to define meaning; and
4. realism—colour is used *qua* colour (water is blue, etc.).

All four are common components of GIS applications, and it is useful for the designer to be aware of the use to which a particular colour is being put. In use of colour for aesthetic reasons, the work of Johannes Itten (1961) of the Bauhaus is interesting. Itten suggests that 'two or more colours are mutually harmonious if their mixture yields a neutral grey' and he proposes that aesthetically pleasing displays will consist mainly of these mutually harmonious colour groupings, or chords. The neutral grey so desired by Itten is of course not obtainable by the normal additive process of colour mixing (where the mixing of red and green would result in yellow) but by the subtractive process as used by printers to mix inks (where red and green would result in grey). The colour wheel (Figure 24.1) shows how these groupings are derived. Any two opposing colours (e.g. yellow and violet) are considered harmonious, as are any three colours that may be connected by rotating the triangle (e.g. yellow, red and blue), or any four connected by rotation of the square (red–orange, blue–green, violet and yellow). The suggestions of Itten may seem simplistic and typical of the reductionist stance of the Bauhaus, but other authors see the value of this neutral grey; Edward Tufte expresses the first rule of Eduard Imhof as (Tufte, 1990: 83, emphasis added)

> color spots against a light gray or muted field highlight and italicize data, *and also help to weave an overall harmony.*

In other words, a display will be at its clearest if the majority of the image is of a muted colour (white or black also appear to work, although not so well, but are the usual backgrounds for paper and VDUs, respectively, and therefore are commonly used), with only the key points highlighted to stand out from the background. The maps generated for the YAA pilot study, for example, use colour sparingly to merely

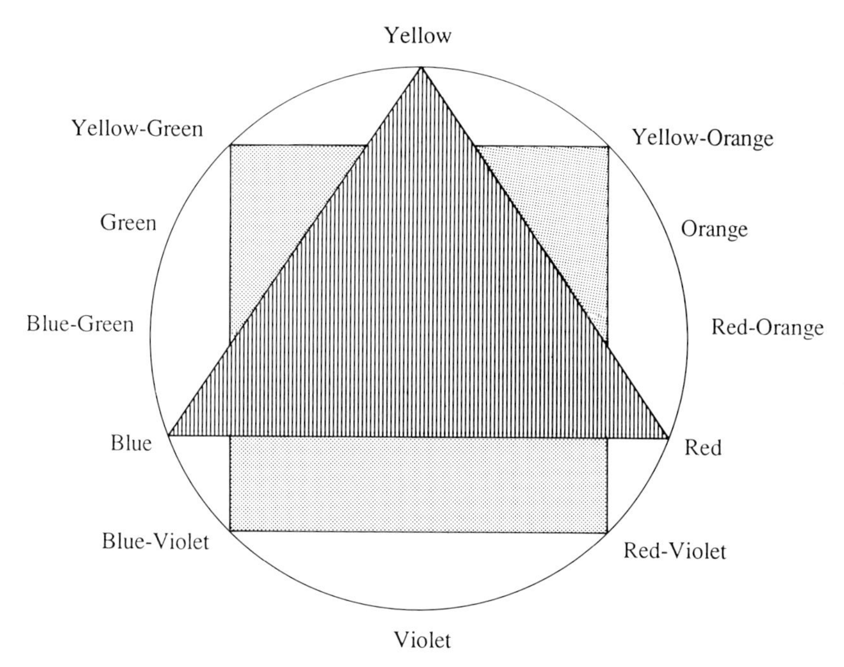

Figure 24.1 Itten's colour wheel (after Travis, 1991).

highlight the areas of significance against a largely subdued background. Where colour is used to code data, Travis defines six principles that should be adhered to (1990: 123–4). Colours should be:

1. **discriminable**: Colours used should be easily distinguishable from each other, even by the visually impaired. They should also be nameable (i.e. green, rather than greeny-browny yellow, with a hint of pink).
2. **detectable**: The colours should be visible under the anticipated lighting conditions.
3. **perceptually equal steps**: The discrimination steps between the chosen colours should appear approximately the same. If a set of five colours is chosen, and two of those colours are very similar, those colours will be associated. The user will think they are linked or related in some way.
4. **meaningful**: Where possible, colour coding should be meaningful. Code water blue! Misleading colour coding can be a serious problem. On a map, for example, blue lines are expected to be water, but if rivers are green and roads blue, unnecessary confusion may arise.
5. **consistent**: Code similar items in the same way across multiple screens. Place consistently present items (such as a scale) in similar (or, even better, the same) locations on all screens.
6. **aesthetically pleasing**: Use as few colours as possible, and ensure that the chosen colours do not clash or vibrate.

A colour is defined in colourspace by reference to the three dimensions of hue,[1] saturation[1] and value (often a measure of lightness or brightness). Variation in one or more of these dimensions may be used in coding different forms of data. Nominal[1] data are best coded by simple variations in hue. Hue exists as a circle in colourspace and thus has no beginning or end, while both saturation and brightness exist as continua (e.g. light to dark) and thus imply relative values not present in nominal data. With nominal data, no value is assigned to the colours used.

Quantitative data, such as temperature, may also be coded by variation in hue, but here small, regular changes should be used to represent a flow or continuity; blue (cold) to red (hot) including all parts of the spectrum in between. Using hue to record this type of information can lead to unintended perceptual problems, as it is possible for underlying variation in hue to destroy the relationship between different levels in the data themselves (Travis, 1991). This effect is most noticeable where large variations in hue are displayed, as the brain has far more difficulty in identifying a sequence for three widely separated hues (orange, green, indigo, for example) than for three similar hues (red, red–orange, orange). It becomes increasingly difficult to perceive the trend represented by the colour scale as the separation between hues increases.

Saturation may usefully be used when trying to visualize deviation of data about a norm. In the YAA, for example, deposits of average thickness may be coded white. Deeper deposits are gradually turned redder along a saturation scale from white (average thickness) to red (greatest thickness), while shallower deposits may progress in a similar fashion from white to green. This method allows trends to be visualized very clearly, but makes the measurement of specific values at a given point quite difficult. Variation in luminosity or brightness may be used in a similar fashion to saturation, but these changes are often difficult to reproduce in hardcopy without expensive colour printers.

24.5 *The visualization engine (VE)*

Colour is merely one aspect of the consideration that should go into the construction of the VE (Figure 24.2). At the heart of the YAA VE is the integration of a series of structured Arc macro language (AML) scripts with a reasoned rule set which contains information on colours, archaeological conventions, etc. The implementation of the VE in this form allows the easy production of multiple images, all conforming to a house style, while leaving flexibility to alter aspects of the display where this may be required by the specific circumstances surrounding an individual image. The VE control program (VECP) is the core of the computer based element of the VE (Figure 24.3), and it is here that parameters may be altered to meet specific circumstances. Input to the VECP may be automated, in the form of a further AML script, or interactive, with the user supplying parameters as requested by the system. The VECP produces output by running a variety of further scripts which request parameters from the VECP where necessary. Following the same drive towards modular structure as advocated within the YAA database design (Chartrand and Miller, 1994) the scripts accessed by the VECP are all small, and designed for the completion of specific tasks. A script exists, for example, simply to generate the north arrow that appears on most YAA output. This script requests (x, y) location and circle radius from the VECP. In the future, the interactive mode of controlling the VECP may well have a menu-driven interface constructed for it using ArcTools, but for the foreseeable future this remains command line driven.

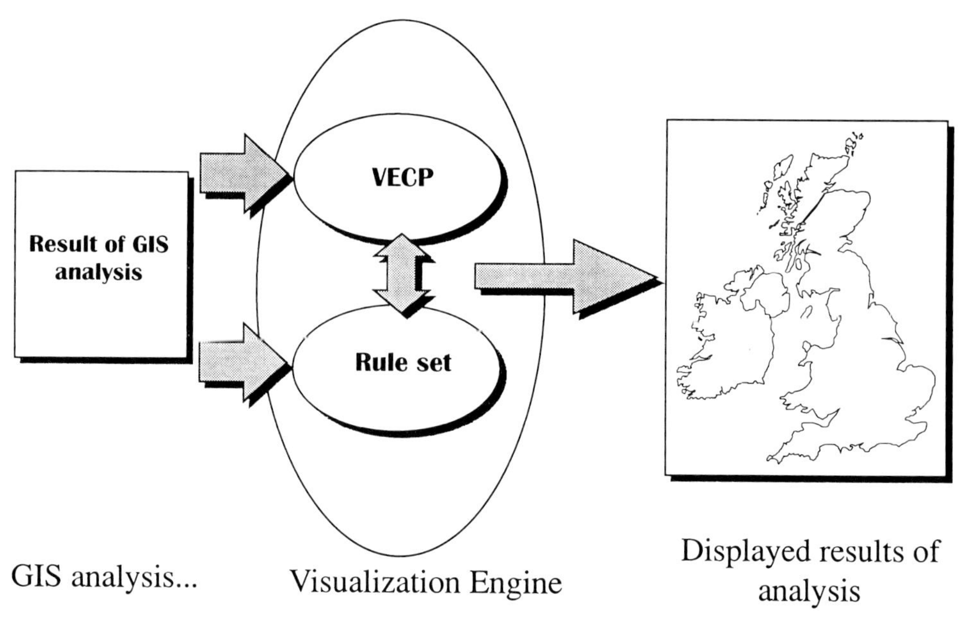

Figure 24.2 The visualization engine.

The rule set within the VE is not known to the software, but is applied by the designer when instructing the VECP. Rules include the guidelines above, such as Travis's (1991) six principles for colour coding, the necessity for an indication of north on any map, and the requirement for a measure of scale on all maps. On most YAA maps, there are two indications of scale; a scale bar showing the scale at which the map is being displayed and a written statement of scale (e.g. 1 : 1250) recording

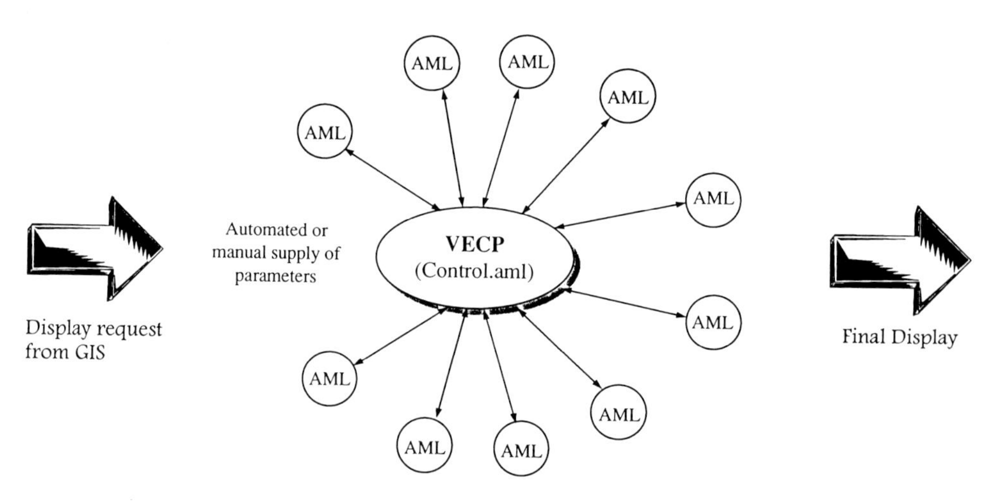

Figure 24.3 The visualization engine control program (VECP).

the scale at which the digital map was digitized or provided by the Ordnance Survey. In practical terms, this is the lowest scale at which the map should be used (see earlier comments).

Most of the rules are to a large degree considered common sense by the majority of GIS users. Since they are so obvious, it may seem of little value to actually state them as rules, but the number of GIS applications that fail to follow even the most obvious of cartographic conventions (scales, north arrows, etc.) make it clear that a formal checklist is in fact necessary. The following is an excerpt from the YAA rule set, indicating the very basic nature of many of the rules.

1. Maps should not be used for analysis at scales greater than that at which they were digitized.
2. Where possible, maps should not be displayed at scales greater than that at which they were digitized.
3. All maps should include an indication of the scale at which they are being displayed.
4. All maps should include an indication of the scale at which they were digitized (or the resolution at which they were obtained for raster images).
5. All maps should include an indication of north.
6. While not necessarily displayed, all maps should have a modification history outlining their source(s), changes made, etc.
7. Where symbolic or colour-based coding are used, a key should be included on the map.
8. Scale bars, keys, etc. should, where possible, be consistently placed to avoid unnecessary confusion.

In creating a display, the designer is forced to compromise between stark informative images and more visually exciting creations which may ultimately confuse the viewer, or obscure the message conveyed by the graphic. Tufte is unimpressed by the 'interior decoration of graphics' (1983: 107) he so scathingly labels chartjunk,[1] and he illustrates his argument with examples of truly awful graphic design. Images just as awful are produced within archaeology, too, and perhaps we can learn from his writing without necessarily adopting the Spartan austerity implied by his data ink[1] approach when taken to extremes.

Any graphic is created primarily to convey information to an audience. In archaeology, that audience may or may not be comfortable with a graphical representation and it is therefore important for archaeological graphics to be clear and relevant, leaving little room for confusion on the part of the viewer. While remaining clear, a little effort put into designing an appealing graphic is often rewarded by greatly increased understanding for the viewer. It is to be hoped that the VE implemented by the YAA has resulted in graphics that, while easily legible, are pleasing to look at and use.

Figure 24.4, for example, shows one result of the VE, where iconic representation has been used in place of a more traditional key to simplify understanding of the map. Rather than being required to relate the line of a more traditional key with the shape on the map, it is merely necessary to relate congruent shapes, thus allowing more time to be spent analysing the meaning of the map rather than decoding unnecessary symbology.

Figure 24.5 represents another display produced through the VE. This display type is frequently used for portraying subsets of the city wide database, such as the

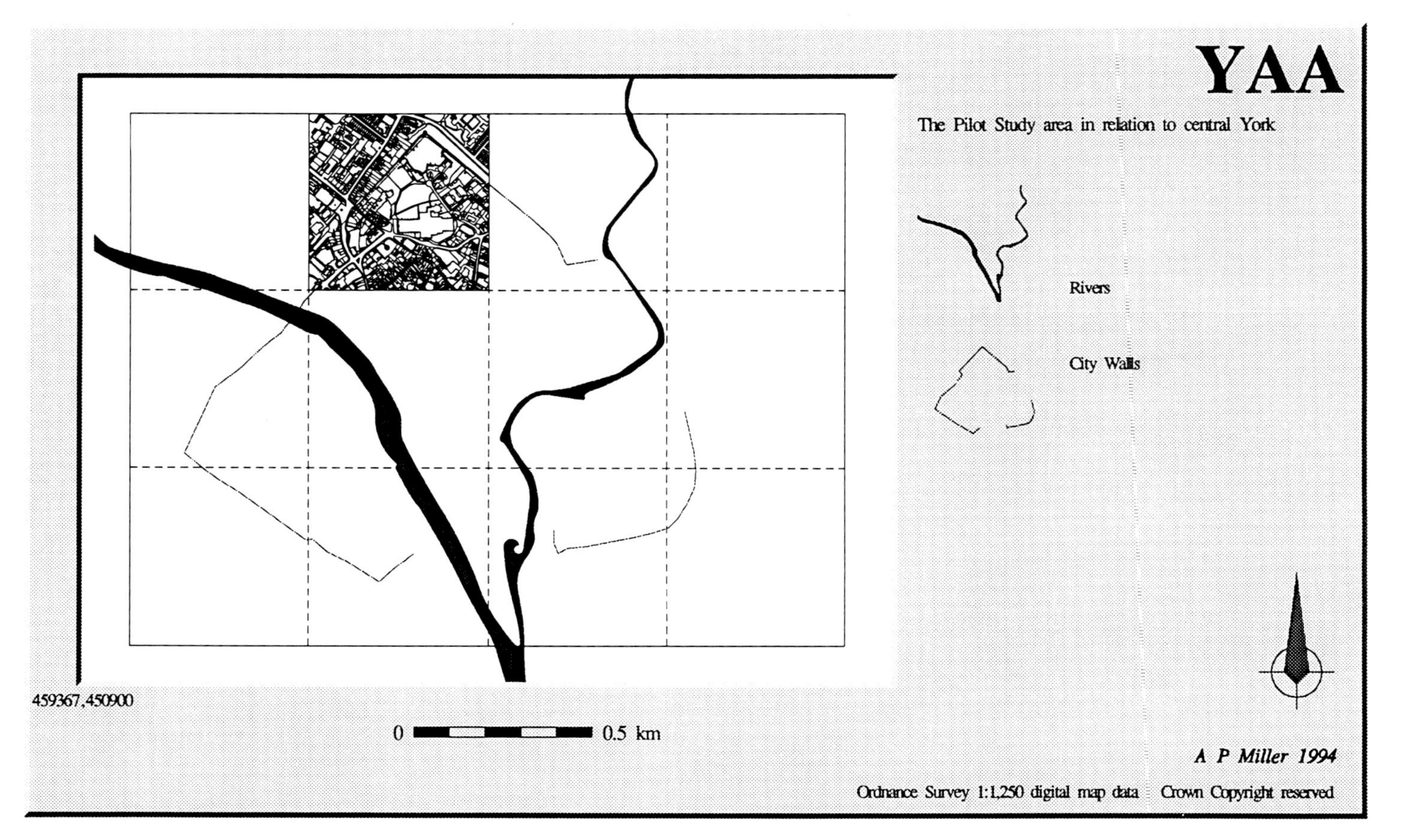

Figure 24.4 The YAA pilot area in relation to central York.

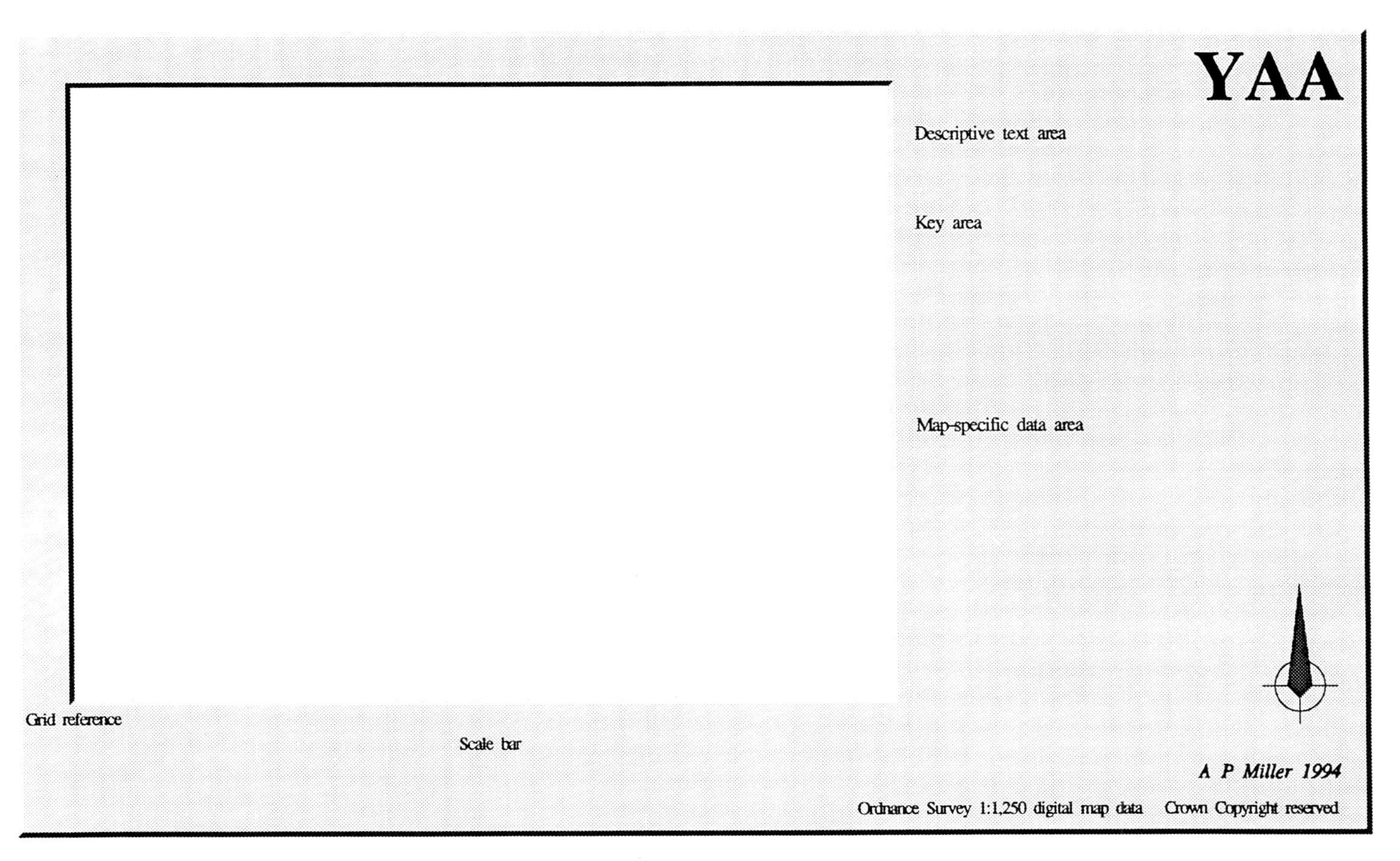

Figure 24.5 A model for the layout of map displays.

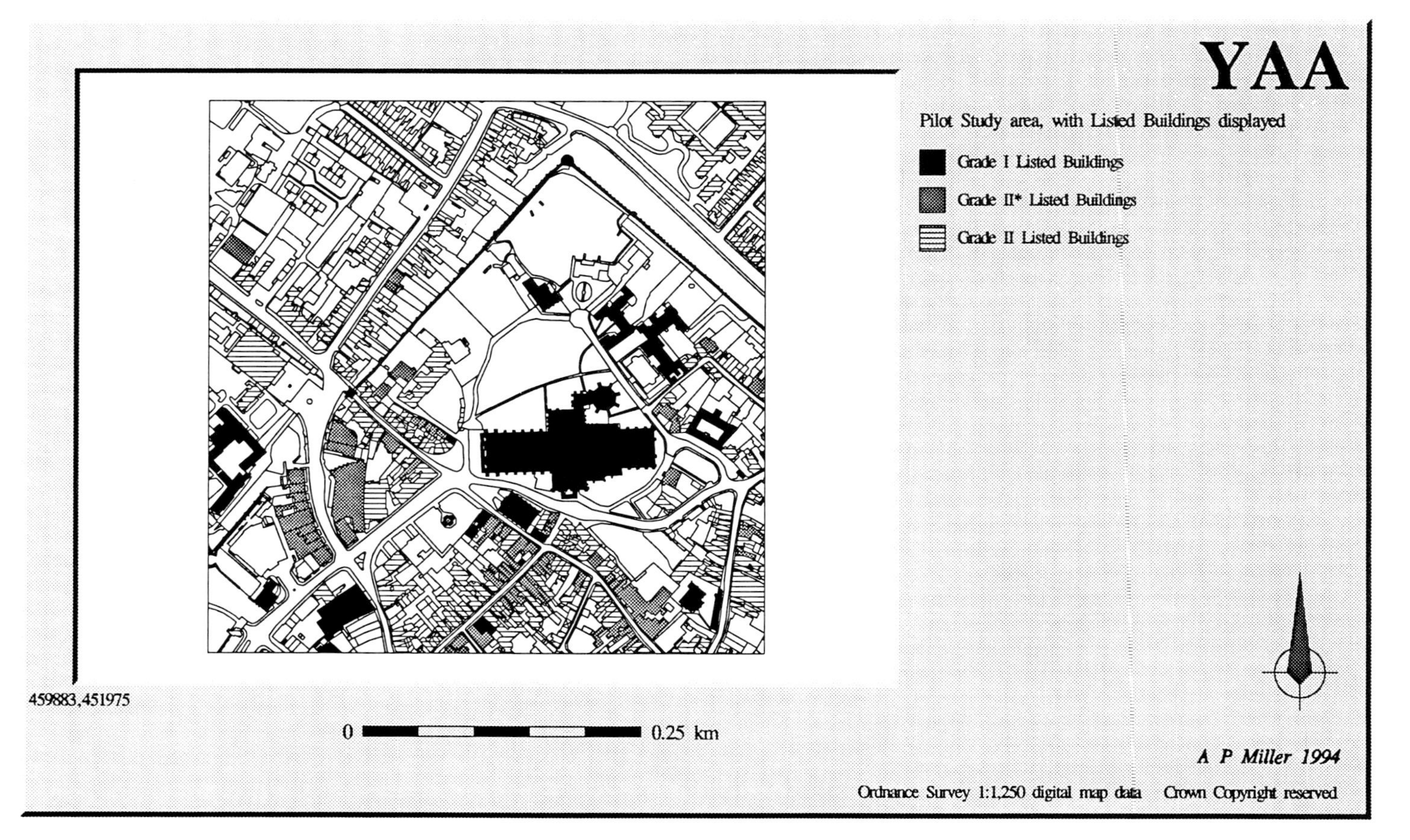

Figure 24.6 The Y A A pilot study area.

Pilot Study area (Miller, 1994) and scripts exists to provide the VECP with data to produce many variations upon this basic model. Figure 24.6 shows the model applied to the Pilot Study area, with the results of a search through the database for listed buildings displayed.

24.6 Conclusions

While most archaeologists cannot afford the time or computing power to produce images such as those of Paul Reilly (1988; 1992) at IBM, and, indeed, many do not need such images, a little care and thought can render far less visually impressive graphics just as informative. Whether graphics are produced on state of the art workstations or on a humble PC, they must still overcome the fundamental problems inherent in archaeological data and the findings of other disciplines regarding the formulation of informative displays (Monk *et al.*, 1993) should be recognized. It is to be hoped that archaeologists are now ready to move *en masse* towards a more reasoned approach to storing, displaying and interpreting all computer-based data.

Acknowledgements

The writing of this chapter involved forays into fields far from my normal work, and I am indebted to several people for their comments and support. Firstly, Anna Cox from the University of Bradford (UK) provided useful pointers into the Psychology literature, and encouraged some of my early thoughts on the subject. Dr Andrew Monk of York University Psychology Department, and Dr Rob Fletcher and Peter Halls, both of York University Computing Service also provided invaluable comments and advice early in the preparation of this chapter. Finally, Dr Julian Richards of York University Archaeology Department, who oversees the York Archaeological Assessment itself, as my MPhil/DPhil supervisor and kindly read and commented upon my earlier drafts.

The project is jointly funded by the Science and Engineering Research Council and York Archaeological Trust, and much of the digital data were provided by York Archaeological Trust.

References

Allen, K. M. S., Green, S. W. and Zubrow, E. B. W. (Eds), 1990, *Interpreting Space: GIS and archaeology*, London: Taylor & Francis.

Andresen, J., Madsen, T. and Scollar, I. (Eds), 1993, *Computing the Past: computer applications and quantitative methods in archaeology 1992*, Århus: Århus University Press.

Boynton, R. M., Fargo, L., Olson, C. X. and Smallman, H. S., 1989, Category effects in color memory, *Color research and application*, **14**, 229–34.

Brown, P. E. and Rubin, B. H., 1982, Patterns of desert resource use: an integrated approach to settlement analysis, in Brown, P. E. and Stone, C. L. (Eds), *Granite*

Reef: a study in desert archaeology, pp. 267–305, Arizona: Arizona State University Press.
Burrough, P. A., 1986, *Principles of Geographical Information Systems for land resources assessment*, Monographs on soil and resources survey 12, Oxford: Clarendon Press.
Chartrand, J. A. H. and Miller, A. P., 1994, Concordance in rural and urban database structure: the York experience, *Archeologia E Calcolatori*, **5**, 203–17.
Foley, J. D. and Van Dam, A., 1982, *Fundamentals of interactive computer graphics*, The systems Programming series, London: Addison–Wesley.
Itten, J., 1961, *The art of colour: the subjective experience and objective rationale of colour*, translated by E. van Haagen, New York: Van Nostrand Reinhold.
Lock, G. and Moffett, J. (Eds), 1992, *Computer applications and quantitative methods in archaeology 1991*. British Archaeological Reports International Series S577, Oxford: Tempus Reparatum.
Miller, A. P., forthcoming, 'The York Archaeological Assessment: computer modelling of urban deposit in the city of York', paper presented at Computer Applications in Archaeology 1993, Stoke on Trent.
Monk, A., Wright, P., Haber, J. and Davenport, L., 1993, *Improving your human–computer interface: a practical technique*, Hemel Hempstead: Prentice-Hall.
Ove Arup, 1991, *York development and archaeology study*, London: English Heritage.
Raper, J. (Ed.), 1989, *Three dimensional applications in Geographic Information Systems*, London: Taylor & Francis.
Reilly, P., 1988, *Data Visualisation: Recent advances in the application of graphic systems to archaeology*, UKSC report **185**, March 1988, Winchester: IBM.
Reilly, P., 1992, Three-dimensional modelling and primary archaeological data, in Reilly, P. and Rahtz, S. P. Q. (Eds), *Archaeology and the information age: a global perspective*, pp. 147–76, One World Archaeology 21, London: Routledge.
Richards, J. D., 1990, Terrain modelling, deposit survival and urban archaeology, *Science and Archaeology*, **32**, 32–8.
Rowell, J., 1988, *Picture perfect: color output for computer graphics*, Oregon: Textronix.
Travis, D., 1991, *Effective colour displays: theory and practice*, London: Academic Press.
Tufte, E. R., 1983, *The visual display of quantitative information*, Cheshire, CO: Graphics Press.
Tufte, E. R., 1990, *Envisioning information*, Cheshire, CO: Graphics Press.

Notes

[1] The following definitions apply.

- **Chartjunk**—the extras added to graphics, sometimes to improve them aesthetically but occasionally to disguise or hide data. While regarded by some as unnecessary at *all* times, these frills can be useful in making data more accessible.
- **Data ink**—a measure of the direct relevance of the whole image to the data being displayed—'the non-erasable core of a graphic, the non-redundant ink' (Tufte, 1983: 93).

- **Flatland**—the limiting restrictions enforced by attempting to display multi-dimensional data on two-dimensional display media.
- **Hue**—is the most basic component of colour, and is the attribute which actually tells us what colour something is.
- **Nominal**—data of a sort where order does not matter—types of pottery on a graph, for example.
- **Quantitative**—data of a sort that may be measured, and quantified.
- **Saturation**—how coloured a colour is—red and pink are of the same hue, but red is more saturated.
- **Threespace**—more than merely three-dimensional display, threespace encompasses the actual display and the conceptual framework behind true multi-dimensional analysis.
- **Visualization engine**—a model defining the means by which geographic data are displayed.

25

Future enhancements to GIS: implications for archaeological theory

J. B. Claxton

25.1 Introduction

Geographical information systems (GIS) are seen by an increasing number of archaeologists as being the optimum information technology to be adopted for locational analysis, spatial data management and spatial modelling (Kvamme, 1992). The extent of this interest is evidenced by the growing number of papers, conference presentations and volumes which have become available over the last five years. Such widespread interest in GIS raises questions regarding the current ability of the technology to meet the expectations placed upon it (Zubrow, 1990).

To date, many researchers have concentrated solely upon the limitations of GIS, rather than seeking complementary technologies to strengthen it (Aangeenbrug, 1991; Maffini *et al.*, 1989; Maguire and Dangermond, 1991; Shepherd, 1991; Zubrow, 1990). This chapter addresses this imbalance by briefly explaining the potential contribution of neural networks, artificial intelligence (AI) and virtual reality (VR) to overcome the limitations faced by the current technological status of GIS. Neural network architectures allied with classic 'strong' (AI) techniques offer the possibility of overcoming the most intractable and error-prone aspects of GIS data collection. The integration of GIS with a VR interface offers the user a more direct experience of differing temporal processes, spatial relationships and the rapid assimilation of multi-dimensional real-world data generated by the interaction of interpersonal and natural forces in the evolution of the landscape. Furthermore, the integration of these technologies offers the prospect of significantly advancing the methods of simulating ancient socio-economic behaviour, raising important implications for the role of GIS as a medium of archaeological discourse.

25.2 Limitations of GIS

There are three significant limitations or problem areas associated with GIS and these concern the process of data input, the modelling of temporal or evolutionary sequences and the inability of most of GIS to handle real-world multi-media data sources satisfactorily.

25.2.1 Data input

GIS data input procedures are error prone and time consuming, and the data are often derived from a variety of different and incompatible sources. For example, these may include aerial photographs, satellite remote sensing images, historical and contemporary maps as well as site surveys. Often these contain a wide range of differing thematic contents, scales and methods of projection (Maguire and Dangermond, 1991). Even if all the data sources are of the same media, e.g. maps, and are drawn to the same scale and using the same method of projection, the user still faces a vast amount of data to be selected and accurately input. Error is inevitable (Maffini *et al.*, 1989; Openshaw, 1989).

25.2.2 Temporal processing

Time is of the essence to the archaeologist and, given the importance of the time dimension, it is perhaps surprising to find that archaeologists have hitherto paid scant attention to an understanding of the different concepts of time and the implications this may have for the understanding of prehistoric peoples (Castleford, 1992). Attention focuses upon the establishment of temporal sequences (understanding stratigraphy, sedimentary succession, and so on) and ranks equal in importance to the understanding of spatial relationships. GIS excels in the processing of spatially related data sets but little has been done so far to model the action of physical processes as they change over time. In fact, most GIS are at present only able to give static views of the data and would require much effort to generate a sequential view of (for instance) the evolution of a river valley or equally long and potentially complex environmental processes.

25.2.3 Multi-media nature of real-world data

The real world is perceived by our senses in many ways; we see things in three dimensions, natural forces have associated sounds and the by-products of their interactions have many differing shapes, sizes, textures, colours, volumes and distributions. When we use instruments to take measurements in wavelengths of the electromagnetic spectrum beyond the visible range (such as infra-red, X-ray, magnetic survey), then the variety of formats in which data about the real world can be collected becomes great indeed. Most GIS have the capacity to display topographical relationships in three dimensions (usually as digital terrain models) yet all lose or filter out much of the non-spatial quality of the data. This is because it is necessary to reduce multi-dimensional and multi-media data sets to two-dimensional representations (maps) in order to work with them. If GIS need to develop their temporal processing capabilities, they equally require ways of incorporating media and formats of data which go beyond the map and the vector line survey.

25.3 Neural network systems

Where data sets are to be transformed, re-formatted and co-registered to common scales, it is important that some expert help should be built into GIS. What cannot be automated should at least be intelligent enough to signal problems and conflicts to the unwary and aid the correct processing decisions to be made. One area which offers great possibilities in this respect is that of hybrid neural network expert systems.

Neural networks draw their architectural inspiration and name from biological models of the structure and function of the brain. Using simple, almost primitive processing units or 'nodes', they avoid the complexity of the more conventional 'Von Neumann' architectures which implement algorithms through the iterative, selective and conditional flow of control instructions through the central processing unit (CPU) and as such have dominated the design of computing architectures to date (Willis and Kerridge, 1983; Dayhoff, 1990). Neural nets consist of layers of simple processing nodes with each node connected to all members of its layer and to all members of the layers immediately preceding and following it. The function of the node is to compare values fed to it from other nodes with its own 'threshold value' (Dayhoff, 1990).

25.3.1 The perceptron

Depending on the level of stimulation from its neighbours, a neural net node will either begin contributing output stimuli of its own or will remain inactive. We can see how simple the basic node is by looking briefly at the earliest such device, the perceptron (Rosenblatt, 1959; 1962).

Figure 25.1 shows the basic node and its configuration in a simple two-layered perceptron. The perceptron consists of two layers of nodes whose interconnections are adjustable values or 'weights'. The threshold value is compared to the value of the inputs by means of a 'transform function' which mediates the input and output stimuli. A network is trained by subjecting it to a series of input patterns and comparing the output pattern to the desired training target pattern (McClelland and Rumelhart, 1988). Weights and threshold values are adjusted until the network gives the exact response required to the training input pattern. A significant but negative assessment of the potential of the perceptron was carried out (Minsky and Papert, 1969) which led to a severe curtailment of work being done in this area in favour of what has now become established as the 'traditional', logicist, rule-based approach to artificial intelligence.

25.3.2 Back-error propagation

The main objection of Minsky and Papert to the potential of the perceptron—or, more accurately, multi-layered perceptrons—concerned the difficulty of training the interconnections of hidden or deep layers within the network. It was argued that for such layers (occurring between otherwise accessible input and output layers) it would be impossible to predict or estimate the errors involved and alter the necessary interconnection weights. The problem was finally solved (McClelland and Rumelhart, 1988) by the method known as back-error propagation. This is a powerful technique for training neural networks and nets designed using it are commonly known as 'back-prop' nets.

25.3.3 Interactive activation

Important to the correct configuration of neural nets is the principle of interactive activation competition (IAC) and the process of lateral inhibition (Rumelhart and McClelland, 1988).

In IAC nets, nodes are configured in layers. Each node is connected to every other node in the same layer. Connections are also arranged between each node and

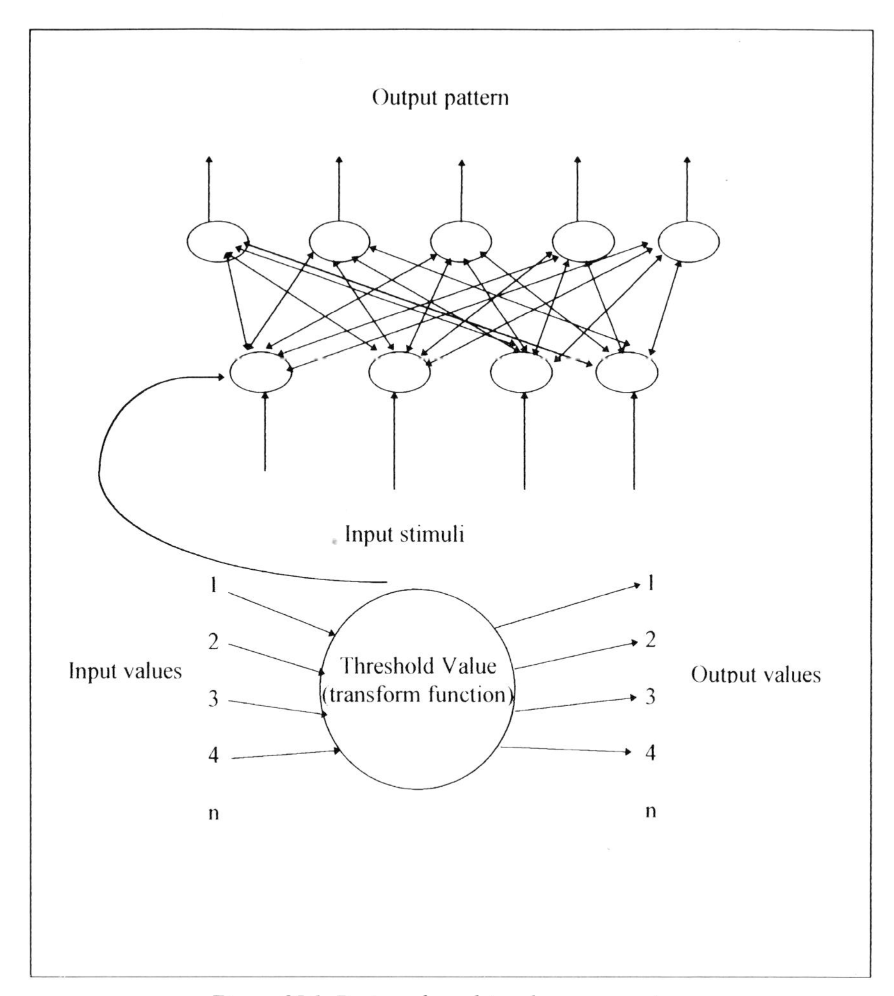

Figure 25.1 Basic node and two-layer perceptron.

the nodes belonging to the layers occurring immediately before and after the current layer. Interconnection is signified by a given value or 'weight' and each node has a 'threshold value' which signifies the stimulation level of that node. There is a fundamental difference in the nature of the interconnections obtaining between nodes of the same layer and those between nodes of differing layers. Interconnections between nodes of the same layer have the effect of suppressing the stimulation of nodes they are attached to by emitting negative or suppressive values. The opposite is the case for connections between layers. Thus connections within layers are known as 'inhibitory' connections, and connections between layers are known as 'excitatory' connections (Rumelhart and McClelland, 1988).

25.3.4 Lateral inhibition

The fact that nodes of the same layer suppress each other by their outputs means that only the node with the strongest stimuli from the layer below will overcome the sum of suppression signals received and be able to issue excitatory signals to the next layer. Nodes of the same layer compete with each other to stimulate nodes in the next layer. In this way, nets can be configured to 'recognize' patterns. Furthermore, when presented with only part of a pattern or shape, a properly configured (trained) network will still respond with outputs corresponding to the amount of stimulation experienced and will approximate an answer in proportion to the amount of the pattern available.

The reverse is also true if a part of the network suddenly becomes damaged, removed or in some other way inoperative, then the part of the net remaining is still able to function. This is known as the principle of 'graceful degradation' and compares favourably to a conventional computer architecture which would stop functioning immediately if a piece of its hardware or executing software were suddenly damaged or removed from the system (Dayhoff, 1990).

25.3.5 Levels of abstraction

An observation to be made at this point concerns the progressive increase in levels of abstraction that a fully configured net moves through. The first layers of a pattern recognition network, designed to read words for instance, are concerned with recognizing the primitive bars and hoops that form the shape of letters. Further layers are concerned with identifying whole letters and even more layers with identifying meaningful patterns of letters, i.e. words. Taking advantage of this aspect, there is no reason why archaeological applications of neural networks should not include the identification of artefact assemblages, pottery sherd classification and the preliminary classification of finds.

25.3.6 Neural nets and GIS

A useful application of neural nets in GIS would be the creation of map-reading devices to overcome the problem of scanning or digitizing maps for GIS data input. Industrial pattern recognition neural nets have already gone far beyond the recognition of points, lines and polygons, the basic building blocks of the vector-based GIS (Dayhoff, 1990). The scope for such map readers includes the reading and interpretation of thematic maps, rescaling, legend translation from foreign languages and the masking or filtering out of unwanted map elements such as awkwardly placed political boundaries or place names. Also, recent work carried out at the Joint Research Centre of the European Community (JRC) on the use of neural networks alongside GIS has seen the successful application of Kohonen map neural nets to the analysis of Landsat-5 remote sensing satellite imagery to extract 'land-cover information with high classification accuracies' (Wilkinson, 1993: 39).

25.4 Extraction of production rules from neural networks

There are significant problems in archaeology such as the classification of artefact types or the interpretation of data in terms of political, trade and cultural boundaries which can only be solved by the application of archaeological expertise gained through practical experience over many years. The process of arriving at sets of production rules which adequately describe and structure the 'knowledge domain' of archaeology is neither simple nor straightforward when using conventional AI techniques. Experts may not remember all they know or may not understand why they know something because of the intuitive nature of learning from experience. Other knowledge domain experts may be dead or simply unavailable. An additional problem concerns discovering production rules capable of dealing with situations which are unexpected or are being encountered for the first time.

The difficulty of capturing knowledge domain expertise and representing it in structured production rule systems is known as 'the knowledge acquisition bottleneck'. Given that neural nets also learn from experience (they have to be trained), can production rules be extracted from neural nets? Encouraging work carried out in the field of medical diagnostic systems (Saito and Nakanono, 1988) suggests that they can. The method is as follows.

First, a back-prop net is created and trained as outlined above. Once the network achieves an acceptable level of accuracy it is considered ready for rule-extraction. Suppose this has been done and the rules to be extracted concern the relationship of observed attributes to object types in a given set of artefacts. It is necessary to establish the general form of positive and negative rules for identifying which attributes refer to which object types and so on. These can be stated (Saito and Nakanono, 1988) as being

- Positive rule (i)
  IF OBJECT (object type)
  THEN (affirmative attributes(1) . . . affirmative attributes(*n*))
- Negative rule (ii)
  IF OBJECT (object type)
  THEN (affirmative attributes(*n*) . . . negative attributes(*n*))

The network is now considered in general systems terminology as a 'black box', which is to say that we are only concerned with its attribute inputs and object type outputs. All input stimuli are switched off and thus the network has no output values. The next step is to turn on individual and combinations of inputs, noting which outputs become activated. In this way, the relationship factor between inputs or attributes, and outputs or object types, can be established. It is from these relationship factors that systems of formal production rules can be built (Saito and Nakanono, 1988).

25.4.1 Hybrid neural expert systems

Having overcome the 'knowledge acquisition bottleneck' in this way, the rules generated can then be used to construct a conventional expert system. However, by combining the strengths of both neural networks and traditional expert systems it is possible to build systems which can learn from experience, generate rules and then apply them. Such systems are called 'hybrid neural expert systems' or neural net knowledge processor systems. An example of one such system, 'Plato/Aristotle', offers a fully working system featuring a fault tolerant, automatic-rule-finding, fully adaptive multi-lingual architecture (Voevodsky, 1987). Hybrid neural expert systems of this kind have great potential for application within the context of computational archaeology and GIS. One archaeological researcher (Gibson, 1993) describes the potential application of such a tool in developing models for archaeofaunal ageing and interpretation. Also, if we agree that models are 'a vital element in all archaeological attempts at hypothesis, theory, explanation, experiment and classification' (Clarke, 1972: 3) then there are significant implications for archaeological theory in the inclusion of hybrid expert systems within a GIS. These arise from their potential contribution as model extraction tools. That is, if neural nets can recognize or identify patterns or structures at different levels of abstraction, then the possibility also arises of extracting models (sets of rules) directly from the raw data of our investigations and those of others.

An immediate implication of this concerns the way in which archaeological data is archived and controlled. It could be effectively argued that the publication of raw data sets (using fractal file compression and CD-ROM storage, for instance) of archaeological interventions becomes not just desirable but actually necessary to the process of the constant re-evaluation of the structure of the archaeological record. Included in this issue is the question of ownership and access to primary data collections and the need for all bodies involved in the collection of archaeological data to establish and adhere to practical and reliable data archiving policies and standards (Cumberpatch, 1993; Hunter and Ralston, 1993).

25.5 The virtual reality user interface

So far, strong argument has been made for the benefits of integrating GIS with neural networks for key tasks such as pattern recognition, expert map reading and satellite image analysis. It is important at this point to consider possible improvements to the user interface. Real-world information systems are highly complex and parallel in structure. For example, wind and rainfall cause erosion while continental drift builds up lands masses and warring populations cause the mass movement of refugees. These are but a few forces affecting the landscape. Considerations such as these confirm our discussion above that GIS technology with its emphasis upon maps as a primary data-input source and the mainly two-dimensional medium of (visual) communication, limits the processing and presentation of multi-dimensional multi-media real-world data (Raper and Kelk, 1991; Shepherd, 1991).

Clearly, what is needed is a user interface which allows the multi-media nature of real-world data to be more readily perceived. In this respect VR systems represent a useful step forward. That is to say, the use of head-mounted visual displays, tactile data manipulation peripherals and three-dimensional auditory localization would allow the user the experience of actually inhabiting the real-world data being simulated (Kalawsky, 1993). Also, VR interfaces imply the notion of a 'virtual archaeology' (Reilly, 1991; Wheatley, 1991) but without such graphical representations being supported by accurate and intelligent spatial databases, the concept stands little chance of gaining further ground.

25.6 Artificial life

People are missing from the majority of traditional archaeological reconstructions, indeed much experimental archaeology seems aimed at eradicating the human factor and ways should be sought to model and understand the difference between past socio-economic structures and our own (Hill, 1992). One way to address these problems, and avoid such pitfalls as economic and technological determinism, would be to simulate the activities of past societies within the context of reconstructed landscapes. In addition, if artificially intelligent simulations were available to model the individual as well as generalized populations of them we might gain some access to the hitherto inaccessible belief structures of ancient non-literate groups. This concept is especially interesting bearing in mind the comments made above regarding the extraction of knowledge rules from experiences encountered.

Following the work of Pollock (1989), whose automata (Oscarites) have survival skills, environmental monitoring and a sense of identity based upon the supervenient re-identification of themselves from second-order internal sensor monitoring and 'De Se' referencing—our aim would be the creation of such automata within the context of the GIS. There are three strands to this approach (Pollock, 1989, ix):

> Token materialism is the thesis that mental acts are physical acts. Agent materialism is the thesis that persons are physical objects having a suitable structure. Strong AI is the thesis that one can construct a person (a thing that literally thinks, feels and is conscious) by building a physical system endowed with 'appropriate artificial intelligence'.

Using these concepts Pollock proceeds to define a strategy for implementing an automaton capable of responding to its environment in a manner which could suggest that it 'is conscious in precisely the same way we are' (Pollock, 1989: 29). Traditional AI is characterized by the production rule-based systems mentioned earlier and this leads to fundamentally logic-driven designs. A number of workers see no other way (Gardin, 1988; Fisher, 1989; Smith and Yang, 1991). Regrettably, the scope of this chapter does not allow us to question whether logic-driven systems are the best way of modelling often irrational, non-logical human behaviour but the point should be regarded as a moot one. However, it should be noted that the methods advocated here allow the possibility that models (logical or otherwise) could be extracted without the intervention of the knowledge domain expert because they involve systems which learn through experience.

The real and exciting possibility is therefore raised of 'recovering' and modelling the expertise of past societal groups through the simulation of automata within the context of intelligent GIS. In this respect, Doran is entirely vindicated in saying, 'If real insight can be achieved into the relationship between dynamic modelling and knowledgeable reasoning systems, then the impact on archaeology is likely to be as much at the conceptual and theoretical level as at the level of practical interpretation' (Doran, 1977: 453).

From a technological perspective, it is now a perfectly legitimate aim to simulate human activity interacting with the reconstructed landscape through the creation of intelligent automata. It is possible to go significantly further than this by seeking to extract the knowledge and belief structures of such simulated individuals as a catalyst to archaeological theory by using the methods described above.

25.7 The real-world simulacrum

In the real world, which we are endeavouring to model using GIS technology, many systems operate in parallel, at different geographical scales and different time periods and yet interact with each other. In archaeological terms, the ancient landscapes we seek to reconstruct are complex palimpsests of that interaction (Aston and Rowley, 1974). Much has been said about production rules and models but what of the information systems needed to handle such complexity? It is important to visualize how these, themselves, might be modelled, particularly in the light of the

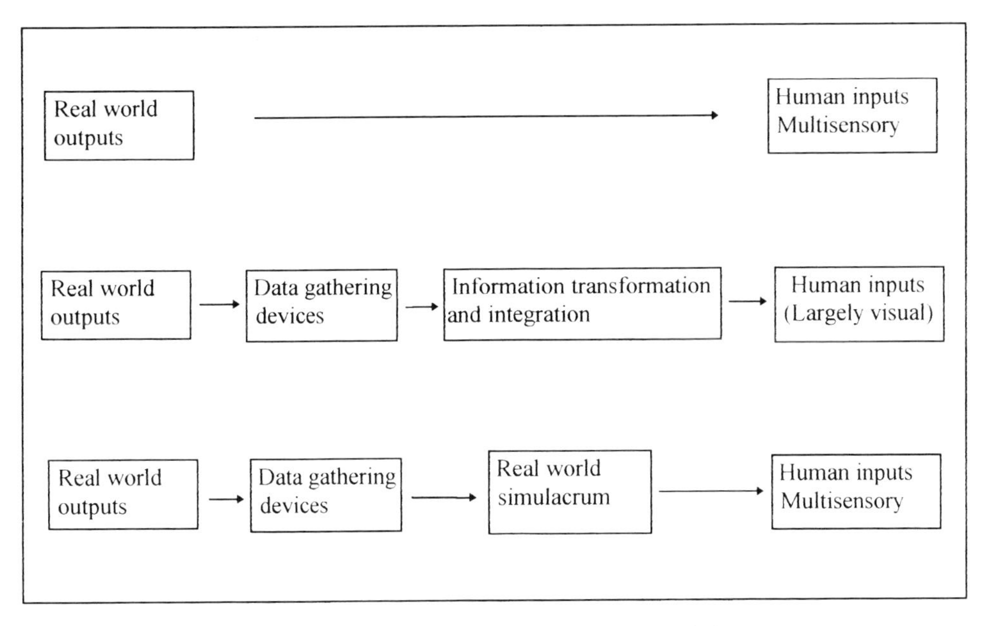

Figure 25.2 Information flows between the real world and the geoscientist (Shepherd, 1991: 355).

current discussion on information integration within GIS (Shepherd, 1991). Shepherd offers a model of information flows between the real world and the geoscientist; see Figure 25.2. The box entitled 'real-world simulacrum' performs the task of taking the multi-sensory, multi-dimensional data flows and presenting them to the human user and yet is not further elaborated by Shepherd. This chapter proposes a way of achieving this by suggesting a strategy of adopting GIS as the integrating information technology with hybrid neural expert systems giving support in key areas such as data input and interpretation and by adopting a virtual reality user interface. Given that there are many computational requirements here, these could be broken down and processed on fast, massively parallel, distributed hardware architectures whose physical location to each other is immaterial given effective computer communication links.

By implementing GIS technology as advocated above, and in the context of fully connected and distributed hardware and software architectures, the full potential of GIS as the integrating medium of the many types of real-world data may be achieved. However, as the technologies discussed continue to merge in the ways outlined, it becomes highly necessary to explore the likely implications for archaeological theory.

25.8 Discussion

It has been noted that the various technologies previously discussed in relation to GIS have contributions to make in overcoming its current limitations and offer better functionality with regard to archaeological requirements. It has been shown that contemporary GIS are strong on spatial modelling and analysis but weak on temporal processing. Also, it has been observed that the nature of real-world data goes beyond the map in forms of representation and that a fundamental requirement of future GIS will be the ability to include multi-media data sets. At the present time, therefore, GIS can be seen as a useful tool for archaeologists to use providing expectations of it do not outstrip what is currently possible.

However, if GIS is seen purely as a tool by archaeologists then its role in the development of archaeological theory will be misunderstood and underestimated. The potential contribution of GIS as 'real-world simulacrum' to archaeological theory can be examined by considering Tilley (1990: 333–4) and his analysis of Foucault. Tilley (1990: 33) suggests we might write material culture as follows:

> First, . . . we need to 'abandon the longing for' and striving after 'a basic originary and primary objectivity . . .'

It is inevitable that the attempt to participate in archaeological discourse using the written word should involve an attempt at concise explanation and the transmission of meaning. The training, examinations and eventual awarding of professional status for the archaeologist depends on the adoption of such skills. Examiners mark papers according to criteria of mutually accepted 'truths', fellow professionals concur more readily when definitive statements are clearly made and transmitted, and public funding bodies require results of an unambiguous nature to justify their continued support.

In the light of these pressures, it is easy to understand why the quest for unambiguous, precise or definitive interpretations of material culture might be sought. However, alternative models of explanation are needed (Tilley, 1990; Hill, 1992) and there is a further problem to consider in that our models should also be made (as far as possible) explicit and testable (Clarke, 1972).

The writer attempting to meet these aims is faced with a very difficult task indeed. The tendency is either to seek a single solution or to attack proposed solutions with all the effort it requires to replace one alternative with another. Either way, the process is very difficult to share with others given the time lags and practicalities of the publishing process and the difficulties of keeping, maintaining and sharing large data sets. As long as the written word remains the sole medium of archaeological discourse, in practice theory will become increasingly divorced from data as methods of examining and extracting the latter from material culture continue to increase and the volume of available data increases exponentially. Tilley (1990: 333) continues:

> Second, . . . we should perhaps write material culture in a manner that has more similarities to novels or poetry than to a treatise on constant acceleration equations.

Technical discourse is available through examination of the models inherent in the design of a GIS, and those generated at run time by its simulated automata. The requirement for literary creation is evident in the fact that such a real-word simulacrum could (with the aid of the virtual reality interface) be directly experienced by the user. As with literary creations where individuals see or understand different aspects of the author's writing, each traveller in a virtual archaeological world would be left with purely personal interpretations and impressions. Is this not the point of literary creation?

Tilley's (1990: 334) third point is that:

> we need to pay much greater attention to detail, being empirical while avoiding empiricism . . . By paying attention to detail and specificity, we impugn any notion that material culture can be adequately understood by reducing it to tables of measurements and statistical correlations of various traits.

If detail is derived from data, as surely it must be, the sheer volume and variety of the data generated by empirical investigations into the nature of the archaeological record (environmental, archaeofloral, archaeozoological, landscape formation) must overwhelm the attempt to synthesize and interpret the mass of multi-media data involved. A medium which allows the assimilation and generation of a multiplicity of possible interpretations of this material more directly is clearly needed. To use GIS as the preparatory stage for verbal synthesis and filtering out of alternative explanations and models is to miss the point entirely. By adopting multi-media intelligent GIS as a medium, there is no longer any need to try to reduce material culture to 'tables of measurement and statistical correlations of various traits'.

Tilley (1990: 334) continues:

> Fourth, data should be used as a means of clarification for an argument rather than as its sole basis. In other words, our texts should incorporate a genuine dialectic between theory, data and practice . . . they should be understood as mediating each other.

In a situation where the models are generated by the system rather than by any one particular author, or biased towards any one particular data set, then the relationships which obtain between the differing systems and subsystems of the real-world simulacrum become much more subtle.

A simulated automaton might, for instance, behave in ways totally unpredictable to the expectations of theories of economic or technological determinism. It may simply not behave rationally as the term 'rational' is normally understood to mean. Would this be called 'ritual behaviour' and what would it mean? Would researchers seek to introduce artificial constraints so that they might more readily recognize what is happening? Would it be possible to examine the 'difference' (Hill, 1992) more readily? In such situations theory, data, and practice become alternative aspects or dimensions of each other.

The final point from Tilley (1990: 334) to consider is that:

> texts should be produced that open out a field of meanings for material culture rather than trying to pin it down as precisely as possible to one meaning: plurality rather than unidimensionality.

GIS, enhanced as suggested above, readily offers itself as a medium of archaeological discourse. It is a medium which is capable of generating many models of explanation which are explicit and testable and yet not governed by a single, predetermined design choice.

Seen as a medium of communication rather than as a tool which eventually aids the production of a written text, GIS offers the real possibility of closing the gap between data and theory. It can achieve this by making data sets readily accessible and interpretations as many as the different routes one may care to explore while walking through the real-world simulacrum.

25.9 Conclusions

GIS are limited in their present form but still represent useful tools for the analysis of spatial relationships. Furthermore, as the technology matures and multi-media, temporal processing systems become more widely available they are likely to become increasingly adopted, particularly in areas such as field archaeology and archaeological resource management.

The scope of the contribution of AI techniques to GIS has yet to be fully explored and the question of how far these should develop in the direction of simulation studies has yet to be fully debated and resolved. It is clear though that significant advances are to be gained from the incorporation of neural networks into GIS.

Finally, until GIS is recognized as a medium of theoretical discourse (which is inevitable as the technologies discussed above converge and are integrated with it) there remains the great danger that the perceived divisions between data, theory and practice will continue to be reinforced to the overall detriment of each.

References

Aangeenbrug, R. T., 1991, A critique of GIS, in Maguire, D. J., Goodchild, M. F. and Rhind, D. W. (Eds), *Geographical Information Systems. Principles and applications*, Vol. 1, pp. 101–7, New York: Longman.

Aston, M. and Rowley, T., 1974, *Landscape archaeology, an introduction to field techniques in post-Roman landscapes*, Newton Abbot: David and Charles.

Castleford, J., 1992, Archaeology, GIS and the time dimension—an overview, in Lock, G and Moffett, J. (Eds), *Computer Applications and Quantitative methods in Archaeology. CAA91*, BAR Intl. S577, pp. 95–104, Oxford: Tempus Reparatum.

Clarke, D. L., 1972, *Models in Archaeology*, London: Methuen.

Cumberpatch, C. G., 1993, An archives policy for the South Yorkshire Archaeology Field and Research Unit, Sheffield: S.Y.A.U.

Dayhoff, J. E., 1990, *Neural Network Architectures—An Introduction*, New York: Von Nostrand.

Doran, J., 1977, Knowledge representation for archaeological inference, *Machine Intelligence*, **8**, 433–54.

Fisher, P. F., 1989, Knowledge-based approaches to determining and correcting areas of unreliability in geographic databases, in Goodchild, M. and Gopal, S., (Eds), *Accuracy of Spatial Databases*, pp. 45–54, London: Taylor & Francis.

Gardin, J. C., 1988, *Artificial Intelligence and Expert systems: case studies in the knowledge domain of archaeology*, New York: Ellis Horwood.

Gibson, P. M., 1993, The potentials of hybrid neural network models for archaeofaunal ageing and interpretation, in Andresen, J., Madsen, T. and Scoller, I. (Eds), *Computing the past. Proceedings of the Conference on Computer Applications and Quantitative Methods in Archaeology CAA92*, Aarhus: Aarhus University Press.

Hill, J. D., 1992, Can we recognise a different past? A contrastive archaeology of later prehistoric settlements in Southern England, *Journal of European Archaeology*, **1**, 57–75.

Hunter, J. and Ralston, I (Eds), 1993, *Archaeological Resource Management in the UK. An Introduction*, Institute of Field Archaeologists. Stroud: Alan Sutton.

Kalawsky, R. S. 1993, *The Science of Virtual Reality and Virtual Environments*, Wokingham: Addison–Wesley.

Kvamme, K., 1992, Geographic Information Systems and Archaeology, in Lock, G. and Moffett, J. (Eds), in *Computer Applications and Quantitative methods in Archaeology CAA91*. BAR Intl. S577, pp. 77–84, Oxford: Tempus Reparatum.

Maffini, G., Arno, M. and Bitterlich, W., 1989, Observations and comments on the generation and treatment of error in digital GIS data, in Goodchild, M. and Gopal, S. (Eds), *Accuracy of Spatial Databases*, pp. 55–68, London: Taylor & Francis.

Maguire, D. J. and Dangermond, J., 1991, The functionality of GIS, In Maguire, D. J., Goodchild, M. F. and Rhind, D. W. (Eds), *Geographic Information Systems*, Vol. 1, pp. 319–35, New York: Longman.

McClelland, J. L. and Rumelhart, D. E., 1988, *Parallel Distributed Processing—Explorations in the Microstructure of Cognition*, Vols 1 and 2, Cambridge, MA: MIT Press.

Minsky, M. and Papert, S., 1969, *Perceptrons*, Cambridge, MA: MIT Press.

Openshaw, S., 1989, Learning to live with errors in spatial databases, in Goodchild, M. and Gopal, S. (Eds), *Accuracy of Spatial Databases*, pp. 45–54, London: Taylor & Francis.

Pollock, J. L., 1989, *How to build a person*, Cambridge, MA: MIT Press.

Raper, J. F. and Kelk, B., 1991, 3-dimensional GIS, in Maguire, D. J., Goodchild, M. F. and Rhind, D. W. (Eds), *Geographic Information Systems*, Vol. 1, pp. 299–317, New York: Longman.

Reilly, P., 1991, Towards a virtual archaeology, in Lockyear, K. and Rahtz, S. (Eds), *Computer Applications and Quantitative methods in Archaeology CAA90*, BAR Intl. S565, pp. 133–6, Oxford: Tempus Reparatum.

Rosenblatt, F., 1959, 'Two theorems of statistical separability in the Perceptron', in Mechanisation of thought processes: Proceedings of a symposium held at the National Physical Laboratory, Nov. 1958, Vol. 1, pp. 421–56, London: HMSO.

Rosenblatt, F., 1962, *Principles of neurodynamics*, New York: Spartan.

Saito, K. and Nakanono, R., 1988, Medical Diagnostic Expert System Based on a PDP Model, *IEEE Int. Conf. on Neural Networks*, **1**, pp. 255–62.

Shepherd, I. D. H., 1991, Information integration and GIS, in Maguire, D. J., Goodchild, M. F. and Rhind, D. W. (Eds), *Geographic Information Systems*, Vol. 1, pp. 337–60, New York: Longman.

Smith, T. R. and Yiang, J. E., 1991, Knowledge-Based Approaches in GIS, in Maguire, D. J., Goodchild, M. F. and Rhind, D. W. (Eds), *Geographic Information Systems*, Vol. 1, pp. 413–25, New York: Longman.

Tilley, C., 1990, *Reading Material Culture*, London: Basil Blackwell Ltd.

Voevodsky, J., 1987, Plato/Aristotle: A Neural Net Knowledge Processor, in Caudhill, M. and Butler, C. (Eds), *IEEE 1st International Conference on Neural Networks*, Vol. 2, 399–410.

Wheatley, D., 1991, Sygraf—resource based teaching with graphics, in Lockyear, K. and Rahtz, S. (Eds), *Computer Applications and Quantitative methods in Archaeology CAA90*, BAR Intl. S565, pp. 9–12, Oxford: Tempus Reparatum.

Wilkinson, G., 1993, Neural Networks: A new analysis tool. *GIS Europe*, Vol. 2(8), pp. 38–40, Cambridge: Longman Geoinformation.

Willis, N. and Kerridge, J., 1983, *Introduction to Computer Architecture*, London: Pitman Books.

Zubrow, E. B. W., 1990, The fantasies of GIS software, in Allen, K. M. S., Green, S. W. and Zubrow, E. B. W. (Eds), *Interpreting Space: GIS and Archaeology*, pp. 184–93. London: Taylor & Francis.

26

Toward an evaluation of GIS in European archaeology: the past, present and future of theory and applications

T. M. Harris and G. R. Lock

26.1 Introduction

In recent years, GIS has experienced explosive growth and archaeology has been one of many disciplines to have been caught up in its technological vortex. The GIS multiplier (Cowen, 1993) in archaeology is in full swing. The rapid diffusion of GIS technology within archaeology, tentatively forecast just a few years ago (Harris and Lock, 1990), has been remarkable. Indeed, such has been the increase in awareness that movement from the 'innovator' to the 'early majority' stage of adoption would seem inevitable. A number of trends in the application of GIS in archaeology are becoming apparent which both pre- and post-date the publication of *Interpreting Space: GIS and Archaeology* (Allen *et al.*, 1990). These trends, as discussed later, indicate profitable early avenues of enquiry. At the same time, the archaeological community is beginning to recognize functional limitations to the technology, and these likewise indicate potential research futures. The purpose of this chapter is threefold: first, to identify the main trends and themes which are developing in the application of GIS in archaeology in the European and North American contexts; second, a number of issues are identified as they impact the adoption of this technology for archaeological purposes; and third, future directions are indicated whereby the specific needs of archaeologists might be tailored to developments in GIS technology itself.

26.2 Past and present: European and North American perspectives

In many respects, the use of GIS in archaeology reflects Crain and MacDonald's (1984) model of GIS maturation. This model essentially proposes a progressive sophistication of GIS use from inventory tool, to analysis, to integrated decision-making system. The need to record or convert archaeological site data from analogue or digital sources into a GIS format is readily understood and represents an obvious first step in the use of GIS. Such inventorying activities, for example, found early adherents in agencies with statutory and regulatory responsibilities for cultural resource management (CRM). In particular, the drive to generate seamless national archaeological coverages and to focus on regional inter-site analyses has been indicative of the early uptake and focus of GIS in archaeology. To date, the application of GIS at the intra-site level is less well-documented although Biró and Fejes, Biswell *et al.* and Meffert (all this volume) show that this deficiency is currently being addressed. Part of the reason for the dearth of studies at the intra-site level is linked to deficiencies and difficulties in handling multi-dimensional data in a GIS environment.

This early focus on inventorying has a number of implications for the archaeological community. The need to establish a GIS database of archaeological site locations and related attributes is an accepted necessity that few would argue against. Such a process is tedious, time consuming, and invariably very expensive. But, as with other disciplines where GIS has made significant inroads, such an early emphasis on taking inventories and theory and only minimal emphasis on applications could have subsequent repercussions for the diffusion and uptake of GIS in the archaeological community as a whole. Although reports of archaeological GIS applications emerged in the early 1980s (Brown and Rubin, 1982), it is only with hindsight that *Interpreting Space* (Allen *et al.*, 1990) can be viewed as a watershed. As a record of achievement of the early innovators within the field of archaeological GIS and as an indicator and enabler of future directions, *Interpreting Space* stands alone in importance. The volume accurately reflects the heavy imbalance in early applications in favour of the USA with only one of the contributions being from European authors (Harris and Lock, 1990) although two applications use European data (Green, 1990; Madry and Crumley, 1990).

The early development of GIS in North American archaeology is detailed by Savage (1990) and can be compared to the sparse pre-1990 developments in Europe (Harris and Lock, 1990: 35). The proceedings of two European meetings of this period reinforce this imbalance. Both proceedings are of UISPP Commission IV, a forum of international importance for the publication of computer and quantitative methods in archaeological research. The 1987 meeting contained one paper concerned with GIS (Kvamme, 1990) which detailed the predictive modelling of archaeological sites, a theme which features strongly in North American GIS applications and an issue to which we return below. The subsequent meeting in 1989 (published 1990) again featured a single paper on archaeological GIS, which was again North American in origin (Williams *et al.* 1990).

The increase in the number of publications describing GIS applications in European archaeology since 1990 has been steady. It is obviously difficult to summarize and quantify research over such a large and diverse area as Europe. While the annual Computer Applications and Quantitative Methods in Archaeology (CAA) conference is heavily biased toward a UK audience and is not representative of the whole continent, it is at least an annual constant that can be used to identify trends within this research domain. Harris (1986; 1988) and Wansleeben (1988) laid the early foundations which, even by the 1990 conference, had attracted little interest with only a single paper (Lock and Harris, 1991) detailing the potential of GIS in integrating spatial information within UK cultural resource management systems and another mentioning the future use of GIS (Arroyo-Bishop, 1991). By the 1991 conference, just fourteen months after the publication of *Interpreting Space*, no less than eight of the thirty papers published by CAA were significantly GIS oriented (Lock and Moffett, 1992). This innovative group is notable not only for its size but also for the breadth of substantive interests displayed by European authors. These interests ranged from ambitious GIS-based national archaeological information systems, through theoretical considerations of data structures, to a consideration of temporality and GIS within archaeology. The same year saw the first publication of an actual European analytical application (Gaffney and Stančič, 1991) which is important for its contribution to theoretical approaches to landscape archaeology. Such a European approach to GIS and landscape archaeology tends to reflect a quite different approach to archaeological GIS than that pursued by North American archaeologists. This blossoming of European interest in all aspects of GIS applications continued the following year with a total of eleven papers in CAA92 (Andresen *et al.*, 1993).

Despite the high profile of GIS within the CAA conferences, however, European archaeological GIS publications have had what can only be described as a disappointing impact elsewhere. *Archaeology and the Information Age* (Reilly and Rahtz, 1992), which is a product of a World Archaeological Congress meeting and represents a global view of information technology in archaeology, contains very few references to GIS. Considering the importance of spatial data within archaeology, the chance to identify GIS as a unifying global theme was missed (Lock and Harris, 1992). An issue of *World Archaeology* dedicated to analytical survey contains two papers which are GIS-based (Brandt *et al.*, 1992; Hunt, 1992). Cox (1992), in the same volume, describes the use of Landsat images within wetlands archaeology in England together with the future use and potential of GIS. In the same year, the publication of the proceedings of a European conference on National Archaeological Records (Larsen, 1992) documents the present and future policies of national and local CRM agencies in many European countries. Preparing an inventory of archaeological sites is obviously a major concern and of the nine countries represented at this conference, only two make no mention of GIS while others are at various stages of evaluating and assessing GIS potential with a minority actually building GIS-based national and local systems.

Other archaeological GIS papers published between 1992 and 1993 were dominated by authors either based in the USA or with contacts there. Kvamme (1992) expanded on the theme of predictive modelling of site locations, while Peterman (1992) applied the same technology and methodology to the more distant location of Jordan. Different theoretical concerns are addressed by Haas and Creamer (1993) who used the concepts of terrain form, viewsheds and visibility to establish the defensibility of sites, while within Europe, Stead (1993) gives an overview of GIS research in archaeology. It is important to mention two US conferences which have yet to be published as these will make substantial contributions to the archaeological GIS literature. The first took place in Santa Barbara in January 1992 (Aldenderfer and Maschner, forthcoming) and presented a wide range of papers with three from the total of twenty-two being from Europe. The second, in Southern Illinois (Maschner, forthcoming), had a proportionally larger European presence accounting for six of the nineteen papers.

26.2.1 The diffusion of GIS in European archaeology

In an earlier paper (Harris and Lock, 1990) we discussed the diffusion of GIS through UK archaeology and it is of interest to return to this theme within a European setting. Diffusion theory identifies a sequence of adopters whereby initial innovators give rise to an early majority who are followed by a late majority and the inevitable laggards who may never adopt the new technology. The published literature since *Interpreting Space*, together with conference contributions yet to be published, suggest that while US archaeology has moved into the early majority phase, European applications are still infrequent enough to be considered at the early innovator stage. It is the nature of the diffusion process that exemplars are quickly duplicated and modified to meet particular circumstances and needs. North American archaeology is exhibiting very rapid adoption of GIS, particularly within CRM where systems and underlying theory are well established. This includes Canada where a great deal of CRM in connection with the forestry industry is GIS-based (Della Bona, 1993). The critical difference between North America and Europe in this instance is that in the latter very few, if any, exemplar systems exist in the literature that are not still in the early research and development stage. Several European projects are now approaching the decisive point whereby they could soon emerge from relative obscurity to become operational systems integrated into the daily workings of local or regional archaeological management and decision making. Projects in Norway (Mikkelsen and Larsen, 1992), The Netherlands (Roorda and Weimer, 1992a; 1992b), France (Guillot, 1992 and this volume; Arroyo-Bishop, 1992 and this volume), Spain (Bosqued *et al.*, forthcoming) and the UK (Robinson, 1993; Chartrand *et al.*, 1993) are all approaching this pivotal point.

Crossing this threshold will be critical in the diffusion process. Whereas technological inhibitors such as the cost and availability of hardware, software and digital data are being reduced by processes external to archaeology, the role and value of GIS within archaeological endeavour is something that can only be judged from within the discipline itself. While individual and institutional research projects are important in that they add to the corpus of knowledge and experience, it is the nature of GIS research that much of what is published remains a minority interest within the discipline as a whole. If GIS is to move beyond minority interest within archaeology and fulfil some of the potential as outlined then projects such as those above must become mainstream. Awareness of GIS among the archaeological community, in general terms even if not in specific terms of functionality, is high. There is almost an air of expectation and waiting, as if the potential has been accepted and now it requires the substantive exemplar applications to initiate the next phase of adoption.

Having reached its current position within European archaeology it is impossible to consider a rejection of GIS by that community. Parallels could be drawn with the expectations and excitement about artificial intelligence and the application of expert systems in archaeology in the mid-1980s. Despite the potential and the number of early research publications, AI has failed to make a major impact on archaeology, primarily because its role and value in terms of underlying archaeological theory were not clarified and justified sufficiently within the wider archaeological community. A simplistic comparison can also be made with relational database (RDMS) technology which is probably the most widely adopted aspect of IT within archaeology and this is not solely because it performs essential archaeological tasks. Relational databases have also had a bearing on the development of archaeological theory and practice at a fundamental level (see Andresen and Madsen, 1992, for an example concerning excavation recording). They have influenced thought on the acquisition and structure of data and on the nature of questions asked of that data. RDMS technology now plays a central role in many archaeological projects from their conceptual design to their everyday application. Quite clearly AI and RDMS provided two possible technology adoption models that GIS might follow. However, it is likely that GIS automation most closely mirrors the way in which archaeologists have traditionally handled and integrated archaeological information. For this reason, GIS is likely to become mainstream and embedded within the archaeological discipline in the future. The pace at which that adoption occurs, however, is open to question.

26.2.2 Trends in archaeological GIS applications

At this point, the diffusion of archaeological GIS is such that certain trends and directions can already be discerned. Three such trends are identified as they reflect the early adoption of GIS within existing traditions of archaeology: cultural resource management (CRM); spatial statistics; and landscape archaeology. At the same time, it is possible to isolate theoretical developments which underlie these themes and which are rooted in either or both of the North American or European traditions.

26.2.2.1 Cultural resource management (CRM)

There is obviously considerable overlap between the three themes suggested here, in terms of theoretical advancement and application. It is the concepts of CRM that unite them within the evolution of GIS-based research in archaeology although the theoretical emphasis in North America and Europe has been quite different. Because CRM, both regional and national, tends to be financed from the public purse, its goals are determined by the need to discharge statutory functions. These are centred on the inventory and management of archaeological information and require extensive data-handling facilities which, in longer established systems, have evolved from manual indices through various database formats (some with digital mapping capabilities), to GIS. While it would be incorrect to claim that GIS development within CRM has been devoid of a theoretical basis in archaeology, the theory developed has been restricted to the development of systems to perform and enhance these obligatory duties.

In many ways, the concerns of GIS-based CRM in North America and in Europe represents a continuation of pre-existing areas of interest. The initial adoption of GIS in archaeology has been to create and perform existing, often manual, operations. North American CRM has a long-standing focus on the predictive modelling of site location. The early thrust of much GIS work in the US has thus been to emulate and pursue this long-standing research theme. European concerns within CRM have centred more on modelling the structure of the cultural landscape and with the related issues of data structure and spatial and temporal definitions of sites (Lang and Stead, 1992). This difference between European and North American archaeological traditions has its origins in the physical characteristics of the archaeological record in the two continents. Compared to North America, European archaeology is temporally and spatially rich and creates a much denser and more complex cultural landscape to record and manage. Within the current discussion, these differing areas of emphasis result in North American CRM having closer links with the spatial statistics tradition whereas European CRM is associated more with the landscape archaeology tradition. These areas of emphasis are replicated in the way in which GIS has been incorporated within the respective archaeological traditions.

26.2.2.2 The spatial statistics tradition

While it has long been recognized that the GIS environment is ideal for the development of new approaches to spatial analysis, there are very few new formal statistical methods which have been widely adopted within GIS (Openshaw, 1991; Fotheringham and Rogerson, 1994). The spatial statistics tradition within archaeological GIS is firmly rooted in the USA and has evolved from the importance of the predictive modelling of site location; see Kohler and Parker (1986) for the 'pre-GIS' importance attached to such methods. The compatibility of predictive modelling with point data, raster data structures and cell-based statistics was quickly and exhaustively exploited (e.g. Kvamme and Kohler, 1988; Kvamme, 1990; Warren, 1990), again being driven mainly by prevailing CRM requirements. Little other work has been done on the integration of spatial statistics within GIS, although an emphasis on cell-based manipulation necessarily prevails. Kvamme has developed new methods of spatial auto-correlation (Kvamme, 1993) and shows the potential of combining statistics and simulation approaches (Kvamme, forthcoming). The integration of spatial exploratory data analysis within GIS promises a number of positive future directions for analysis.

26.2.2.3 The landscape archaeology tradition

The theory and methods of predictive modelling and related statistical approaches have never sat comfortably within the landscape analysis approach that dominates part of European archaeology. Predictive modelling is rooted in the quantitative methods of the processual school of archaeological theory and is inherently deterministic and reductionist. Because environmental variables describing the topography, lithology and hydrology of an area are relatively simple to map, the trap of environmental determinism is difficult to avoid; see Gaffney (Chapter 15) and van Leusen (Chapter 3), and following sections in this chapter. Also, to represent an archaeological site as a single point may be acceptable for a lithic scatter in the US Southwest, but for a typical landscape palimpsest of Europe, which may be continuous over many hectares, it is unacceptably reductionist. The richness of many European landscapes has encouraged an analytical approach which concentrates on the cultural meaning of the spatial relationships identified between different landscape elements such as field systems, trackways, ritual sites and different types of settlements. Attempts to apply social, economic, and political models to gain an explanation of spatial structure rather than just identifying structure have developed a different emphasis to GIS applications. It is also worth noting here that the spatial complexity of the cultural landscape over much of Europe encourages the use of vector-based systems with their greater precision and abilities in representing and analysing non-point sites.

Of course, we are not suggesting a rigid differentiation between North America and Europe. Not all North American GIS applications are concerned with predictive modelling. Allen (1990) and Zubrow (1990) are just two examples of a wider analytical approach to landscape analysis. Conversely, predictive modelling has been practised in Europe although it is interesting that this is concentrated in The Netherlands (for example Brandt *et al.*, 1992) where contact and exchange with North American archaeologists using GIS is particularly evident. What is beginning to emerge as a strong theme within Europe, however, are the efforts to move beyond environmental determinism by aspiring towards representations of social landscapes and associated analyses. The seeds of this approach were sown by the publication of the first European GIS landscape analysis (Gaffney and Stančič, 1991) and reactions to it. Attempts at developing new analyses centre around the use of viewshed analysis and multiple viewsheds (Ruggles *et al.*, 1993; Wheatley, 1993; Lock and Harris, forthcoming (b); Gaffney *et al.*, forthcoming; plus Chapters 15 and 27 in this volume), various types of cost or effort surfaces (Stead, Chapter 23) and combinations of these. We must be careful not to be technologically deterministic here as the importance lies not in the methods used but in the underlying archaeological theory involving explanation and meaning inherent within spatial relationships (e.g. Boaz and Uleberg, Chapter 18). It is worth emphasizing that GIS is not an objective, value neutral, unbiased technology. Data is likewise not value neutral. GIS thus represents the social reproduction of knowledge and, as such, the development of a GIS methodology cannot be divorced from the development of the theory needed to sustain it. Nor can archaeological applications of GIS be undertaken without acknowledgement of the pitfalls which await the unwary. These issues are discussed more fully in the following sections but the development of these linkages is crucial if the technology is to find general acceptance within the broader archaeological community.

26.3 GIS futures and archaeology

26.3.1 Multi-dimensionality and temporality

Many of the themes discussed above regarding the application of GIS to archaeological applications will necessarily continue to be an important part of archaeological research in the short to medium term. This section identifies a number of other areas in which we envisage greater archaeological focus in the future.

Perhaps one of the most obvious and immediate deficiencies in current GIS revolves around its enforced uni-dimensional representation of the world. Archaeologists operate in a multi-dimensional world in which the addition of a third dimension, in the form of time, or depth, or height, are fundamentally interwoven within the archaeological analysis. Scientific visualization has provided some important capabilities for the display of data but it lacks the capability provided by three independent axes. With few exceptions, GIS applications in archaeology to date have been firmly rooted in two-dimensional abstractions of reality. In part, this represents the continuation of traditional manual or CAD approaches to handling archaeological information. It also represents severe limitations in GIS functionality for the handling of spatial information in more than two dimensions. Traditionally, where an application area has called for the handling of a third dimension, such as depth or time, then the approach has been to construct, integrate, and analyse within a stacked vertical series of two-dimensional geographies (Jones, 1989; Raper, 1989; Turner, 1991). In many instances, two and a half dimensional graphics is achieved via the draping of two-dimensional coverages over a wire frame representation of a landform or other surface. This facility adds to the visual interpretation of the archaeological phenomenon but the approach is still limited by the restriction of using only two independent dimensions. Two and a half dimensional graphics, or quasi three-dimensional, should not be confused with true three-dimensional capabilities in which multiple attribute data may be recorded for any unique combination of three-dimensional space represented along three independent axes of *x*, *y*, and *z*.

It is perhaps not surprising therefore that the majority of GIS applications in archaeology have occurred at the inter site, regional scale. It is here that GIS functionality is at its strongest, in identifying distribution patterns and exploring latent relationships between sites and their environs. At the intra-site level, however, the recording of excavation data in three-dimensional space produces a number of very real problems when using current GIS. At this point, conventional GIS approaches are inadequate and limited in their ability to record and explore relationships in three-dimensional space. Only in a few instances of the dollar rich commercial world of petroleum and gas exploration has GIS been developed and implemented which possess true three-dimensional functionality, and even here implementation falls short of what can reasonably be envisioned. In this context, the demands of archaeologists would clearly indicate a fundamental need for three-dimensional GIS capability.

There are a variety of ways in which three-dimensional data models can be constructed, including: volumetric or geocellular approaches; surface patches welded by parametric polynomial functions; and triangulated tessellations (Jones, 1989; Belcher and Paradis, 1991; Fisher and Wales, 1991). It is likely that GIS capability to handle three-dimensional data is not far from being a reality. The availability of three-dimensional capability in GIS promises to unleash many exciting and innovative avenues of enquiry for archaeologists. Three potential application areas in archaeology become immediately apparent. The first relates to the recording and analysis of archaeological phenomena *in situ* at the intra-site scale. This would entail the replacement of traditional hand-drawn or CAD plans with the digital recording of artefacts and features in three-dimensional GIS space; although, beware the precision of recording necessary for this and the potential conflict with resource pressures on modern rescue excavation as shown by Biswell *et al.* (Chapter 20). One immediate outcome of such a system would be the coupling of GIS to the identification of standard stratigraphic relationships using the Harris matrix. The development of such linkages would be particularly valuable to the process of integrating GIS and existing archaeological methods.

A second area of enquiry is related to the investigation of change through time at both the intra- and inter-site scales. Current GIS are effectively atemporal in nature and historical or causal explanation is poorly represented. As with the issue of incorporating a third dimension representing depth, time is invariably incorporated within the GIS in the form of a stacked series of temporal 'snapshots'. Most human activity, however, is fundamentally dependent on time. Current GIS thus fossilizes geographies and restricts consideration of time as being a continuum. The representation of archaeological time on a continuous axis would permit the construction of dynamic models based on the combined temporal attributes of different archaeological elements within the same site or landscape. Thus the 'time-cube' would enable the pursuit of a variety of sophisticated spatio–temporal questions.

Furthermore, the exploration of temporal questions in which the rate of change between archaeological sequences may be conjectured as being linear (a regular rate of change), catastrophic (a dramatic rate of change), or non-linear (an increasing or decreasing rate of change) may be examined. This is particularly useful when the temporal quality of the archaeological data varies. The possibility of using topology, traditionally used to explore spatial relationships, to explore spatio–temporal patterns is particularly attractive. Thus the temporal sequencing of a context relative to other contexts which may occur on top (later), underneath (earlier), or to the side (contemporaneous) may be explored using the topological relationships between objects. This situation is invariably more complex in that one period may intrude on another, as in the excavation of negative features, but object–object relations could be specified to reflect this. The ability to address issues associated with fuzzy space and fuzzy time raises further potentially rewarding avenues of enquiry. We develop these arguments further in Lock and Harris (forthcoming (a)) by detailing a three-dimensional probability model which would enable the combination of different pieces of dating information for sites within a landscape. The product is a series of 'columns' representing the third, or temporal, axis which show the probability of use for each site at any point in time and thus change through time.

Thirdly, three-dimensional capability would enable the development of spatial statistics capable of heuristically seeking adjacency and other spatial relationships in three dimensions simultaneously. Such a development would greatly benefit the advancement of more powerful statistical algorithms which would draw upon the full data and functionality available within such systems; the potential is shown by several papers in Fotheringham and Rogerson (1994).

26.3.2 Issues of data, knowledge distortion, behaviouralism and fuzziness

Two themes related to the issue of data are discussed here as they impact contemporary and future developments in archaeological GIS. The first theme is drawn from an ongoing debate currently engaging the GIS and geography communities, which revolves around issues of knowledge distortion and the incorporation of behavioural data within GIS. Given the lack of critical evaluation of GIS by other disciplines, this debate is especially important for disciplines beyond geography including the archaeological and GIS communities. Taylor's concern for knowledge distortion and bias in knowledge production (Taylor, 1991; Taylor and Overton, 1991) brings to the fore the selective nature of data, information, and group participation in knowledge production within a GIS. He, and others, have argued that GIS are not value neutral nor objective but incorporate a number of biases generated by such factors as data selection, classification, and the type of GIS analyses undertaken (see also Chrisman, 1987; Harris *et al.*, forthcoming; Pickles, 1991; Edney, 1991).

In the archaeological community, this knowledge distortion finds outlets in the data collected, and the coverages created and stored within the GIS. In particular, it appears to be giving rise to the return of a GIS-driven environmental determinism. There can be little doubt that some of the easiest coverages to incorporate within an archaeological GIS are derived from generalizations of the physical environment. Thus from elevation data may be derived slope, aspect, and sun-intensity. Combined with soil information and hydrological features, and supported by the buffering and overlay capabilities of GIS, these coverages provide an obvious starting point from which to explore the decision-making processes of earlier societies. However, if these coverages provide the bulk of the explanatory evidence then one should be wary of allowing them to attain the status of key explanatory determinants. Clearly, this represents another instance of Taylor's knowledge distortion and the lack of complementary, yet more difficult to acquire, information. There is growing evidence that the diffusion of GIS within archaeology is engendering an environmentally deterministic interpretation of the archaeological past (again, see Gaffney and van Leusen, Chapter 27). One way of combating this environmental emphasis is to incorporate complementary behavioural data within the GIS and seek to reconstruct alternative perspectives on prehistoric cultural landscapes. The use of viewshed analysis and cultural friction surfaces represent attempts to redress social deficiencies in archaeological GIS. Identifying alternative perspectives represents a significant challenge to archaeologists and the potential focus of some of the more demanding and yet rewarding aspects of archaeological GIS.

A second, and related theme, concerns the nature of archaeological data itself and the requirements for incorporation within a GIS. In effect, GIS at the present time are spatially deterministic. The GIS industry is driven by the need for accurate, geographically precise data. However, the type of data generated by archaeologists is often geographically imprecise and not easily expressed within a Cartesian point/line/polygon paradigm. An emphasis on fuzziness also runs counter to the GIS focus on accuracy and error minimization or management, although it does reflect some current research areas within the field. There can be little doubt that GIS represents progress in the advancement of computer-based spatial accuracy with regard to recording and interrelating archaeological phenomena. What becomes apparent, however, is that the demands of the digital technology to record site boundary information 'accurately' can be somewhat spurious in terms of the implied accuracy. At one level, perceived or affected accuracy can arise simply from the representation of an archaeological feature in the computer or on a hard-copy plot. This aspect is already acknowledged within the GIS world for the mere process of encoding a feature can in itself contribute to a misleading perception of accuracy which the source data and encoding accuracy does not actually support. In this respect, the issue of scale, accuracy, and error in data capture and reproduction becomes critical. The temptation to push data interpretation beyond the limits which the scale or accuracy of data capture will permit is unfortunately all too prevalent in GIS applications and represents a gross misuse of GIS technology. The issue of multiplicative cascading errors which arise from the combination of data layers captured at varying levels of accuracy is also a well-known, if little understood, element of GIS.

What archaeology brings to this discussion is that the boundaries of archaeological sites are often in themselves fuzzy in nature and do not conveniently lend themselves to digital capture in either vector or raster form. The need to deal with imprecise or fuzzy archaeological information parallels similar questions raised in geography; for an early discussion of this debate see Gale (1972). In the recording of soil distribution maps, for example, the soil type rarely changes as abruptly as the line demarcating the distributions on a map would suggest. There is invariably a transition zone of varying width between the respective soil types, but because of the crudity of both the survey information and current computer spatial data handling technology, this transition is only represented at present as a linear division (Fisher, 1987).

The issue of fuzzy space is equally applicable in the definition and recording of archaeological site boundaries. This concern for recording either the 'cartographic' representation of archaeological phenomena or a 'symbolic' representation of sites has important implications for GIS applications in archaeology. For the purpose of a site retrieval system, for example, a case can be made that the detailed cartographic representation of an archaeological site is neither required nor warranted. Thus complex site features would be grouped into a single entity. It would not be intended as a true record of an actual site description. The more accurate descriptions of sites which accompany excavation or detailed site surveys would represent the other extreme for archaeologists. Thus, the issue of the scale of data capture, between inter-site and intra-site, becomes significant for the design and construction of an archaeological GIS database. Furthermore, the detailed recording of a building, for example, clearly does not represent the known limits of human use of the space surrounding such a construction. Consequently, GIS in this respect forces archaeologists to return to earlier archaeological issues concerning the definition of site boundaries. The issue as to what level of accuracy should site information be captured and for what purposes is evident. Defining an optimal scale of data capture, identifying the level of accuracy or error involved, and the qualifications which this level of data accuracy imposes on archaeological research questions is something that will continue to become of importance to archaeologists as GIS is adopted.

26.3.3 Multi-media in archaeological GIS

Related to these issues of data accuracy and representation, and the incorporation of behavioural data within a GIS, is the question of how archaeological source materials from disparate origins and media may be incorporated within an archaeological GIS. Once again this represents an area in which archaeology may act to focus a number of needs which may drive the development of GIS. While GIS takes us some way down the road toward an integrated information system, it is still relatively crude when compared with the rich breadth of media sources upon which archaeologists traditionally draw. GIS are designed to handle data which have a spatial component, and yet there are many sources which archaeologists regularly call upon which would represent major problems if integration were sought within current GIS. For example, an industrial archaeologist may wish to use old or present-day survey plans of an industrial site; or utilize oral histories passed on by those who may have worked in a building; or incorporate old and recent photographs or paintings of a structure and its surroundings; or peruse documents, ledgers, or letters concerning everyday life in a particular building; or examine other studies written about a site. There may even be moving images and film which could illustrate the operational nature of an industrial site far better than any text or GIS attribute could impart. Much of this information would clearly provide an understanding of a site from the perspective of those who lived and worked there when it was in operation.

This contextual perspective could be invaluable to those seeking to understand the human experience of a relict structure and the landscape in which it stood. Because of the difficulties encountered in incorporating such diverse media into a GIS, there has been a noted tendency for archaeological explanation to focus on seeking causal explanation from data which are more readily available. Invariably these data pertain to the physical environment and great care should be taken not to move toward the discredited notions of environmental determinism because of limitations in the handling of non-standard information. For these reasons, GIS must seek additional means for recording and displaying material which can aid and facilitate explanation. The development of multi-media applications is only now becoming available and yet if integrated within a GIS tremendous possibilities exist to utilize the full power of a GIS linked to the enormous explanatory power of multi-media presentation. This powerful combination would add another 'dimension' to the exploratory and explanatory power of a GIS.

26.4 Conclusions

In this chapter, we have sought to evaluate some of the directions in which archaeological GIS is progressing. The rapid diffusion of GIS integration within archaeology in Europe and North America is traced and differences in the pace of adoption are identified. Importantly, three traditional areas of archaeological analysis comprising cultural resource management (CRM), spatial analysis, and landscape archaeology are examined as they have shaped the early use of GIS in archaeology, and account for differing archaeological GIS emphases in North America and Europe.

A number of functional limitations in the technology are also identified from the perspective of the archaeologist which, in turn, point to possible future developments in the linkage of GIS and archaeology. Perhaps two of the greatest GIS limitations confronting archaeologists are those encountered in handling multi-dimensional data and in dealing with the temporal aspect of archaeological data. Current GIS software represents a two-dimensional abstraction of the world. In archaeological terms, such a view is crude and far removed from reality. Quite how archaeologists deal with the addition of the depth of an object or site using current GIS functionality is problematic. A stacked series of two-dimensional coverages is indicative of these limitations. Similarly, an archaeological interpretation devoid of the temporal dimension is meaningless. And yet current GIS functionality restricts our ability to handle the full archaeological record or to address a number of fundamental questions which face the archaeologist in seeking to identify relationships within the archaeological record. The need for multi-dimensional capability in GIS is clear.

Similarly, with its primary focus on geographical location a GIS could stand accused of spatial determinism. Data that are locationally fuzzy, or do not easily convert to contemporary notions of distance, are difficult to incorporate within a GIS. Likewise an archaeological information system based on GIS has significant deficiencies in its ability to display the full archaeological record. Such data sources are not easily incorporated within existing GIS. However, the development of multi-media applications within GIS would suggest one very promising avenue of exploration toward an integrated archaeological database. In identifying present trends and recognizing where archaeological questions remained unanswered, GIS technology can more fully be adapted to the needs of the archaeologist.

References

Aldenderfer, M. and Maschner, H. (Eds), forthcoming, *The Anthropology of Human Behavior through Geographic Information and Analysis*, New York: Oxford University Press.

Allen, K. M. S., 1990, Modelling early historic trade in the eastern Great Lakes using geographic information systems, in Allen, K. M. S., Green, S. W., and Zubrow, E. B. W. (Eds), *Interpreting Space: GIS and Archaeology*, pp. 319–29, London: Taylor & Francis.

Allen, K. M. S., Green, S. W. and Zubrow, E. B. W. (Eds), 1990, *Interpreting Space: GIS and archaeology*, London: Taylor & Francis.

Andresen, J. and Madsen, T., 1992 Data structures for excavation recording. A case of complex information management, in Larsen, C. U. (Ed.), *Sites and Monuments: National Archaeological Records*, pp. 49–67, Copenhagen: The National Museum of Denmark, DKC.

Andresen, J., Madsen, T. and Scollar, I. (Eds), 1993), *Computing the Past: Computer Applications and Quantitative Methods in Archaeology CAA92*, Aarhus: Aarhus University Press.

Arroyo-Bishop, D., 1991, The ArcheoDATA System: towards a European archaeological document, in Lockyear, K. and Rahtz, S. (Eds), *Computer Applications and Quantitative Methods in Archaeology 1990*, British Archaeological Reports International Series, 565, pp. 61–70, Oxford.

Arroyo-Bishop, D., 1992, Further structuring of the ArcheoDATA System, in Lock, G. R. and Moffett, J. (Eds), *Computer Applications and Quantitative Methods in Archaeology 1991*, BAR International Series S577, pp. 89–94. British Archaeological Reports, Oxford: Tempus Reparatum.

Belcher, R. C. and Paradis, A., 1991, A mapping approach to three dimensional modelling, in Turner, A. K. (Ed.), *Three dimensional modelling with Geoscientific Information Systems*, pp. 107–122, Dordrecht: Kluwer Academic Publ.

Bosqued, C. B., Espiago, J. and Preysler, J. B., forthcoming, The role of GIS in the management of archaeological data: an example of an application to the Spanish administration, in Aldenderfer, M. and Maschner, H. (Eds), *The Anthropology of Human Behaviour through Geographic Information and Analysis*, New York: Oxford University Press.

Brandt, R., Groenewoudt, B. J. and Kvamme, K. L., 1992, An experiment in archaeological site location: modelling in the Netherlands using GIS techniques, *World Archaeology*, **24**, 268–82.

Brown, P. E. and Rubin, B. H., 1982, Patterns of desert resource use: an integrated approach to settlement analysis, in Brown, P. E. and Stone, C. L. (Eds), 1982, *Granite Reef: A Study in Desert Archaeology*, pp. 267–305, Anthropological Research Papers No. 28, Tempe: Arizona State University.

Chartrand, J., Richards, J. and Vyner, B., 1993, Bridging the urban–rural gap: GIS and the York Environs Project, in Andresen, J., Madsen, T. and Scollar, I. (Eds), *Computing the Past: Computer Applications and Quantitative Methods in Archaeology, CAA92*, pp. 159–66, Aarhus: Aarhus University Press.

Chrisman, N. R., 1987, Design of geographic information systems based on social and cultural goals, *Photogrammetric Engineering and Remote Sensing*, **53**(10) 1367–70.

Cowen, D. J., 1993, The GIS multiplier model, *Proceedings URISA*, **6**, 155–64.

Cox, C., 1992, Satellite imagery, aerial photography and wetland archaeology, *World Archaeology*, **24**(2), 249–67.

Crain, I. K. and MacDonald, C. L., 1984, From land inventory to land management, *Cartographica*, **21**, 40–46.

Della Bona, L., 1993, A preliminary predictive model of prehistoric activity location for the western Lake Nipigon watershed, *Archaeological Computing Newsletter*, **37**, 11–19.

Edney, M. H., 1991, Strategies for maintaining the democratic nature of geographic information systems, *Papers and Proceedings of the Applied Geography Conferences*, **14**, 100–108.

Fisher, P. F., 1987, The nature of soil data in GIS: error or uncertainty, in Aangeenbrug, R. T. and Schiffman, Y. M. (Eds), *Proceedings of International Geographic Information Systems (IGIS) Symposium: the research agenda, NASA, Washington DC*, **3**, 307–18.

Fisher, T. R. and Wales, R. Q., 1991, Three dimensional solid modelling of geo-objects using non-uniform rational B-splines (NURBS), in Turner, A. K. (Ed.), *Three dimensional modelling with Geoscientific Information Systems*, pp. 85–105, Dordrecht: Kluwer Academic Publ.

Fotheringham, S. and Rogerson, P. (Eds), 1994, *Spatial Analysis and GIS*, London: Taylor & Francis.

Gaffney, V. and Stančič, Z., 1991, *GIS approaches to regional analysis: a case study of the island of Hvar*, Ljubljana: Znanstveni institut Filozofske fakultete.

Gaffney, V., Stančič, Z. and Watson, H., forthcoming, Moving from catchments to cognition: tentative steps towards a larger archaeological context for GIS, in Aldenderfer, M. and Maschner, H. (Eds), *The Anthropology of Human Behavior through Geographic Information and Analysis*, New York: Oxford University Press.

Gale, S., 1972, Inexactness, fuzzy sets and the foundation of behavioral geography, *Geographical Analysis*, **4**, 337–49.

Green, S. W., 1990, Sorting out settlement in southeastern Ireland: landscape archaeology and geographic information systems, in Allen, K. M. S., Green, S. W. and Zubrow, E. B. W. (Eds), *Interpreting Space: GIS and archaeology*, pp. 356–63, London: Taylor & Francis.

Guillot, D., 1992, The National Archaeological Record of France: advances in computerization, in Larsen, C. (Ed.), *Sites and Monuments: National Archaeological Records*, pp. 125–32, Copenhagen: The National Museum of Denmark.

Harris, T., 1986, Geographic information system design for archaeological site information retrieval, *Computer Applications in Archaeology 1986*, pp. 148–61, Birmingham: Centre for Computing and Computer Science, University of Birmingham.

Harris, T., 1988, Digital terrain modelling and three-dimensional surface graphics for landscape and site analysis in archaeology, in Ruggles, C. L. N. and Rahtz, S. P. Q. (Eds), *Computer Applications and Quantitative Methods in Archaeology 1987*, pp. 161–70, British Archaeological Reports International Series 393, Oxford: Tempus Reparatum.

Harris, T. M. and Lock, G. R., 1990, The diffusion of a new technology: a perspective on the adoption of geographic information systems within UK archaeology, in Allen, K. M. S., Green, S. W. and Zubrow, E. B. W. (Eds), *Interpreting Space: GIS and archaeology*, pp. 33–53, London: Taylor & Francis.

Harris, T. M., Weiner, D., Warner, T. and Levin, R., forthcoming, pursuing social goals through participatory GIS?: redressing South Africa's historical political ecology, in Pickles, J. (Ed.), *Representations in an Electronic Age: GIS and Geography*, New York: Guilford Press.

Haas, J. and Creamer, W., 1993, Stress and warfare among the Kayenta Anasazi of the thirteenth century AD, *Fieldiana Anthropology* (new series), **21** (1450), Field Museum of Natural History, Chicago.

Hunt, E. D., 1992, Upgrading site–catchment analyses with the use of GIS: investigating the settlement patterns of horticulturalists, *World Archaeology*, **24**(2), 283–309.

Jones, E. R., 1989, Data structures for 3D spatial information systems, *International Journal of Geographical Information Systems*, **3**(1), 15–31.

Kohler, T. A. and Parker, S. C., 1986, Predictive models for archaeological resource location, in Schiffer, M. B. (Ed.), *Advances in Archaeological Method and Theory*, Vol. 9, pp. 397–452, New York: Academic Press.

Kvamme, K. L., 1990, 'The fundamental principles and practice of predictive modelling', in Voorrips, A. (Ed.), Mathematics and Information Science in Archaeology: A Flexible Framework, pp. 257–95, *Studies in Modern Archaeology* Vol. 3, Bonn: Holos–Verlag.

Kvamme, K. L., 1992, A predictive site location model on the high plains: an example with an independent test, *Plains Anthropologist*, 56(2), 19–40.

Kvamme, K. L., 1993, Spatial statistics and GIS: an integrated approach, in Andresen, J., Madsen, T. and Schollar, I. (Eds), *Computing the Past: Computer Applications and Quantitative Methods in Archaeology, CAA92*, pp. 91–103, Aarhus: Aarhus University Press.

Kvamme, K. L., forthcoming, Understanding within-site spatial complexity: linking visualization, simulation and spatial statistical approaches within GIS, in Maschner, H. (Ed.), *Geographic Information Systems and the Advancement of Archaeological Methods and Theory*, Center for Archaeological Investigations, Carbondale, IL: University of Southern Illinois Press.

Kvamme, K. L. and Kohler, T. W., 1988, Geographic information systems: technical aids for data collection, analysis and display, in Judge, J. W. and Sebastian, L. (Eds), *Quantifying the Present and Preserving the Past: Theory, method and Application of Archaeological Predictive Modelling*, pp. 493–548, Denver: US Department of the Interior, Bureau of Land Management.

Lang, N. and Stead, S., 1992, Sites and monuments records in England—theory and practice, in Lock, G. R. and Moffett, J. (Eds), *Computer Applications and Quantitative Methods in Archaeology 1991*, BAR International Series S577, pp. 69–76, British Archaeological Reports, Oxford: Tempus Reparatum.

Larsen, C. U. (Ed.), 1992, *Sites and Monuments: National Archaeological Records*, Copenhagen: The National Museum of Denmark.

Lock, G. R. and Harris, T. M., 1991, Integrating spatial information in computerised SMRs: meeting archaeological requirements in the 1990s, in Lockyear, K. and Rahtz, S. (Eds), *Computers and Quantitative Methods in Archaeology 1990*, British Archaeological Reports International Series, 565, pp. 165–173, Oxford.

Lock, G. R. and Harris, T. M., 1992, Visualising spatial data through geographic information systems: a new perspective on traditional approaches to spatial analysis, in Reilly, P. and Rhats, S. P. Q. (Eds), *Archaeology and the information age: a global perspective*, pp. 81–96, London: Routledge.

Lock, G. R. and Harris, T. M., forthcoming (a), Analyzing change through time within a cultural landscape: conceptual and functional limitations of a GIS approach, in Sinclair, P. (Ed.), *Urban origins in eastern Africa*, One World Series (Archaeology), London: Routledge.

Lock, G. R. and Harris, T. M., forthcoming (b), Danebury revisited: an Iron Age hillfort in a digital landscape, in Aldenderfer, M. and Maschner, H. (Eds), *The Anthropology of Human Behavior through Geographic Information and Analysis*, New York: Oxford University Press.

Lock, G. R. and Moffett, J. (Eds), 1992, *Computer Applications and Quantitative Methods in Archaeology 1991*, BAR International Series S577, British Archaeological Reports, Oxford: Tempus Reparatum.

Maschner, H. (Ed.), forthcoming, *Geographic Information Systems and the Advancement of Archaeological Methods and Theory*, Center for Archaeological Investigations, Carbondale, IL: University of Southern Illinois Press.

Madry, S. L. H. and Crumley, C. L., 1990, An application of remote sensing and GIS in a regional archaeological settlement pattern analysis: the Arroux River valley, Burgundy, France, in Allen, K. M. S., Green, S. W. and Zubrow, E. B. W. (Eds), *Interpreting Space: Geographical Information Systems and Archaeology*, pp. 365–380, New York: Taylor & Francis.

Mikkelsen, E. and Larsen, J. H., 1992, Recording archaeological sites in Norway, in Larsen, C. U. (Ed.), *Sites and Monuments: National Archaeological Records*, pp. 71–78, Copenhagen: The National Museum of Denmark, DKC.

Openshaw, S., 1991, Developing appropriate spatial analysis methods for GIS, in Maguire, D. J., Goodchild, M. F. and Rhind, D. W. (Eds), *Geographical Information Systems*, Vol. 2, 389–402.

Peterman, G. L., 1992, Geographic information systems: archaeology's latest tool, *Biblical Archaeologist*, **55**(3), 162–67.

Pickles, J., 1991, Geography, GIS, and the surveillant society, *Papers and Proceedings of Applied Geography Conferences*, **14**, 80–91.

Raper, J. F. (Ed.), 1989, *Three dimensional applications in Geographic Information Systems*, London: Taylor & Francis.

Reilly, P. and Rahtz, S., 1992, *Archaeology and the Information Age: a global perspective*, One World Archaeology Series, London: Routledge.

Robinson, H., 1993, The archaeological implications of a computerised integrated National Heritage Information system, in Andresen, J., Madsen, T. and Scollar, I. (Eds), *Computing the Past: Computer Applications and Quantitative Methods in Archaeology CAA92*, pp. 139–50, Aarhus: Aarhus University Press.

Roorda, I. M. and Weimer, R., 1992a, Towards a new archaeological information system in The Netherlands, in Lock, G. and Moffett, J. (Eds), *CAA91 Computer Applications and Quantitative Methods in Archaeology 1991*, British Archaeological Reports International Series S577, pp. 85–88, Oxford: Tempus Reparatum.

Roorda, I. M. and Weimer, R., 1992b, The ARCHIS project: towards a new national archaeological record in The Netherlands, in Larsen, C. U. (Ed.), *Sites and Monuments: National Archaeological Records*, pp. 117–22, Copenhagen: The National Museum of Denmark.

Ruggles, C. L. N., Medyckyj-Scott, D. J. and Gruffydd, A., 1993, Multiple viewshed analysis using GIS and its archaeological application: a case study in northern Mull, in Andresen, J., Madsen, T. and Scollar, I. (Eds), *Computing the past: CAA92*, Aarhus: Aarhus University Press.

Savage, S. H., 1990, GIS in archaeological research, in Allen, K. M. S., Green, S. W. and Zubrow, E. B. W. (Eds), *Interpreting Space: Geographical Information Systems and Archaeology*, pp. 22–32, New York: Taylor & Francis.

Stead, S., 1993, GIS in archaeology: a research summary, *Mapping Awareness and GIS in Europe*, **7**(3), 41–3.

Taylor, P. J., 1991, A distorted world of knowledge, *Journal of Geography in Higher Education*, **15**, 85–90.

Taylor, P. J. and Overton, M., 1991, Further thoughts on Geography and GIS, *Environment and Planning A*, **23**, 1087–94.

Turner, A. K. (Ed.), 1991, *Three dimensional modelling with Geoscientific Information Systems*, Dordrecht: Kluwer Academic Publications.

Wansleeben, M., 1988, Geographical Information Systems in archaeological research, in Rahtz, S. P. Q. (Ed.), *Computer and Quantitative Methods in Archaeology 1988*, British Archaeological Reports International Series 446, pp. 435–51, Oxford: Tempus Reparatum.

Warren, R. E., 1990, Predictive modelling of archaeological site location, in Allen, K. M., Green, S. W. and Zubrow, E. B. W. (Eds), *Interpreting space: GIS and archaeology*, pp. 201–15, London: Taylor & Francis.

Wheatley, D., 1993, Going over old ground: GIS, archaeological theory and the act of perception, in Andresen, J., Madsen, T. and Scollar, I. (Eds), *Computing the Past: Computer Applications and Quantitative Methods in Archaeology, CAA92*, pp. 133–8, Aarhus: Aarhus University Press.

Williams, G. I., Parker, S. Jnr., Limp, W. F. and Farley, J. A., 1990, The integration of GRASS-GIS, S, and relational database management: a comprehensive interaction environment for spatial analysis, in Voorrips, A. and Ottaway, B. S. (Eds), *New Tools from Mathematical Archaeology*, Warsaw: Polish Academy of Sciences.

Zubrow, E. B. W., 1990, Modelling and prediction with geographic information systems: a demographic example from prehistoric and historic New York, in Allen, K. M. S., Green, S. W. and Zubrow, E. B. W. (Eds), *Intepreting Space: GIS and Archaeology*, pp. 307–18, London: Taylor & Francis.

27

Postscript—GIS, environmental determinism and archaeology: a parallel text

V. Gaffney and M. van Leusen

27.1 Introduction

The impact of a new technology on methodology and theoretical approaches in archaeology can be profound. The relatively recent introduction of geographic information systems (GIS) within archaeology is a case in point. In what ways is GIS technology influencing our research? The question has recently become a subject of discussion among archaeologists working with GIS (Wheatley, 1993). In particular, the observation that researchers using GIS are prone to using determinist (functionalist) approaches to archaeological explanation will be explored in this chapter. This tendency, which became apparent among geographers some years earlier, has led to concerns of conceptual poverty of a kind that is referred to as environmental determinism (ED). In papers and discussions at the Ravello conference, it became apparent that considerable differences of opinion and confusion exist among European archaeologists as to the exact nature and consequences of this problem.

In the hope of providing a foundation for future discussion, the authors have attempted to represent some of this debate here. The arguments are presented as two parallel texts with the two authors asserting their own perspective on the situation separately. Whilst this format may appear somewhat clumsy, it does attempt to represent the present situation among GIS practitioners within European archaeology.

27.2 A defence of environmental determinism

Martijn van Leusen

27.2.1 Introduction

In its most restricted sense, ED is a theoretical approach to archaeology that regards past (and present) cultures as somehow functions of, or shaped by, environmental pressures. It relies to an extreme extent on the pre-eminence of human economic behaviour as opposed to cultural behaviour, and can be historically traced, via the proponents and adherents of New Archaeology, to the early Marxist archaeologists. This approach translates into the explicit and exclusive use of environmental variables in GIS models in order to explain patterns of past behaviour.

In a less restricted sense, ED is the (often involuntary) result of the influence of two factors: the limited availability of geographically anchored datasets; and the limited functionality of the GIS. First, a GIS is a computerized system for the display and analysis of geographically anchored information. Since data have to be geographically defined before they can be put into the GIS and used (for instance) for archaeological modelling, there are restrictions to the type of data that can be used in such models. Second, since data development is an extremely time-consuming activity, GIS modellers have tended to restrict themselves to acquiring datasets that are more or less readily available. Thus soils, topography, geology, hydrology, the digital terrain model and its derivatives, and all environmental variables mapped by national agencies at standard scales (1 : 25 000 or 1 : 50 000), tend to form the reservoir of 'base maps' out of which current archaeological models are built. These two factors have tended to push archaeological researchers using GIS into producing models that focus on relationships between regional distribution patterns and mappable components (or variables) of the environment.

We suppose that ED in its restricted sense has few or no adherents among contemporary archaeologists. However, there is no doubt that many recent applications of GIS, particularly in regional settlement location studies, reflect an ED approach to archaeological explanation. The next section examines the charges against the ED approach in such cases, arguments in its favour, and ways of circumventing the problems raised by ED. In section 27.2.3 the proposed alternative to ED models in GIS, cognitive models, will be discussed.

27.2.2 ED models in GIS

If the use of a GIS in archaeological research tends to push the researcher towards ED, leading to neglect of ritual and cognitive aspects of (for instance) site location, then there is cause for concern that archaeological theory will be impoverished. Gaffney *et al.* (1995, see also Chapter 15) argue that the sterility of the purely quantitative approach was recognized in the late 1970s and early 1980s and that the use of GIS threatens to revive that approach:

> Culture and belief systems are increasingly interpreted as being able to order the physical environment within absolute limits. The need to develop 'cognised models' which incorporate the belief systems and perceptions of past societies has become an imperative.
>
> . . . the aptitude of GIS to analyse such (mappable environmental) data and the relative sophistication of the results do not legitimate such pursuits archaeologically. There are good reasons to suggest that the application of GIS techniques in such a way could ultimately prove to be restrictive to the general development of archaeological thought. In its least harmful form, the indiscriminate use of GIS solely in conjunction with mapped physical data may result in the slick, but repetitious, confirmation of otherwise obvious relationships. In the worst case, it might involve the unwitting exposition of an environmentally or functionally determinist analytical viewpoint of a type which has largely been rejected by most archaeologists.

And again:

> . . . the nature of most geographical information systems is such that they are most readily applied to data which is most conveniently stored in map format

> [which] may ultimately be restrictive to the natural development of archaeological analysis.

I feel this puts too much stress on the negative aspects of ED. On the one hand, responsibility for the (mis-) use of GIS in this way (as with any tool) must ultimately lie with the archaeologists themselves. Sloppy research has always existed and GIS is merely a new toolbox with which to conduct it. On the other hand, valid reasons may well exist to want to take the ED approach to GIS modelling. Three of these will be briefly discussed below.

27.2.2.1. The ED approach to cultural resource management (CRM)

Much of the early impetus for the use of GIS in archaeology came from CRM needs. The main CRM application of GIS has been and will be in predictive models of cultural resource location, to be used in the creation of protective zoning maps by planning departments and in the preparatory stages of development projects. Because such models are aimed at the effective protection of the cultural (archaeological) heritage rather than its understanding, I have argued that a different set of rules should apply (van Leusen, Chapter 3), essentially sanctioning the ED approach for practical reasons. In particular, it should be perfectly valid to try to hunt down environmental correlates of settlement location, using methods such as those described by Kvamme (1988) and Warren (1990), as long as no simplistic causal 'explanation' is attached to these correlations. The aim of a CRM-type model is to detect and describe patterns in the known cultural resource data and to use these in predicting the likely presence of as yet unknown resources. No 'story' is needed to justify the predictions.

I want to stress here that the criticism by Gaffney *et al.*, quoted above, that this leads to 'the repetitious confirmation of otherwise obvious relationships' may well apply here—but also that, in the context of planning and development requirements, practical limits on time and money available, and the difficulty of reaching a consensus on relationships that are less than obvious, this is perhaps the only practical approach to GIS in CRM.

27.2.2.2 The ED approach to scale

'Scale' is a key word in discussions of GIS theory and methodology. In the case of ED and GIS, both spatial and temporal scales are relevant to the validity of taking the ED approach.

The relative importance of cultural factors in influencing the archaeological record diminishes with smaller spatial scales (larger regions). On a global scale, human settlement patterns are largely determined by climate, topography and other aspects of the physical environment. On a regional scale, the influence of cultural factors may become noticeable while the environment still largely determines patterns of settlement and exploitation. It would seem that with larger spatial scales (smaller regions) the charge of ED becomes more important.

With temporal scales, a similar relation to ED holds. Cultural behaviour changes at a much faster pace than does the physical environment. Unless the datasets used are of exceptional quality, it will often not be possible to 'see' periods short enough to exhibit definite cultural traits. A typical archaeological 'culture' has a time-depth of perhaps a few hundred years; even if cultural behaviour remained static during that time, any resultant patterning of the dataset would probably be statistically insignificant because of the small size of the dataset. Conversely, with larger time slices, shifts in cultural behaviour would result in random noise.

27.2.2.3. The ED approach to patterning
By using a GIS to detect patterns in the archaeological record, the archaeologist has the means to detect also the non-patterned, exceptional or 'random' part of the record. By applying an ED model to a dataset, one can eliminate environmental patterning in the data, leaving a clearer view of whatever cultural factors may have influenced the data. In this approach, ED is used as a 'data cleaning' or pre-processing operation in its own right. I am not aware of any applications that have already taken this approach; if not, a proper methodology will have to be established before such applications can be evaluated.

27.2.3 Cognitive models in GIS

How should all this influence the way in which we use GIS to model and understand past cultural landscapes? Taking a backward look, Boaz and Uleberg (Chapter 18) argue that the validity of recent GIS-based cultural landscape studies must be examined. This should reveal whether ED influences have indeed crept in and whether they have been handled properly by the researchers. Looking forward, Gaffney *et al.* (Chapter 15) stress the need to develop 'cognitive' models, i.e. models that incorporate human cultural behaviour.

Many kinds of cultural variables are either spatially based or cause spatial patterning. The patterns that are present in the archaeological record, be they the result of environmental or cultural factors, can be analysed in a GIS. For example, locational preferences can be seen as reflecting important aspects of past social and symbolic, as well as economic behaviour. Gaffney *et al.* put it like this:

> In attempting such models we are encouraged by Renfrew's (1982: 11) suggestion that 'if people's actions are systematically patterned by their beliefs, the patterning (if not their beliefs, as such) can become embodied in the archaeological record'. The suggestion that cognitive information on the way communities perceive and interpret their environment should be patterned, indicates that such qualities will be measurable and potentially mappable.

To this I would add that even if some cultural variables prove to be unmappable, hypotheses about cultural behaviour can still be tested in a GIS if they have spatial consequences.

Harris and Lock (Chapter 26) see a number of fundamental problems as to how behavioural and cognitive perspectives can be incorporated into, or derived from, an archaeological GIS. In their view, the early focus on territoriality and viewshed analysis (visible in the nascent archaeological GIS literature) represents attempts to address this factor. In discussing some of these early applications of 'cognitive' models, it became apparent that they were seen by some (Kvamme, van Leusen) as being essentially no different from the ED approach, in that they still involve measurable properties of the environment (visibility analyses being based on a derivative of the DTM, friction surfaces on evaluation rules for environmental variables such as slope, vegetation cover, etc.). Such models are therefore limited in exactly the same ways that ED models are limited.

So the question becomes: which parts of archaeological analysis lie outside those limits? The short answer is, analysis of the non-patterned part of the archaeological record. The GIS can help to establish which part of the dataset is non-patterned, but it cannot analyse that part.

27.2.4 Conclusions

GIS were not designed for archaeological uses; therefore their functionality is not optimal. It is useless to blame the GIS for being unable to do things it was not designed for. Responsibility for the (mis-)use of GIS rests with the archaeologist, not with the toolbox.

For the reasons given already, I feel that the ED approach to modelling with a GIS should not be abandoned, and that an effort should be made to make plain its advantages as well as its disadvantages. It is possible to make use of the limits of a GIS by using an ED model as a 'first approach' or 'data cleaning' instrument.

In section 27.2.3 I have argued that 'cognitive' models so far produced are in an important sense very similar to ED models and that perhaps we should not be concerned that 'cognition' is disregarded by GIS users but rather that small-scale and non-patterned phenomena (both environmental and cognitive) will be disregarded. It is not yet apparent that archaeologists using GIS will be able (or should even try) to surmount the limitations of this type of modelling, but in the meantime there is quite enough to explore.

27.3 GIS and ED: a trap for the unwary?

Vincent Gaffney

27.3.1 Introduction

The final discussion period at the 1993 Ravello conference was the most heated session within the event. It was at this point that an important division among the participants relating to a fundamental relationship between GIS and archaeology emerged. It was unfortunate, however, that the debate took place largely without reference to theoretical developments elsewhere within the archaeological community (Hodder, 1991; Wheatley, 1993). Indeed, I would suggest that this isolation is central to some of the current problems facing practitioners of GIS and will ultimately emerge as a major obstacle to their ambition to assert the role of GIS as, potentially, one of the most important analytical tools to become available to the archaeologist for decades (van Leusen, 1993).

In making such an assertion it is necessary that some explicit definition of the problem under discussion is attempted. This is not a simple task as I believe that the major problem with the application of GIS in archaeology is archaeology itself—and here I understand archaeology to mean the study of past societies in their entirety, from the analysis of their cultural and environmental remains, and through the inferences which may legitimately be deduced from such remains. Such studies are not passive descriptions of material culture; they attempt explanation relating to the genesis and transformation of archaeological entities through time. The success or failure of attempts by archaeologists to carry out such studies rests on the adequacy of their results in explaining the evidence that remains for study. Advocates of GIS within archaeology carry out analysis within the same intellectual framework.

GIS will therefore be judged by its success in 'doing' archaeology. Within the context of its own abilities and capabilities, GIS must contribute to the wider archaeological debate. If archaeological GIS applications are seen to be simply the vehicles for individuals who seek merely to use archaeological data to demonstrate the merits of various GIS hardware or software, the technology will have a limited impact.

In the context of the current debate, the question is whether the charges levelled at GIS of being environmentally or functionally deterministic are valid or, following from van Leusen's contribution above, that the negative emphasis placed on environmental determinism by some archaeologists, including myself, is misplaced and that such analyses either have a place within archaeology or their overall benefits outweigh any minor drawbacks that may be incurred.

I intend to argue that not only is GIS, as used by many of its adherents, environmentally and functionally deterministic, and that such stances will ultimately be unproductive, but also that this situation results largely from the restricted theoretical perspectives of all too many archaeologists currently using GIS as an analytical tool. I will also suggest that these limitations are compounded by the repeated use of data types which most easily fit the prevalent GIS data model and that too little consideration is paid to whether these data sets allow valid descriptions of past societies or even of settlement systems. From this position, we can begin to question the adequacy of such studies from an archaeological perspective. I will also, however, argue that GIS has a major place within archaeology if it can overcome these problems (Gaffney *et al.*, Chapter 15; Wheatley, 1993).

27.3.2 CRM models in archaeology

Paradoxically, I intend to start this process by agreeing with van Leusen with respect to some of his views on the role of GIS within CRM. I have no doubts that different rules apply when GIS is utilized for the management of known archaeological resources rather than to explain such resources. GIS is in its element when asked where something is and what its spatial relationship is with another object. I begin to diverge from van Leusen, however, when he begins to discuss cultural resource management models. He states that:

> . . . it should be perfectly valid to try to hunt down environmental correlates of settlement location . . . as long as no simplistic causal 'explanation' is attached to these correlations . . . the aim of a CRM-type model is to detect and describe patterns in the known cultural resource data and to use these in predicting the likely presence of as yet unknown resources. No 'story' is needed to justify the predictions.

This position is debatable, and the assertion that we can produce models which can detect and describe patterns without the attachment of a 'simplistic' explanation, should be questioned. Descriptions of patterns are not necessarily predictive. If a pattern is held to be the result of an environmental variable, this evidence should be presented. Alternatively, if the significance of a casual variable within the model can not be proven, we must question the adequacy of the model for predictive purposes. Otherwise we risk creating models which are inadequate descriptions of the resource and whose application risks the destruction of archaeological data that are not included in the model. For a CRM model to be successful, it must therefore explain the archaeological data, including the original cultural and environmental factors which produced the resource, and the post-depositional factors which have allowed its survival or caused its destruction. It is not possible for any archaeological model, GIS-based or otherwise, to stand aside from these complex matters and somehow to provide solutions 'objectively'. At a pragmatic level, I accept that there may be problems in fitting the more abstruse arguments of archaeological epistemology into the day-to-day operations of a CRM organization. However, I see little or no value in the utilization of inadequate or misleading models simply because they are easy to understand or easy to use.

27.3.3 Archaeological data in GIS applications: starting from a false premise?

In considering the adequacy of GIS approaches within archaeological analysis, van Leusen's exposition on CRM models is instructive for another reason. He notes that 'it should be perfectly valid to try to hunt down environmental correlates of settlement location'. This unwitting emphasis on settlement location is a significant theme within published GIS applications and archaeological models (Judge and Sebastian, 1988, Kvamme and Kohler, 1988). Even where this is not explicit it may be observed that there is also a tendency for many GIS models to explain the location of 'sites' without any qualitative description of what a site means. At the risk of generalization, I would suggest that, in most applications, it is frequently assumed that these sites are in fact 'settlements', i.e. habitation sites where most day-to-day economic activities are carried out and analysis is carried out on that premise.

If such an observation is correct, it becomes pertinent to study the GIS data structure as frequently demonstrated in many archaeological GIS applications. In the past, the debate related to the ED structure of GIS-based archaeological analyses has usually revolved around the apparent necessity to incorporate mappable environmental data into any study. Clearly, this is important. However, it must be equally important to question whether GIS data layers containing archaeological data are adequate descriptions of the human systems they seek to represent.

The vexed question of how we should map and display archaeological data has been with us for decades. Fortunately, we are helped by the existence of a considerable archaeological and ethnographic literature relating to the nature of human systems and the structuring of their residues. These studies suggest that the concept of the archaeological site is an inadequate description of human remains and that settlement sites rarely reflect the full range of human activities, many of which are differentially distributed in space (Foley, 1981). 'Sites', as such, only come to prominence because of the intensity of activities which occur at these loci and the concomitant chance of increased material residue discard and subsequent discovery (Gaffney and Tingle, 1984). This complexity is rarely reflected within GIS models, which all too frequently represents human activity as a series of isolated points (sites/settlements) without reference to the continuous activity that occurred across the landscape (e.g. van Leusen, 1993).

There is another, related point. Having simplified a vastly complex landscape record into a few spatially discrete, unrepresentative points, many GIS practitioners then ignore the qualitative data which come from excavation and survey of many of these sites/settlements. Despite the voluminous evidence which suggests significant social and economic differences within and between archaeological sites of virtually every period, many GIS applications represent the data as a series of bland egalitarian symbols.

The overall result of this situation is a significant coarsening of the resolution of the archaeological data under study and, I would suggest, a resulting database that will only be susceptible to very crude analyses. This is reflected in the emphasis on the locational and point analysis of sites within many GIS studies. It is to be wondered whether the databases used in such studies are capable of anything else. The fact that such crude data appear to fit into environmentally deterministic modelling is, probably, largely irrelevant. I doubt whether most studies could isolate non-environmental factors even if they wanted to.

Despite such observations, I believe that most archaeological data are amenable to GIS-based spatial analysis and that data structures which reflect the archaeological record more realistically are possible.

27.3.4 Pulling it all together: is environmental determinism enough?

Having considered the archaeological database—and found it, on occasion, wanting—we must now consider the interpretative regime to be applied to archaeological data in GIS-based applications. Van Leusen suggests that the criticisms of environmental determinism within archaeological interpretation is at least partially a reflection of scale, with the influence of cultural factors becoming most apparent at larger scales. He further suggests that given the poor temporal resolution of archaeological data there will be a difficulty in isolating cultural variation. However, some hope is offered in that it is suggested that ED models have a role in cleaning data and that after having isolated environmental variation with data we may somehow be left with the 'cultural'.

Several points are suggested from van Leusen's comments. The first is the apparent primacy of environmental patterning within the archaeological record and thus the dominance of such variables in archaeological explanation. There is also an implicit scale of inference in which the economic basis of human activities is clearly attainable, but that the more obtuse factors in human decision making are both less tangible or amenable to analysis. Even where such analysis is attempted, for example through the application of viewshed data to ritual monuments (Gaffney *et al.*, Chapter 15; Wheatley, 1993), van Leusen asserts that such work is 'essentially no different from the ED approach, in that they still involve measurable properties of the environment'.

The world view suggested by van Leusen is one in which human action is essentially a passive adaptation or by-product of the economic system (Hodder, 1991: 32–4). This attitude is closely linked with systems theory and, as Wheatley (1993) has already commented, such theoretical perspectives tend to be almost inevitably deterministic and characterized by analyses which emphasize cross-cultural rules and metrical testing. Social action is clearly seen as secondary and dependent upon the environment and economy. In such a situation there is no other context for explanation other than within the environmental system, and abstract models provided through an environmentally deterministic viewpoint appear perfectly adequate descriptions of cultural remains. In all too many analyses, spatial relationships defined from such studies are subjected to batteries of descriptive statistics and the results held to have been vindicated (Wansleeben, Chapter 12). Does this not suggest a confusion between the abstraction of patterns and their explanation?

Perhaps the clearest indication of such a situation is provided through the confusion displayed by van Leusen and Kvamme at Ravallo concerning models which use physical measurements and those which view physical measurement as an explanation in itself. The important point about the use of viewpoint analysis in relation to ritual monuments was not that such monuments could either be seen or not, but that human communities imbued such positions with some value. The landscape is given significance by human perceptions, it is not formed geologically with such qualities. Consequently, perceived physical relationships are not to be confused with archaeological explanations, but have to be seen within the context of our interpretation of belief systems.

Such observations are, of course, not of any great significance if the ordering of such data is such that can we use van Leusen's idea of using ED models to 'clean' the archaeological record and isolate the 'cultural'. This apparently neat partition between the environment/economy and the arcane represents a clinical, Western view of cultural organization and is once again frequently representative of systems-based approaches to archaeological analysis. There are good reasons to suggest that such attitudes are misplaced when dealing with cultures past or present (Layton, 1989a, b). Such models tend to ignore the pervasive nature of belief systems or the impact of human agency within past societies and minimize their impact in ordering all levels of human action (Barrett, 1993).

The alternative view demands that we acknowledge that human actions to material situations are culturally mediated and that we need to approach regularities in the archaeological record, not from a perspective which views pattern as a direct behavioural response to stimuli, but which interprets change and patterns within a historically specific context (Barrett, 1993: 165). As Colin Renfrew (1979: 3) has said, 'To know what happened in the past is not enough: the aim is to understand why it happened'.

There is a need, therefore, for a 'contextual' analysis which incorporates all levels of human activity and discourse in an interpretational framework and does not simply rely upon the environment. Unfortunately, the lack of such a context within most GIS analyses is assured because of the almost total abstraction, noted above, of a vastly complex archaeological database, and the loss of qualitative information in favour of the quantification of coarse resolution data which simply circumscribes the limits of human existence, rather than the reasons for cultural development.

One of the problems of not adopting such an interpretational stance can be illustrated through several GIS analyses that have been published and which are based upon the Neolithic linearbandkeramic (LBK) culture of central and north-western Europe (van Leusen, 1993; Wansleeben, Chapter 12). Both van Leusen and Wansleeben emphasized the strong relationship between settlement and environmental factors, particularly the relationship to loess soils. However, van Leusen (1993: 107) also states that:

> LBK settlement is determined by economic and social factors. Since the latter cannot be reliably reconstructed, settlement locations can only be related to what we know about the economic activities of the inhabitants.

On this basis, van Leusen (1993: 107) carries out a point analysis of settlement data in relation to four variables which are 'important and amenable to cartography: the availability of loess soils, the slope of the terrain, the amount of relief, and the distance to permanent streams'.

Van Leusen has suggested that this approach is legitimate on the grounds that it is simply a methodological example: a GIS duplication of a pre-existing archaeological model which demonstrates 'the archaeological possibilities of a GIS'.[1] While I am prepared, up to a point, to accept this, one should also be aware that it also illustrates the archaeological inadequacies of a GIS study which does not incorporate the social 'story'. Consequently, it is still valid to explore the possible implications of the absence of any input other than environmental spatial data.

The well-known stability of the LBK tradition and its clear relationship between settlement and agricultural resource might make an environmentally deterministic analysis of such data appear attractive. There is, however, a large amount of important research concerning the nature and spread of the LBK which would not be reflected within such a study. Even in 1985, Whittle (1985: 95) observed that:

> . . . there was a pause in settlement expansion at the limits of the loess. This may reflect the predilection of the LBK for loess soils in preference to the sandy and morainic soils to the north, but it may also be a further indication that the LBK was not simply a question of the growth and spread of incoming population but may have heavily involved the indigenous population.

The significance of social relations within the LBK and between the LBK and indigenous inhabitants has also been raised recently by Keeley (1992) and Bradley (1993). In particular, it is significant to observe that LBK settlements were not a totally uniform phenomenon. Apart from some evidence for social differentiation, the presence of enclosures on a minority of sites in peripheral regions of the LBK may be particularly significant. Although some enclosures may have been interpreted as defensive settlements, others may have been associated with the control of craft activities, and it is possible that some may have developed at the interface between agriculturalist and hunter–gatherer territories (Keeley, 1992; Bradley, 1993: 79). The interpretation of some of these enclosures as monuments, rather than as mundane structural artefacts is particularly significant especially in the light of Bradley's observations that some of these sites appear to have been chosen as settlements after their creation as monuments.

I am not an expert on the northwest European Neolithic, however, the data available to me suggest that while the LBK settlement pattern may appear to conform to broad economic and environmental rules, it may actually develop according to a number of factors including social relations with indigenous groups, control of traded or manufactured objects or perceived relationships to monuments whose significance may have had no connection to the formal economic base. Any attempt to understand the pattern of the LBK must incorporate all these elements.

An analysis of LBK settlement, which is presented in purely economic or environmentally deterministic terms, cannot be regarded as interpretive. It is merely descriptive, and given the very large amount of excavated data for the LBK across Europe, one must question whether such models provide an adequate description of our archaeological knowledge of these societies.

27.3.5 Digging deep: time and cultural transformations in GIS applications

There is another problem frequently associated with GIS analyses of archaeological data—the lack of any context for cultural and historical time. GIS are currently two-dimensional tools and all too frequently the data are presented by time slice and in isolation from what has gone before or what comes afterwards. This is a very artificial situation as few cultures have the luxury of arriving first on any landscape. The apparent isolation of settlement systems in time and space gives the impression that choice is unconstrained, except by environmental or economic factors.

The existence of a landscape history, however, frequently ensures that choice may be prescribed by earlier events. At the crudest environmental level, we might consider the effect of severe land erosion or drainage caused by inadequate farming practices and its effect on later farming practices. More significant, within the terms of this chapter, is the way in which the cultural perception of landscape may determine land use and how this may vary over time.

In attempting this, I will adopt what appears to be a soft option! I have already alluded to the importance of understanding ascribed significance when interpreting monuments. It is equally important to understand how these perceptions structure space. With respect to monuments, we should note that an attempt to understand liturgical movement purely in terms of the physical area of monuments is frequently to miss their point. Monuments may be connected to the landscape physically, visually, or even, through a call to prayer or the ringing of a bell, aurally, and we must attempt to understand their impact on the landscape at every level. However, whatever the original significance of any structure, it should be noted that monuments also have a history. They are modified, incorporated and re-interpreted within landscapes and studying this process is essential to understanding their role (Bradley, 1993).

The potential impact of such monuments has been noted by Bradley (1993) in his study of monuments, *Altering the Earth*. His comments on the division of the landscape by monuments including the Dorset Cursus is particularly instructive. The Cursus, an early Neolithic linear monument over 10 km in length and up to 100 m wide, divided upland areas of settlement and flint extraction, from lowland riverine zones associated with construction of enclosures and henge monuments. The impact of such a massive monument on the landscape must be accepted. As Barrett (1993) and Bradley (1993) observe, the Cursus not only had important direct ritual functions, it also physically separated major land zones, and the effect on the daily life of people using such areas must have been substantial. This is reflected in the way that the monument actually dictated the spatial division of land in later periods.

A similar point may be made using survey data on the Roman occupation on the Berkshire Downs in England. This study has revealed a regular pattern of settlement and land use (Gaffney and Tingle, 1989). Comparison of territories of villas derived from surface data and Thiessen polygon analysis suggest significant similarities between these estates and late Bronze Age linear ditch systems indicating some degree of continuity in land divisions over time. However, archaeologists have also noted that settlement and land-use activity have tended to avoid the area of the Lambourne Seven Barrows Cemetery—a linear barrow group originating with a very early Neolithic earthen long barrow. This pattern of avoidance extends into the Roman period with Roman settlement and intensive Roman discard activity being located within adjacent valleys but avoiding the Seven Barrows valley itself.

The peculiar zonation of land use in the area has prompted the suggestion that the valley of the Seven Barrows Cemetery was a ritual landscape during later prehistory (Richards, 1978). While one might accept such an assertion for part of the period under discussion, the direct extrapolation of belief systems from the early Neolithic to the Roman periods seems unlikely. Despite this, it seems clear that very early patterns of land use, including ritual behaviour, not only ordered contemporary land use, they also influenced succeeding phases. The Romano–British, who frequently seem the most functional of peoples, demonstrated some consciousness relating to the Seven Barrows valley and this is reflected in the discard of artefacts within the area. How such monuments were incorporated into early historic belief systems is uncertain. It may be that, at a day-to-day level, such monuments appeared little more than incidental to the Roman occupants of the Berkshire Downs. However, the existence of a cultural landscape of considerable time depth did have an impact on settlement location at a landscape level in this case, and here we must emphasize the role of human agency to transform and mediate action and acknowledge that these actions structure the archaeological record in a spatial manner.

Clearly, there are difficulties in incorporating the concept of cultural time within GIS-based analyses. However, this does not excuse us from attempting analysis. Indeed, the presence of time-depth is, to a certain extent, the essence of archaeology and, in the context of the general ED debate, one might emphasize the accumulator effect of cultural variables over time. Not to accept the challenge of using such data seems, to me, to be an attempt to side-step some of the most substantial issues in archaeological analysis.

27.3.6 Conclusion

The current preoccupation with environmentally and functionally deterministic models is not a useful development. These models offer an inadequate description of the archaeological record and a slavishly adaptive view of the world that is misleading. The abstraction of settlement data for simple pattern recognition purposes and the use of a variety of descriptive statistics to lend authority to such procedures will not ultimately be profitable.

The alternative to such a situation is the application of a more contextual approach to analysis. This will involve a fuller and more thoughtful use of the available archaeological data. There will be less emphasis on pattern recognition and physical measurement as goals in their own right, and more concern with interpretation of the historical processes which result in such patterns. This will not be easy. However, there is evidence that relevant non-environmental data may be a structured phenomenon, and if this is the case such data can be incorporated within GIS-based analysis. An appreciation of the full complexity of the archaeological record by GIS practitioners may seem a daunting prospect. Without such an overview, it is unlikely that GIS will ever be used to its full potential by archaeologists.

27.4 A reply

Martijn van Leusen

Gaffney's brief discussion of pattern and prediction in CRM models brings to light a recurrent theme in his contribution: the supposed 'unacceptability', 'inadequacy', even 'impossibility' of detecting patterns in an archaeological dataset without at the same time attaching an explanation (causal or otherwise) to the existence of such patterns. In putting my views, as he sees them (section 27.3.4), against a systems theory backdrop while he himself assumes a contextual stance, Gaffney makes plain that he sees the problem of ED in archaeological GIS applications as a problem of archaeological understanding, not of methodology (Trigger 1989: 303–11; 348–57). Since my own original argument discussed the methodological utility of ED in GIS, I feel that Gaffney's lengthy exposition of its rather obvious shortcomings in explaining the past and contextualism in general are beside the point. This is particularly clear in his discussion of LBK settlement, in which he rightly, but irrelevantly, points out various defects in my own supposed ED 'explanation' of LBK settlement in the southern Netherlands.[1] I agree with most of what he says here, but I am afraid it has no bearing on the utility of ED in archaeological GIS, which I have argued lies specifically in the field of pattern detection. I reject Gaffney's insistence on demanding an archaeologically 'adequate' explanation before such methods could be useful. To me, the methodological issue (is it possible and scientifically valid to use a GIS in this manner?) is separate from the archaeological issue (does it get us 'adequate' or 'useful' information about the past?), but I maintain that both questions should be answered in the affirmative. Statistics can be used to describe patterning in archaeological datasets in a rigorous manner without reference to the cause(s) of those patterns, and if the extrapolation of those patterns yields predictions that are useful in CRM (such as restricting fieldwalking areas, or helping to determine strategies for unbiased sampling) the method is validated. In his concluding section, Gaffney argues that archaeologists should be concerned with interpretation of the historical processes which result in patterns in the data. I will not argue with that except to note that this begs the question of how such patterns can be detected in the first place.

Of course, Gaffney is right in his criticisms of the typical GIS database and of current GIS technology. He argues that the representation of archaeological systems in many published GIS models is inadequate because they are reduced to dimensionless 'sites' (locations). He observes that current GIS technology limits archaeologists to representing two-dimensional time slices of the past, and that the apparent spatial and temporal isolation of settlement systems in such time slices makes us forget about the previous history of the landscape and about the fact that it is part of a living social system. He argues that this tends to make us believe that environmental factors are of paramount importance here, because the social/historical context has been swept, as it were, under the carpet. I agree with him that better data structures should and could be designed to improve archaeological modelling in GIS,[2] but again fail to see the relevance of these criticisms to the issue of ED. It is the responsibility of the archaeologist to decide on the quality of the data used in a

GIS model, the hypotheses to be tested, and the methods to be used. You cannot blame the toolbox for being misused. Besides, in practice the location is often just about the only thing that is reliably known about a group of archaeological objects. Should GIS analysis on such a basis (e.g. for CRM purposes) therefore be abandoned? Nor is it clear how Gaffney's (section 27.3.5) two examples of cultural expressions influencing later patterns of settlement and exploitation speak against the ED approach in GIS modelling. The Dorset Cursus and the Lambourne Seven Barrows Cemetery were part of later cultures' contemporary physical and mental environment. As physical objects, they should be part of an ED model. Of course, Gaffney's point here is that the ascribed significance of such monuments may be much more important than their physical presence and dimensions. Yet the only way to reconstruct this ascribed significance, besides studying properties of the monuments themselves (size, amount of work that went into making them, etc.), is to study their relations with the surrounding landscape—including other monuments. This brings us full circle to such techniques as friction surfaces and viewshed analyses that rely on an analysis of measurable properties of the physical landscape—the ED approach.

I must conclude that Gaffney's position and mine are not as different as they may seem. We agree on the tendency of the GIS toolbox to produce ED models, on the utility of ED modelling for CRM, on the inadequacy of ED explanations of the past, and on the possibility of analysing non-environmental patterning in the data. Because 'cognitive' models would (and do) apply the same methodology to cognitive aspects of the past landscape as 'ED' models do to environmental variables, Kvamme and I choose to regard both approaches as essentially identical. Because 'cognitive' models would result in a deeper understanding of past societies, whereas ED models result in statistical descriptions of a dataset, Gaffney chooses to regard both approaches as essentially different. The issue will no doubt be resolved before long by that ultimate of arbiters, the changing practice of archaeology.

References

Bakels, C. C., 1978, Four Linearbandkeramik Settlements and their Environment: a palaeoecological study of Sittard, *Stein, Asloo and Heinheim, Analecta Praehistorica Leidensia*, **11**.

Bakels, C. C., 1982, The settlement system of the Dutch Linearbandkeramik, *Analecta Praehistorica Leidensia* **15**, Leiden.

Barrett, J. C., 1993, *Fragments from Antiquity: an archaeology of social life in Britain, 2900–1200 BC*, Oxford: Blackwell.

Bradley, R., 1993, *Altering the Earth*, Edinburgh: Society of Antiquaries of Scotland Monograph Series Number 8.

Foley, R., 1981, *Off-site Archaeology and Human Adaptation: An Analysis of Regional Artefact Density in the Amboseli, Southern Kenya*, British Archaeological Reports International Series 97, Oxford.

Gaffney, V., Stančič, Z. and Watson, H., 1995, The impact of GIS on archaeology, this volume, Chapter 15.

Gaffney, V. and Tingle, M., 1984, The tyranny of the site: method and theory in archaeological field survey, *Scottish Archaeological Review* **3**, 134–40.

Gaffney, V. and Tingle, M., 1989, *The Maddle Farm Project: an integrated survey of prehistoric and Roman landscapes on the Berkshire Downs*, British Archaeological Reports Series 200, Oxford.

Hodder, I., 1991, *Reading the Past*, Cambridge: Cambridge University Press.

Judge, J. W. and Sebastian, L. 1988, *Quantifying the Present and Predicting the Past: Theory, Method and Application of Archaeological Predictive Modelling*, Denver: US Department of the Interior, Bureau of Land Management.

Keeley, L., 1992, The introduction of agriculture to the western North European plain, in Gebauer, A. B. and Price, T. D. (Eds.), *Transitions to Agriculture in Prehistory*, World Monographs in Archaeology 4, Madison.

Kvamme, K. L., 1988, Developing and Testing of Quantitative Models, in Judge, J. W. and Sebastian, L. (Eds), *Quantifying the Present and Predicting the Past: Theory, Method and Application of Archaeological Predictive Modelling*, pp. 315–428, Denver: US Department of the Interior, Bureau of Land Management.

Kvamme, K. L. and Kohler T. W., 1988, Geographic information systems: technical aids for data collection, analysis and display, in Judge, J. W. and Sebastian, L. (Eds), *Quantifying the Present and Predicting the Past: Theory, Method and Application of Archaeological Predictive Modelling*, pp. 493–548, Denver: US Department of the Interior, Bureau of Land Management.

Layton, R., 1989a, *Conflict in the Archaeology of Living Traditions*, London: Unwin Hyman.

Layton, R., 1989b, *Who Needs the Past? Indigenous Values and Archaeology*, London: Unwin Hyman.

Renfrew, C., 1979, Transformations, in C. Renfrew and K. L. Cooke (eds.), *Transformations: mathematical approaches to culture change*, pp. 3–44, Oxford: Blackwell.

Renfrew, C., 1982, *Towards an Archaeology of Mind*, Inaugural Lecture, Cambridge: Cambridge University Press.

Richards, J. C., 1978, *The Archaeology of the Berkshire Downs*, Berkshire Archaeological Committee Publications, **3**, Reading.

Trigger, B. R., 1989, *A history of archaeological thought*, Cambridge: Cambridge University Press.

van Leusen, P. M., 1993, Cartographic Modelling in a Cell-based GIS, in Andresen, J. Madsen, T. and Scollar, I. (Eds), *Computing the Past: Computer Applications and Quantitative Methods in Archaeology 1992*, pp. 105–24, Aarhus: Aarhus University Press.

van Leusen, P. M., forthcoming, CRM and Academia in the Netherlands: assessing current approaches to locational modelling in archaeology, in Maschner, H. (Ed), *Geographic Information Systems and the Advancement of Archaeological Methods and Theory*, Carbondale, IL: Centre for Archaeological Investigations.

Warren, R. E., 1990, Predictive modelling in archaeology: a primer, in Allen, K. M. S., Green, S. W. and Zubrow, E. B. W. (Eds), *Interpreting Space: GIS and archaeology*, London: Taylor & Francis.

Wheatley, D., 1993, Going over old ground: GIS, archaeological theory and the act of perception. In Andresen, J. Madsen, T. and Scollar, I. (Eds), *Computing the Past: Computer Applications and Quantitative Methods in Archaeology 1992*, pp. 133–8, Aarhus: Aarhus University Press.

Whittle, A., 1985, *Neolithic Europe: a survey*, Cambridge: Cambridge University Press.

Notes

[1] Here and elsewhere (section 27.3.4) Gaffney cites my own 1993 paper on LBK settlement to demonstrate the inadequacy of ED explanations in the typical GIS model. Unfortunately the model I described is atypical, in that it is an explicit duplication of previous (and pre-GIS) research in the New Archaeology tradition by Bakels (1978, 1982). It was created only to demonstrate the methodological possibilities of GIS and cannot, therefore, be used to infer any view of mine regarding its utility or legitimacy.

[2] With regard to the particular criticism by Gaffney that the concept of 'sites' is an inadequate abstraction of what in essence is a spatially and temporally continuous variable, I should point out that discrete map layers of 'sites' could well be replaced by continuous map layers of some density measure or 'activity index'.

Index